Configuring Windows Server Hybrid Advanced Services Exam Ref AZ-801

Configure advanced Windows Server services for on-premises, hybrid, and cloud environments

Chris Gill

BIRMINGHAM—MUMBAI

Configuring Windows Server Hybrid Advanced Services Exam Ref AZ-801

Associate Group Product Manager: Mohd Riyan Khan
Senior Editor: Nihar Kapadia, Divya Vijayan
Technical Editor: Shruthi Shetty
Copy Editor: Safis Editing
Project Coordinator: Ashwin Kharwa
Proofreader: Safis Editing
Indexer: Hemangini Bari
Production Designer: Ponraj Dhandapani
Marketing Coordinator: Marylou De Mello

First published: April 2023

Production reference: 1310323

Published by Packt Publishing Ltd.
Livery Place
35 Livery Street
Birmingham
B3 2PB, UK.

ISBN 978-1-80461-509-6

www.packtpub.com

To my wife, Jenn – thank you for being incredibly supportive of this hustle every step of the way through life's rollercoaster ride. You are my North Star, my driving force, and my inspiration to keep going no matter what the challenge may be! To my children, Coleston and Kennady – you both give me the courage, the patience, and the inspiration to share knowledge and experiences with the community, no matter the generation (or the topic, for that matter). While I realize I borrowed time from our busy schedules to create this resource, know that you both were front of mind throughout! And to my extended family, coworkers, and community friends – your encouragement, support, and check-ins were appreciated and necessary to help ensure that I finished writing this book!

They say that legacy is a shared experience – I hope that you are not only left with valuable insights from this resource, but your own visions and experience that encompass the past, present, and future states of the cloud.

– Christopher "CG" Gill

Foreword

Reports of Windows Server's death have been greatly over-exaggerated. I know companies of all sizes are hyper-focused on the cloud, but Windows Server skills are not the equivalent of knowing Latin (at least not yet anyway). I am excited to see Microsoft's reaffirmed focus on Windows Server skills, especially given how much everyone invested in those skills over the last few decades (me included).

In my previous role, I worked on the Cloud Advocacy team, and I was involved with the initial promotion of the AZ-800 and AZ-801 during the Ignite 2021 timeframe (November 2021). I worked with teams to get content out to a broader audience of folks all across the globe who could benefit from taking the exam. The overwhelming response to having these certifications brought to life was such a positive experience for me to witness. These exams are meant to give more traditional on-premises administrators exposure to the world of Azure!

In thinking through my own journey as a former system administrator, I love that we're getting back to the basics of what made most of us technical: Windows Server. I learned so much about how to configure Active Directory, DNS, file services, PowerShell, DHCP, and so on, just by installing a server operating system onto a random spare computer I was no longer using. In preparing for the exams myself, I re-remembered some of the more complex and complicated parts of Windows Server, which is a must-have if someone needs to validate skills for a job role. Additionally, learning about the Windows Server components that extend into Azure is one awesome way to keep skills current. I'll say it again: infrastructure skills are NOT going away…they're just evolving.

The one thing that has changed is the way in which Windows Server skills mix with cloud-based skills. The fundamentals are still very important and showcasing that knowledge to current and future employers is equally as important. A lot of this is best understood by going through appropriate how-tos, especially because the fundamentals are still the same.

Folks interested in sitting for this exam are usually those responsible for configuring and managing Windows Server on-premises. Additionally, these same folks are sometimes managing hybrid and Infrastructure as a Service (IaaS) platform workloads in Azure. The Windows Server hybrid administrator usually finds themselves integrating Windows Server environments on-premises with Azure services. The learner would also need to manage and maintain Windows Server IaaS workloads in Azure as well as knowing how to both migrate and deploy workloads in Azure.

Everything covered in this book will more than equip students to take and pass the AZ-801. The exam itself can be tricky in parts, so the best bet will be to read through everything, lab up, and dig in! I really like how the labs are outlined with hands-on activities, which is my favorite way of learning (or relearning) tech!

I have known Chris for a few years now and because of the pandemic, everything we have worked on together has been virtual. Hopefully, that will change soon. I was so ecstatic to be selected as a reviewer and the foreword writer!

I hope you find this book enjoyable, practical, and easy to follow! Best of luck if you're going to schedule and take the exam!

Shannon Kuehn

Senior Product Manager – Digital Influence Team – Identity and Network Access, Microsoft

Contributors

About the author

Chris Gill is a director of Microsoft applications at Nixon Peabody LLP, a leading global law firm. He has been in IT, development, infrastructure, security, and architecture for over 20 years. His passion for tech has led to his becoming a **Microsoft Certified Trainer (MCT)** and a Microsoft **Most Valuable Professional (MVP)** in Microsoft Azure.

Chris has a Bachelor of Science degree in computer science from St Bonaventure University and holds over 20 Microsoft certifications including Azure Solutions Architect, Azure Security Engineer, Azure Administrator, and Microsoft 365 Enterprise Administrator. Chris can be found interacting with folks on social networks, mentoring tech and soft skills, speaking with and hosting user groups, and providing insights on his blog.

Originally from Northern Cambria, PA, Chris resides in Penfield, NY with his wife and two incredible children.

About the reviewers

Alex ter Neuzen has been a system engineer on Windows for over 20 years and has spent the last 12 years working on Microsoft 365 with a main focus on on-premises combined with Microsoft Azure services. On-premises services are a specialty that he has developed in those 20 years and are his expertise. Currently, he is working as an IT consultant for a major company that is using Microsoft Azure in a hybrid form. He is responsible for migrating services and devices.

I would like to thank my girlfriend and children for their patience during the review of this book. It took a lot of time that I was not able to give to them. Microsoft Azure and on-premises services are an interesting world and it was a lot of fun to be part of the technical review team for this book.

Shannon Kuehn is a not-so-typical technologist. Her roots are grounded in public speaking and writing. She gravitated toward technology through music, by way of being a DJ and recording demos for club promoters. After taking time to develop initial skills in troubleshooting, she quickly found herself in jobs that graduated her toward deeper levels of technical skill. After receiving training and on-the-job experience, her career arc evolved into building comprehensive technical solutions in Azure. She is currently a senior product manager on the identity and network access team at Microsoft. She focuses on community outreach and content creation for Microsoft's identity platform to get developers and IT pros comfortable with life in the cloud.

Windows Server is STILL widely used in the technical world we live in. The two AZ-8XX exams were something I worked very closely with the Cloud Advocacy team on back in 2021. A year plus into seeing those exams come online and the number of folks taking these exams continues to grow. That alone is quite a testament to Windows being here for the long haul. Thanks to Chris and the Packt team for letting me review the final product!

Steve Miles, aka SMiles, is an MCT, Microsoft Azure, and Hybrid MVP with over 20 years of experience in on-prem security, networking, data center infrastructure, managed hosting, and hybrid cloud solutions. Steve is the author of the Packt book *AZ-800 Exam Guide Administering Windows Server Hybrid Core Infrastructure*, as well as being the author of the Packt book *AZ-900 Exam Guide Azure Fundamentals and Azure Security Cookbook*.

His first Microsoft certification was on Windows NT; he is an MCITP, MCSA, and MCSE for Windows Server and many other Microsoft products. He also holds multiple Microsoft Fundamentals, Associate, Expert, and Specialty certifications in Azure Security, Identity, Network as well as M365, and D365.

This is my contribution to the worldwide technical learning community, and I would like to thank all of you who are investing your valuable time in committing to reading this book and learning these skills.

Jimmy Bandell is an experienced IT professional with a wealth of knowledge and skills honed over 15 years in the industry. Throughout his career, he has developed a deep understanding of the technical complexities involved in managing IT systems and networks, as well as the importance of effective communication and collaboration in ensuring successful outcomes for clients and stakeholders.

Known for his strong work ethic and commitment to excellence, he is a trusted advisor to his clients and colleagues. Whether he is troubleshooting issues or mentoring team members, he approaches each task with a can-do attitude and a focus on achieving results.

Outside of work, he enjoys spending time with his family and his hobbies, such as cooking and building Lego.

Table of Contents

Part 1: Exam Overview and the Current State of On-Premises, Hybrid, and Cloud Workflows

1

Part 2: Secure Windows Server On-Premises and Hybrid Infrastructures

2

Securing the Windows Server Operating System   35

3

Securing a Hybrid Active Directory (AD) Infrastructure 77

4

Identifying and Remediating Windows Server Security Issues Using Azure Services 119

Part 3: Implement and Manage Windows Server High Availability

11

Implementing Disaster Recovery Using Azure Site Recovery   283

12

Protecting Virtual Machines by Using Hyper-V Replicas   309

Part 5: Migrate Servers and Workloads

13

Migrating On-Premises Storage to On-Premises Servers or Azure   329

14

Migrating On-Premises Servers to Azure   353

15

Migrating Workloads from Previous Versions to Windows Server 2022
381

Part 6: Monitor and Troubleshoot Windows Server Environments

22

Final Assessment and Mock Exam/Questions   535

Index   555

Other Books You May Enjoy   572

Preface

With this book, you'll learn everything you need to know to administer core and advanced Windows Server and Infrastructure-as-a-Service workloads utilizing available on-premises, hybrid, and cloud technologies to meet your needs in your private and public cloud-native management and transformation efforts while preparing you to pass the AZ-801: Configuring Windows Server Hybrid Advanced Services exam.

Who this book is for

This book is intended for cloud and data center management administrators and engineers, enterprise architects, Microsoft 365 administrators, network engineers, and anyone seeking to gain additional working knowledge with Windows Server operating systems and managing on-premises, hybrid, and cloud workloads with administrative tools.

What this book covers

Chapter 1, Exam Overview and the Current State of Cloud Workloads, sets you up for success by helping you prepare for a Microsoft exam, helping you to identify additional community resources for training, and establishing a lab and trial Microsoft 365 subscription to reinforce skills with hands-on learning.

Chapter 2, Securing the Windows Server Operating System, covers how to configure and manage settings such as **Exploit Protection** and **SmartScreen**, **Windows Defender Application Control** and **Credential Guard**, and firewall and antimalware features such as **Microsoft Defender for Endpoint**. We will also learn how to implement additional system security configurations using **Group Policies** and configuration baselines.

Chapter 3, Securing a Hybrid Active Directory (AD) Infrastructure, helps you learn how to apply appropriate layers of security to protect Active Directory domain controllers against attack while allowing for continued productivity and secure workloads. You will also learn how to manage protected users, about the delegation of privileges and administrators, how to secure administrative workflows, about the authentication of and to domain controllers, and how to successfully implement and manage Microsoft Defender for Identity.

Chapter 4, Identifying and Remediating Windows Server Security Issues Using Azure Services, helps you learn how to successfully monitor virtual machines running both on-premises and in Azure using Azure Arc, Azure Monitor, and Microsoft Sentinel, allowing for telemetry and metrics insights, analysis, and response. We will also cover how to onboard devices into Microsoft Defender for Cloud so that we can proactively identify and remediate security issues wherever the virtual machine may be running within the infrastructure.

Chapter 5, Secure Windows Server Networking, covers how to configure and manage Windows Defender Firewall, how to successfully plan for and configure domain isolation, and how to implement connection security and authentication request rules for your Windows servers.

Chapter 6, Secure Windows Server Storage, helps you discover how to properly secure Windows Server storage to help protect against data theft, exposure, and ransomware.

Chapter 7, Implementing a Windows Server Failover Cluster, covers how to successfully establish the building blocks for a Windows Server failover cluster, where you will learn how to configure various storage options, and successfully design and configure appropriate network settings for the failover cluster.

Chapter 8, Managing Failover Clustering, covers how to configure components such as cluster-aware updating, how to recover failed cluster nodes, and how to upgrade existing cluster nodes to a newer Windows Server version.

Chapter 9, Implementing and Managing Storage Spaces Direct, covers how to configure **Storage Spaces Direct (S2D)** and then manage S2D within a failover cluster. We will also discuss how to upgrade the S2D node, as well as implementing proper security and networking configurations for S2D, discovering topics such as converged and hyper-converged deployments.

Chapter 10, Managing Backup and Recovery for Windows Server, discusses managing backup and recovery options for Windows Server. We will cover how to install and use Azure Backup Server for general backup and recovery of files and folders. We will then discuss how to configure and use an Azure Recovery Services vault to manage the backup of files and folders using backup policies.

Chapter 11, Implementing Disaster Recovery Using Azure Site Recovery, covers how to configure Azure Site Recovery, how to create and implement a recovery plan, and how to configure recovery policies to ensure your workload remains online in the event of a planned or unplanned outage/failure.

Chapter 12, Protecting Virtual Machines by Using Hyper-V Replicas, covers how to configure Hyper-V hosts for replication, including the management of the replica servers. We will then discuss how to configure VM replication between replica hosts and ultimately perform a failover to learn about the failover orchestration process.

Chapter 13, Migrating On-Premises Storage to On-Premises Servers or Azure, identifies how to successfully transfer data and share configurations from on-premises Windows servers to other available Windows servers running on-premises or in Microsoft Azure. We will learn how to use Windows Admin Center and Storage Migration Service to migrate services from one server to another.

Chapter 14, Migrating On-Premises Servers to Azure, covers how to deploy and configure the Azure Migrate appliance. Working with the Azure Migrate appliance, we will then migrate VM workloads to Microsoft Azure IaaS, migrate physical workloads to Azure IaaS, and finally, identify additional tools that can be used within Azure Migrate to achieve your migration objectives.

Chapter 15, Migrating Workloads from Previous Versions to Windows Server 2022, covers the available tools for migrating various legacy Windows Server workloads to Windows Server 2022. We will dive into how to migrate **Internet Information Services (IIS)** workloads, Hyper-V hosts, **Remote Desktop Services (RDS)** host servers, **Dynamic Host Configuration Protocol (DHCP)**, and print servers from an older Windows Server version to Windows Server 2022.

Chapter 16, Migrating IIS Workloads to Azure, covers the available tools for migrating IIS workloads to Microsoft Azure. We will dive into how to migrate IIS workloads to Azure App Service and Azure Web Apps, and how to migrate IIS workloads to Windows containers by using Dockerfile technology.

Chapter 17, Migrating an Active Directory Domain Services (AD DS) Infrastructure to Windows Server 2022 AD DS, helps you learn how to determine the various approaches for moving domain controllers to Windows Server 2022. You will learn how to use the Active Directory Migration Tool to migrate Active Directory objects such as users, groups, and Group Policy objects. You will also learn how to migrate Active Directory objects to a new Active Directory forest.

Chapter 18, Monitoring Windows Server Using Windows Server Tools and Azure Services, covers how to use performance monitoring and data collector sets, how to monitor servers and event logs using Windows Admin Center, and how to monitor overall server health using System Insights.

Chapter 19, Troubleshooting Windows Server On-Premises and Hybrid Networking, covers how to effectively troubleshoot network connectivity for both on-premises and hybrid networking.

Chapter 20, Troubleshooting Windows Server Virtual Machines in Azure, covers how to effectively troubleshoot Windows Server virtual machine workloads in Microsoft Azure. This includes learning how to troubleshoot deployment and booting failures, VM performance and extension issues, disk encryption, storage, and overall VM connection issues.

Chapter 21, Troubleshooting Active Directory, covers how to enable and use the **Active Directory Recycle Bin** to restore deleted objects, how to use **Directory Services Restore Mode** to recover a corrupt Active Directory database and/or corrupted objects, and how to recover the **SYSVOL** folder and files necessary for running Active Directory services.

Chapter 22, Final Assessment and Mock Exam/Questions, is intended for use as additional reinforcement for the objectives reviewed in this guide, giving you a question-and-answer approach to test and validate your knowledge prior to scheduling your exam. There is also a section for lab environment housekeeping to keep your device and Microsoft Azure costs low.

To get the most out of this book

You should have a basic understanding of how to configure advanced Windows Server services utilizing existing on-premises technology in combination with hybrid and cloud technologies. This should include expertise in the management and implementation of on-premises and hybrid solutions, backup and high availability, security, networking, monitoring, the migration of workloads, and overall troubleshooting. This should also include a basic understanding of technologies such as PowerShell, Windows Admin Center, Azure Security Center, Azure Monitor, and Azure Arc.

If you are using the digital version of this book, we advise you to type the code yourself or access the code via the GitHub repository (link available in the next section). Doing so will help you avoid any potential errors related to the copying and pasting of code.

Before you start, be sure to check *Chapter 1, Exam Overview and the Current State of Cloud Workflows*, to review the technical requirements and ensure you complete the lab setup.

Download the color images

We also provide a PDF file that has color images of the screenshots/diagrams used in this book. You can download it here: https://packt.link/oG707.

Conventions used

There are a number of text conventions used throughout this book.

`Code in text`: Indicates code words in text, database table names, folder names, filenames, file extensions, pathnames, dummy URLs, user input, and Twitter handles. Here is an example: "Mount the downloaded `WebStorm-10*.dmg` disk image file as another disk in your system."

A block of code is set as follows:

```
html, body, #map {
  height: 100%;
  margin: 0;
  padding: 0
}
```

When we wish to draw your attention to a particular part of a code block, the relevant lines or items are set in bold:

```
[default]
exten => s,1,Dial(Zap/1|30)
exten => s,2,Voicemail(u100)
exten => s,102,Voicemail(b100)
exten => i,1,Voicemail(s0)
```

Any command-line input or output is written as follows:

```
$ mkdir css
$ cd css
```

Bold: Indicates a new term, an important word, or words that you see onscreen. For example, words in menus or dialog boxes appear in the text like this. Here is an example: "Select **System info** from the **Administration** panel."

> **Tips or important notes**
> Appear like this.

Get in touch

Feedback from our readers is always welcome.

General feedback: If you have questions about any aspect of this book, mention the book title in the subject of your message and email us at customercare@packtpub.com.

Errata: Although we have taken every care to ensure the accuracy of our content, mistakes do happen. If you have found a mistake in this book, we would be grateful if you would report this to us. Please visit www.packtpub.com/support/errata, selecting your book, clicking on the Errata Submission Form link, and entering the details.

Piracy: If you come across any illegal copies of our works in any form on the Internet, we would be grateful if you would provide us with the location address or website name. Please contact us at copyright@packt.com with a link to the material.

If you are interested in becoming an author: If there is a topic that you have expertise in and you are interested in either writing or contributing to a book, please visit authors.packtpub.com.

Share Your Thoughts

Once you've read *Configuring Windows Server Hybrid Advanced Services Exam Ref AZ-801*, we'd love to hear your thoughts! Scan the QR code below to go straight to the Amazon review page for this book and share your feedback.

https://packt.link/r/1804615099

Your review is important to us and the tech community and will help us make sure we're delivering excellent quality content.

Download a free PDF copy of this book

Thanks for purchasing this book!

Do you like to read on the go but are unable to carry your print books everywhere?

Is your eBook purchase not compatible with the device of your choice?

Don't worry, now with every Packt book you get a DRM-free PDF version of that book at no cost.

Read anywhere, any place, on any device. Search, copy, and paste code from your favorite technical books directly into your application.

The perks don't stop there, you can get exclusive access to discounts, newsletters, and great free content in your inbox daily

Follow these simple steps to get the benefits:

1. Scan the QR code or visit the link below

https://packt.link/free-ebook/9781804615096

2. Submit your proof of purchase
3. That's it! We'll send your free PDF and other benefits to your email directly

Part 1: Exam Overview and the Current State of On-Premises, Hybrid, and Cloud Workflows

This section will focus on the overall exam objectives, including what to expect on exam day, as well as a high-level overview of the current state of the technology stacks encountered on on-premises, hybrid, and cloud platforms and how these services can work collectively together to help you and your business succeed.

This part of the book comprises the following chapter:

- *Chapter 1, Exam Overview and the Current State of Cloud Workloads*

1

Exam Overview and the Current State of Cloud Workloads

You have decided to step up to the plate and take the AZ-801 exam, but what should you do next to achieve your desired Microsoft Certification? The AZ-801 exam focuses on the configuration and management of Windows Server on-premises workloads, hybrid workloads, and **Infrastructure-as-a-Service** (**IaaS**) platform workloads. This chapter will set you up for success by helping you to prepare for your Microsoft exam, helping you to identify additional community resources for training, and establishing a lab and trial Microsoft 365 subscription to reinforce skills with hands-on learning.

Completing this chapter will arm you with the skills and tools to successfully prepare for the AZ-801 exam, complete additional lab exercises throughout this book, and ultimately achieve the Configuring Windows Server Hybrid Advanced Services certification. In this chapter, we will cover the following topics:

- Technical requirements and lab setup
- Preparing for a Microsoft exam
- Resources available and utilizing Microsoft Learn for supplemental experience
- Creating a Microsoft 365 trial subscription and a free Azure account
- Setting up a lab environment using Hyper-V
- Exam objectives
- Who should take the AZ-801 exam?
- The current state of on-premises, hybrid, and cloud workloads

Technical requirements and lab setup

To successfully follow along and complete tasks and exercises throughout this book, you will need to first gain access to an **Azure Active Directory** (**Azure AD**) tenant. Establishing this access can be accomplished by getting a trial subscription of Microsoft 365. Advanced topics throughout this book will require an Azure AD Premium license and may require the purchase of features or services in a Microsoft Azure pay-as-you-go subscription to complete the exercises.

In addition, you will need to establish a Hyper-V lab. This can be accomplished by ensuring your desktop/laptop meets the Hyper-V system requirements, as well as completing setup and configuration for the device. The detailed steps to establish both the Microsoft 365 trial subscription and the Hyper-V lab requirements will be covered later in this chapter.

Preparing for a Microsoft exam

Microsoft certification exams follow a role-based approach, with a focus on areas of tech based on either your current role or a role that you would like to pursue. Two incredible resources that can be used to plan both your certification path as well as your career progression are found at the following links:

- `https://aka.ms/RoleBasedCerts`
- `https://aka.ms/TrainCertPoster`

There is a multitude of ways to prepare for any Microsoft exam and each of us approaches this differently, from hands-on learning, microlearning, and reading books to taking practice evaluations or instructor-led courses to help grow our current skillsets. I recommend the following three-step process for preparing for Microsoft exams, so let's review the steps together.

Step 1 – Reviewing the exam skills outline

When any new Microsoft certification exams are published to the public, one of the first questions I ask myself is *is this exam right for me and my career?* Starting with the exam page for the exam you are pursuing, you will quickly find details at the top of the page indicating what skills and expertise you should have or want to gain. Let's start by reviewing the details for the AZ-801 exam together by navigating to the following link: `https://docs.microsoft.com/learn/certifications/exams/az-801`.

This screenshot shows the details for the AZ-801 Microsoft certification exam:

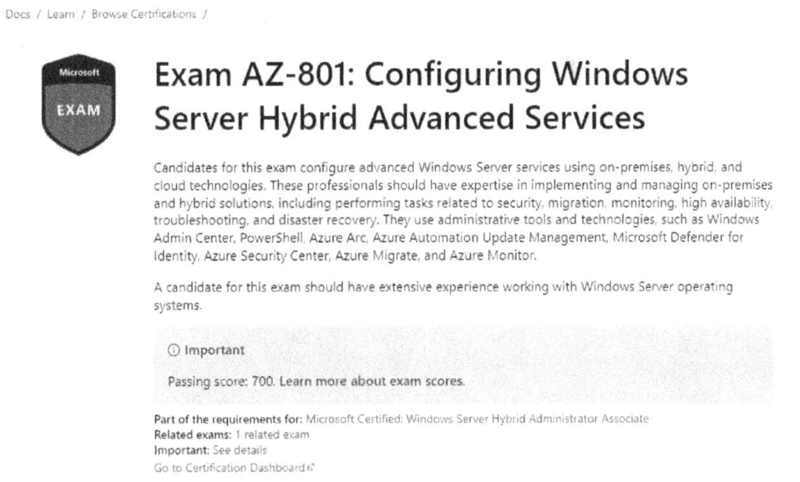

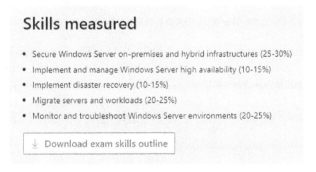

Docs / Learn / Browse Certifications /

Exam AZ-801: Configuring Windows Server Hybrid Advanced Services

Candidates for this exam configure advanced Windows Server services using on-premises, hybrid, and cloud technologies. These professionals should have expertise in implementing and managing on-premises and hybrid solutions, including performing tasks related to security, migration, monitoring, high availability, troubleshooting, and disaster recovery. They use administrative tools and technologies, such as Windows Admin Center, PowerShell, Azure Arc, Azure Automation Update Management, Microsoft Defender for Identity, Azure Security Center, Azure Migrate, and Azure Monitor.

A candidate for this exam should have extensive experience working with Windows Server operating systems.

ⓘ Important

Passing score: 700. Learn more about exam scores.

Part of the requirements for: Microsoft Certified: Windows Server Hybrid Administrator Associate
Related exams: 1 related exam
Important: See details
Go to Certification Dashboard ↗

Figure 1.1 – Exam AZ-801 overview page

This AZ-801 exam is a part of the requirements for the **Microsoft Certified: Windows Server Hybrid Administrator Associate** certification, and it requires a related AZ-800 exam to achieve the Associate certification. Note that we will focus only on the exam objectives of the AZ-801 exam.

Step 2 – Review the skills measured and practice to improve your skills

As you continue to scroll down the exam page, you will find the **Skills Measured** section, giving the exam participant an overall view of what high-level skills are being validated. Each objective shown in the following screenshot is weighted on a scale and, depending on the type of question, may span multiple objectives in the same question:

Skills measured

- Secure Windows Server on-premises and hybrid infrastructures (25-30%)
- Implement and manage Windows Server high availability (10-15%)
- Implement disaster recovery (10-15%)
- Migrate servers and workloads (20-25%)
- Monitor and troubleshoot Windows Server environments (20-25%)

↓ Download exam skills outline

Figure 1.2 – Exam AZ-801 skills measured

Step 3 – Identify additional learning materials

Scrolling further down on the exam page for AZ-801, you will find an additional section that provides two ways to prepare for your exam – online (free) learning through Microsoft Learn and instructor-led (paid) learning. These two separate resources provide great value, depending on the level of effort you are willing to put forth or spend on additional professionally moderated training:

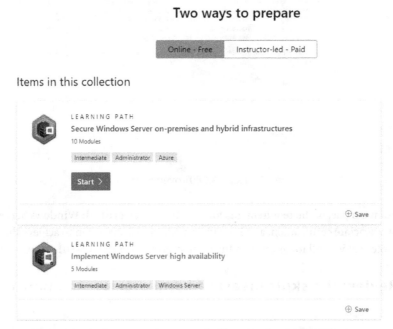

Figure 1.3 – Online versus instructor-led learning for the AZ-801 exam

The next section will provide additional details and guidance on free resources provided by Microsoft to prepare you for your exam experience, as well as any supplemental content.

Exam format and exam demo experience

Microsoft certification exams have recently released an online exam demonstration experience to be more inclusive and help reduce the anxieties of preparing for and sitting an exam. This new online experience gives the exam candidate a real-world look into the type of exam questions that appear and gives the candidate a feel for how the questions work and function. Let's take a closer look by visiting the following URL: `https://aka.ms/examdemo`.

Microsoft certification exams comprise the following types of questions:

- **Multiple choice** questions contain multiple answer choices and have only one correct answer. These types of questions may have several varieties and may appear to you as part of a series or a set.

- **Multiple choice, multiple select** questions require one or more selections and indicate how many answers to select for providing the answer.

- **Drag and drop** questions may include text, graphics, or code. Interacting with these questions requires you to drag or move objects to target locations in the answer area. One, many, or all of the available answers may be used to successfully complete these questions.

- **Build, list, reorder** questions contain a list of movable actions or answer choices that must be arranged in the correct order in the answer area. Again, one, many, or all of the available answers may be used to successfully complete these questions.

- **Active screen** questions contain interactive screen elements such as menus, option buttons, or boxes to select. Several variations of these questions include selecting a drop-down element to complete a statement or section of code, or changing the up/down arrows to increase or decrease the answer value.

- **Hot area** questions contain one or multiple clickable areas where the selections represent the correct answer. Think of these questions as an interactive screen, menu, or configuration page where you actively select the correct actions.

- The preceding question types may be grouped into a **case study**. You will need to answer questions based on case information, which can be reviewed from the left-side menu option, and will typically contain an overview, environment details, and requirements section.

- **Exhibit** questions typically contain a screenshot, design document, or configuration page where you need to review and identify either a problem or a resolution to the question. Exhibit questions may also have multiple exhibit tabs, so you must select each tab to review all the presented material to correctly answer the question.

- **True/false** and **best-answer scenario** questions will really test your understanding and mastery of the exam content. These are typically delivered in a sequence with no option of navigating back to a previous question. The question will be providing you with a specific scenario, asking you whether the recommended answer appropriately solves the question or issue presented.

For an additional review of exam features, visit the following URL: `https://aka.ms/ExamFeatures`. The next section will cover how to best utilize Microsoft Learn for any supplemental experience in addition to this book.

Resources available and utilizing Microsoft Learn for supplemental experience

Microsoft Learn combined with Microsoft Docs provides an enormous amount of free content and value to help you prepare for both the exam and your career progression.

Let's begin by reviewing how to access Microsoft Learn and locate learning paths directed at the AZ-801 exam candidates. Accessing Microsoft Learn can be done through the following link: `https://aka.ms/learn`. Creating a free account can be accomplished by clicking the **Sign in** link, as shown in the following screenshot:

Figure 1.4 – Microsoft Learn website sign in

After selecting **Sign in**, you have the option of creating a new account if you do not already have one, or using an existing account as seen here:

Figure 1.5 – Creating a new Microsoft account or using an existing Microsoft account for Microsoft Learn

Selecting the following link will quickly take you to a curated Microsoft Learn learning path that covers both the AZ-800 and AZ-801 exams: `https://aka.ms/AzureLearn_WindowsServerHybridAdmin`.

Where and when to take the exam

The classic IT answer comes to mind when people ask where and when they should take the exam – *it depends*! When you feel that you have successfully mastered the skills based on the exam outline and have finished the exercises and materials in this book, it would be a good time to consider scheduling the exam. You can take the exam either at a proctored exam site near you or remotely from your home or office.

Both options have their pros and cons, depending on the comfort level of the exam candidate. There are some important preparations that need to occur from a remote proctor's perspective before considering scheduling this exam to be taken at your home or office. Let's review the steps together to prepare for a remote proctor.

Your device must meet the requirements of having a webcam, a microphone, speakers, and a stable internet connection (wired is preferred). You will also need to stop or disable any software that is unnecessarily running on your device (for instance, PowerToys, screen capture tools, VPN software, and other productivity tools). You will need a mobile device to complete a remote check-in process, which involves taking photos of your form of identification, as well as the surrounding area in which you will be taking the exam. Your work area and walls must be relatively free of any pictures, papers, pens, phones, and books. A word of caution – you will need to remain quiet and stay within the frame of the camera throughout the exam so as not to risk disqualification from the exam attempt.

Creating a Microsoft 365 trial subscription and free Azure account

Microsoft offers trial subscriptions for both Microsoft 365 and Microsoft Azure, and this is arguably one of the best ways to get hands-on experience with Microsoft products before implementing them in a customer or business environment. This also gives you the advantage of being more marketable and advances your career by giving you skillsets that prospective employers are seeking in an ever-evolving world.

To follow along with the labs and exercises in this book surrounding Microsoft 365 and Microsoft Azure, we must first establish a subscription to both Microsoft 365 and Microsoft Azure AD Premium to learn about the advanced enterprise-licensed features within the exam objectives. The following steps will help you create 30-day trials to help get you started:

1. Navigate to `https://www.microsoft.com/microsoft-365/enterprise/compare-office-365-plans` and select the **Try for free** option under the Office 365 E5 licensing plan, as shown in the following screenshot:

Figure 1.6 – Selecting an Office 365 E5 trial subscription

2. Complete the walk-through to either create an account or utilize a different email address to create a free trial:

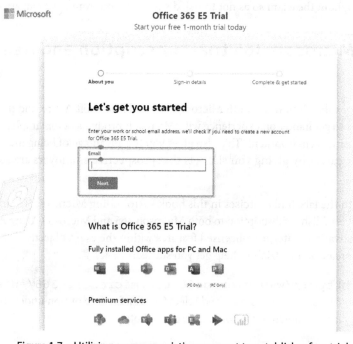

Figure 1.7 – Utilizing a new or existing account to establish a free trial

3. Once you have completed all the necessary steps to create your new Microsoft 365 and Microsoft Azure tenant, you will have access to the initial Microsoft 365 services. We will need to complete additional steps for an additional license to gain all the Microsoft 365 and Microsoft Azure services necessary to follow along with this book and complete the exercises.

    We will also need to access an Azure AD Premium license for all the advanced identity and security features that we are reviewing in the exam objectives and in this book. We will navigate to https://www.microsoft.com/microsoft-365/enterprise-mobility-security/compare-plans-and-pricing:

Figure 1.8 – Signing up for Enterprise Mobility + Security E5

4. As we are adding this license to the Microsoft 365 E5 services we just set up, you will want to use the same email address that you used to establish the Office 365 E5 subscription in the following example:

Enterprise Mobility + Security E5 Trial
Start your free 3-month trial today

○	○	○
About you	Sign-in details	Complete & get started

Let's get you started

Enter your work or school email address, we'll check if you need to create a new account for Enterprise Mobility + Security E5 Trial.

Email

This is required

Next

Figure 1.9 – Signing up for Enterprise Mobility + Security E5, continued

We now have all the Microsoft 365 and Microsoft Azure services necessary to follow along with this book and complete the exercises. The next section will provide steps to build out a local Hyper-V lab so that we have everything we need for our hands-on experience and exam preparation based on the exam objectives.

Setting up a lab environment using Hyper-V

Hyper-V can be enabled in numerous ways on a Windows 10/11 operating system. With Hyper-V features built into Windows, this becomes an easy task to complete with only minor configuration changes needed. We will be running Hyper-V inside of a nested virtualization environment, and deeper details on this approach can be found at `https://docs.microsoft.com/virtualization/` `hyper-v-on-windows/user-guide/nested-virtualization`.

For the most straightforward and repeatable approach, we will be using instructions that help you enable Hyper-V using the Windows PowerShell method. Let's begin the configuration using the following steps:

1. We must first validate that the device we are using meets the following hardware requirements:

    *   The operating system must be running either Windows 10 or Windows 11 Enterprise, Pro, or Education versions; Windows Server (version 2012 R2 or later) can also be used

    *   The installed 64-bit processor (either AMD or Intel) supports **Second Level Address Translation (SLAT)**

    *   The installed CPU supports the VM Monitor Mode Extension (called VT-x on Intel-based CPUs)

    *   A minimum of 4 GB of memory installed

    > **Important note**
    > Virtual machines share memory with the Hyper-V host computing environment, so you will need enough memory to handle the workload for these labs. We recommend at least 8 GB of memory to successfully complete the lab exercises in this book.

    *   The following configurations will need to be enabled in the system BIOS (each manufacturer will have a slightly different way to enter the boot/BIOS configuration menus, so it is best to search for and identify configuration instructions pertaining to your specific device model):

        *   A feature called **Virtualization Technology** (each motherboard manufacturer may have a different label for this, e.g., *Intel ® Virtualization Technology* or *Intel VT-d* where appropriate)

        *   A feature called **Hardware Enforced Data Execution Prevention**

2. Once the above requirements are met, we can continue with the setup of the intended device by opening Windows PowerShell as an administrator. This can be done by opening the Windows start menu, typing `PowerShell`, and then right-clicking on the entry and selecting **Run as Administrator**. You may need to select **Yes** to accept the user account control popup to continue.

3. Enter the `systeminfo` command and hit *Enter* to run the command:

    -   At the bottom of the returned result set, if you receive the message shown in *Figure 1.10* saying `A hypervisor has been detected. Features required for Hyper-V will not be displayed`, then you already have Hyper-V installed and can skip to *Step 6* to continue network switch configuration for our Hyper-V lab:

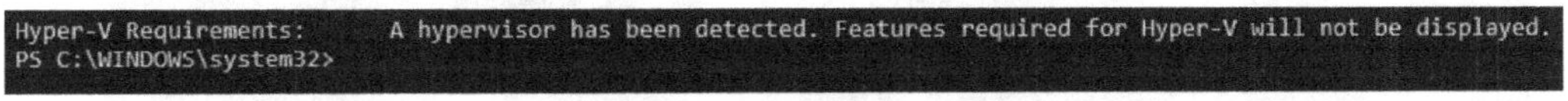

```
Hyper-V Requirements:    A hypervisor has been detected. Features required for Hyper-V will not be displayed.
PS C:\WINDOWS\system32>
```

Figure 1.10 – Hyper-V requirements indicating that a hypervisor has been detected

    -   At the bottom of the result set returned, if you receive a list of four requirements as shown in *Figure 1.11*, confirm that all requirements result in a **Yes** response validating that all requirements for Hyper-V have been met on this device:

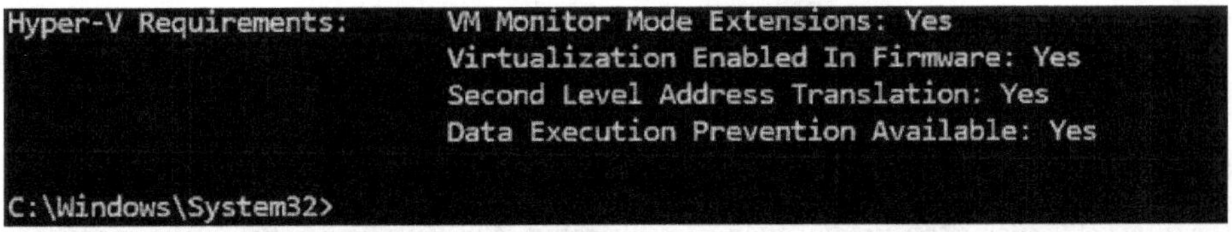

```
Hyper-V Requirements:    VM Monitor Mode Extensions: Yes
                         Virtualization Enabled In Firmware: Yes
                         Second Level Address Translation: Yes
                         Data Execution Prevention Available: Yes

C:\Windows\System32>
```

Figure 1.11 – Hyper-V requirements indicating that a hypervisor has not been detected

4. Reusing the already open administrative Windows PowerShell, enter the following command to enable the Hyper-V components on your device:

```
Enable-WindowsOptionalFeature -Online -FeatureName
Microsoft-Hyper-V -All
```

> **Important – when using Windows Server for your Hyper-V lab**
>
> If you choose to use Window Server as your operating system for this Hyper-V lab environment, you can utilize the following command to install *both* Hyper-V and the Hyper-V Management Console for use on Windows Server:
>
> ```
> Install-WindowsFeature -Name Hyper-V -IncludeManagementTools
> ```

5. When you are prompted to reboot the device, please choose **Yes**. Note that you may be required to reboot the device multiple times to complete the setup and configuration.

6. If this is the first time you have installed Hyper-V on this device, you can proceed to *Step 7*:

    - If you are unsure *or* have an existing virtual switch configured, reuse the already open administrative Windows PowerShell and enter the following command:

    ```
    Get-VMSwitch
    ```

    - Carefully review the results returned, per *Figure 1.12*, to determine whether any of the existing virtual switches have a `SwitchType` value of `External`:

Figure 1.12 – Running the Get-VMSwitch PowerShell command to determine the existing virtual switches

    - Capture the name of the existing virtual switch that has a `SwitchType` value of `External` – we will use this later in multiple important configuration changes in *Step 9* that are specific to your configuration. Please skip to *Step 9* to continue our configuration.

7. Next, we will determine the network adapter for the device that you will utilize for connecting to the internet. Reusing the already open administrative Windows PowerShell, enter the following command and review the results returned in *Figure 1.13*:

    ```
    (Get-NetAdapter | Where-Object {$_.Status -eq "Up" -and
    !$_.Virtual}).Name
    ```

    Issuing the previous `Get-NetAdapter` command will present the results shown in *Figure 1.13*:

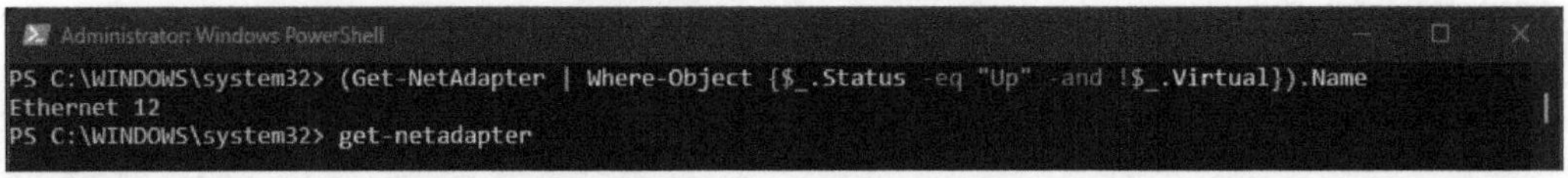

Figure 1.13 – Running a PowerShell command to determine a network adapter connected to the internet

    The results in your open Windows PowerShell instance should successfully identify the name of the current network interface you are using for your internet connection on this device. Verify that this is the interface that you want to use for this lab setup.

Optional – If you intend to use another network interface for this internet connection, enter the following command in Windows PowerShell and review the results returned in *Figure 1.14*:

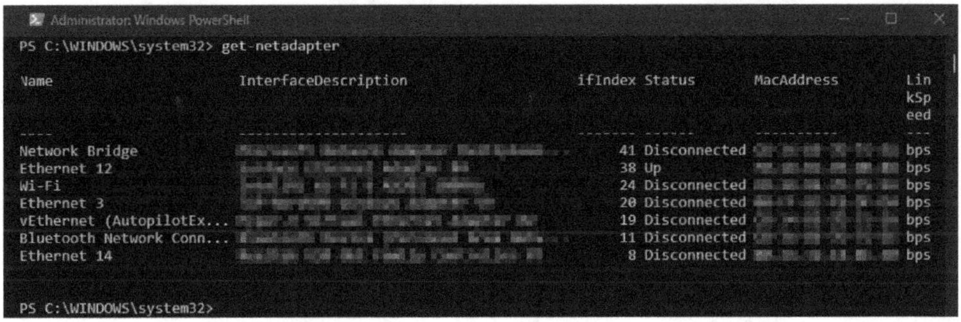

Figure 1.14 – Running a PowerShell command to determine a network adapter connected to the internet

In the preceding instance, `NetAdapterName` was identified as **Ethernet 12**, but we may want to use **Ethernet 14** instead. In this case, we will update the following command to be used later in *Step 9*:

```
New-VMSwitch -Name AZ801PacktLabExternal
 -AllowManagementOS $true -NetAdapterName Ethernet 14
```

8. Visit `https://www.microsoft.com/evalcenter/evaluate-windows-server-2022-preview` to download an evaluation copy of Windows Server 2022. The following steps will help you ensure that the newly downloaded evaluation software is safely unblocked for use on your system:

  *   Once the ISO file has finished downloading, you will need to rename the ISO to `Server2022.iso` for our scripted lab deployment.

  *   Locate and right-click on the newly downloaded file and select **Properties**. In the properties, click the **General** tab near the bottom. There is a security section where you will select the checkbox for **Unblock**. Select **OK** to continue:

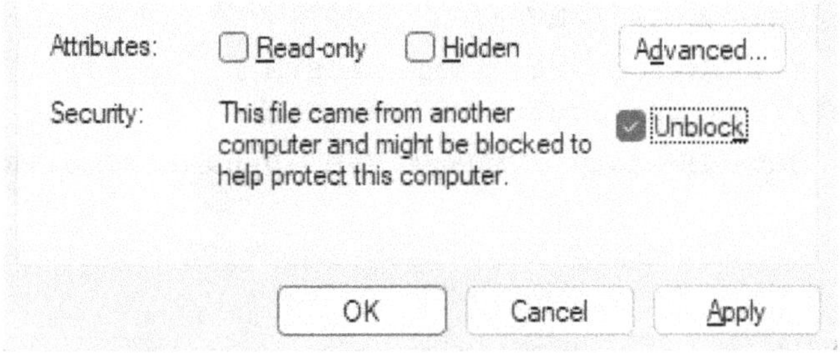

Figure 1.15 – Unblocking a recently downloaded ISO image file for use in our lab

- Copy the downloaded file into `C:\AZ801PacktLab\iso`.

9. Use PowerShell to create a demo VM:

   - We will be utilizing the following PowerShell script to automatically build out all required network, disk, and virtual machine requirements for our AZ-801 local Hyper-V lab (this script is also located at `https://github.com/cgill/Books/blob/main/AZ-801/CH1/CH1-BuildHVLab.ps1` for a copy-paste approach):

```
mkdir C:\AZ801PacktLab
Set-Location C:\AZ801PacktLab


$VMName='AZ801PacktLab-DC-01','AZ801PacktLab-HV-
01','AZ801PacktLab-HV-02','AZ801PacktLab-FS-01'
$VMExternalSwitch='AZ801PacktLabExternal'
$VMInternalSwitch='AZ801PacktLabInternal'
$VMIso='c:\AZ801PacktLab\iso\Server2022.iso'


New-VMSwitch -Name $VMExternalSwitch -AllowManagementOS
$true -NetAdapterName (Get-NetAdapter | Where-Object {$_.
Status -eq "Up" -and !$_.Virtual}).Name


New-VMSwitch -name $VMInternalSwitch -SwitchType Internal


Foreach ($VM in $VMName) {
    New-VM -Name $VM -MemoryStartupBytes 2GB
-BootDevice VHD -NewVHDPath ".\VMs\$VM.vhdx" -Path .\
VMData -NewVHDSizeBytes 40GB -Generation 2 -Switch
$VMInternalSwitch


    if($VM -match '-HV-') {
        Set-VMProcessor -VMName $VM
-ExposeVirtualizationExtensions $true
        Set-VMMemory -VMName $VM -DynamicMemoryEnabled
```

```
$false
        Get-VMNetworkAdapter -VMName $VM |
Set-VMNetworkAdapter -MacAddressSpoofing On
    }


    Add-VMScsiController -VMName $VM
    Add-VMDvdDrive -Path $VMIso -VMName $VM
-ControllerNumber 1 -ControllerLocation 0


    $VMDvd = Get-VMDvdDrive -VMName $VM


    Set-VMFirmware -VMName $VM -FirstBootDevice $VMDvd


    Set-VMKeyProtector -NewLocalKeyProtector -VMName $VM
    Enable-VMTPM -VMName $VM

}
```

> **Hyper-V switch management and pre-existing virtual switches**
>
> Please note that if you have already been using Hyper-V on the device where you intend to build our lab environment, you will want to consider either removing your existing Hyper-V external switch or making the necessary script adjustments for the existing virtual switch name in your configuration.

Now that we have created all of the necessary virtual machines, we will need to install the Windows Server 2022 operating system on all of our virtual machines. Let's begin the configuration using the following steps:

1. Install Windows Server 2022 on the VM:

    -   Open the Windows Start menu and search for Hyper-V Manager, then select **Open**. You will be presented with a list of virtual machines that we have previously created via a script.

- Connect to one of our four virtual machines by right-clicking and selecting **Connect…**, as displayed in the following screenshot:

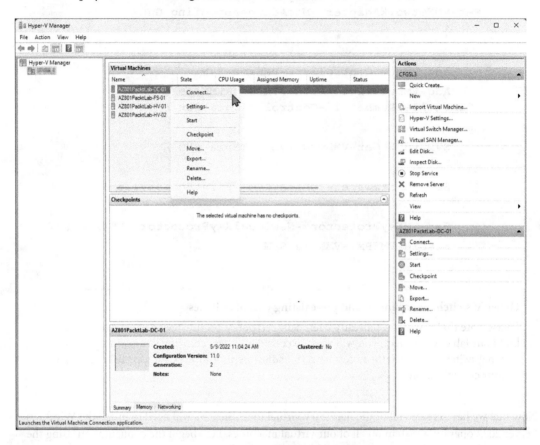

Figure 1.16 – Selecting a virtual machine to start in Hyper-V Manager

- From the **Virtual Machine Connection** window, we can either select the **Start** button in the center of the screen or select the green **Start** button in the taskbar at the top of the screen:

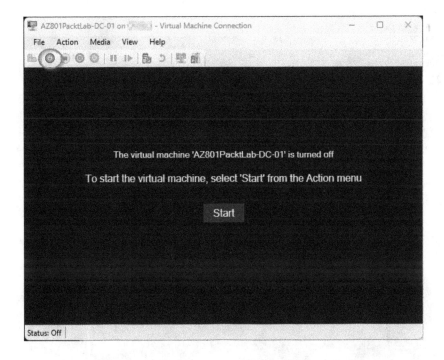

Figure 1.17 – Starting a virtual machine inside of the Virtual Machine Connection console

2. Ensure that you press a key inside **Virtual Machine Connection** to begin the operating system's installation quickly after the virtual machine boots. If you do not receive a **Microsoft Server Operating System Setup** screen, you will need to reset the VM and attempt to boot into the DVD again to begin the installation.

3.   In **Microsoft Server Operating System Setup**, select **Next** to continue:

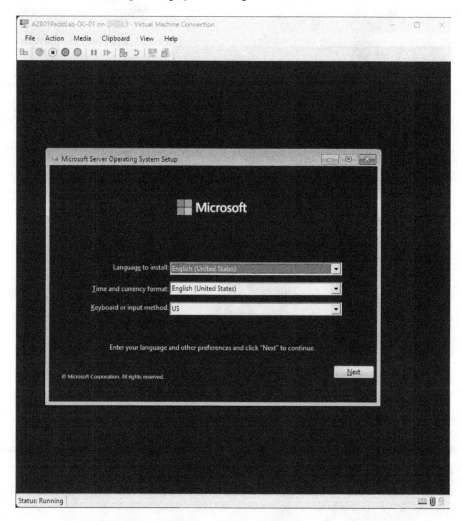

Figure 1.18 – Starting the Microsoft Server 2022 operating system setup

4.   On the following page, select **Install Now**:

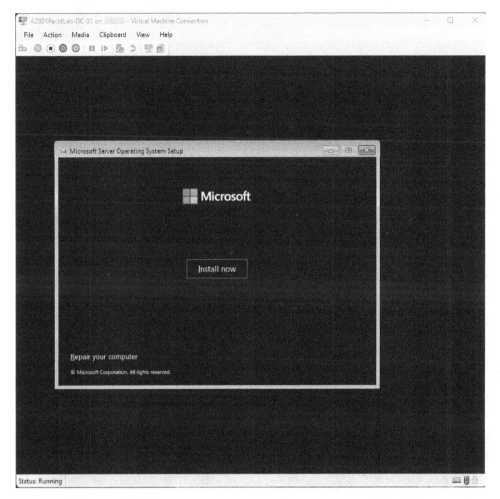

Figure 1.19 – Continuing the Windows Server 2022 operating system installation

5. In **Activate Microsoft Server Operating System Setup**, select **I don't have a product key** to continue our installation as shown in the following screenshot. Please note that these evaluation virtual machines will have 180 days before they are no longer licensed under the evaluation period:

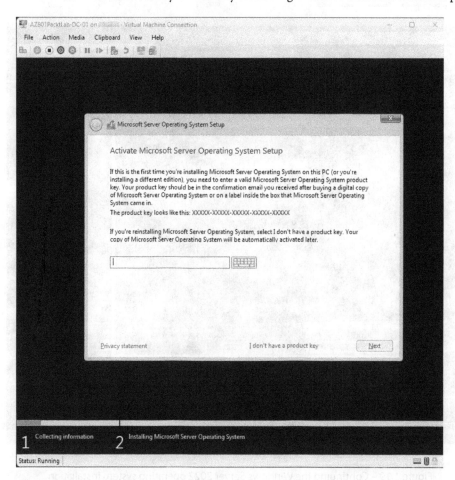

Figure 1.20 – Selecting to use an evaluation version in the Windows Server 2022 installation

6. On the **Select the operating system you want to install** screen, change your selection to **Windows Server 2022 Datacenter (Desktop Experience)**, and select **Next** to continue:

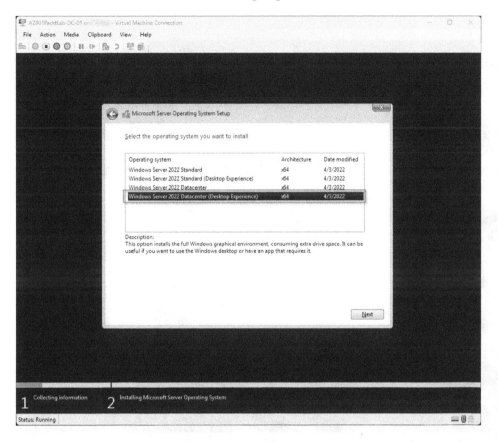

Figure 1.21 – Selecting to install the Windows Server 2022 Datacenter (Desktop Experience)

7. On the **Applicable notices and license terms** page, select **I accept…** and then select **Next** to continue:

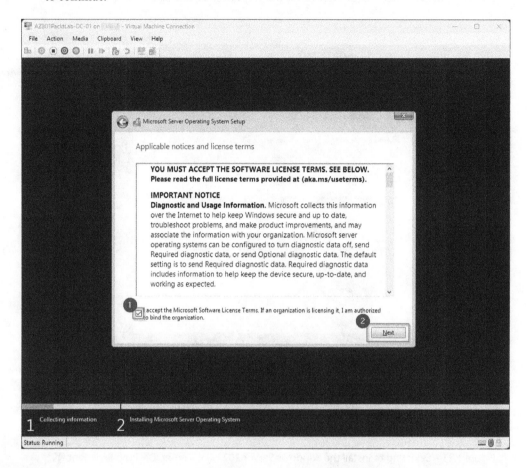

Figure 1.22 – Accepting the Microsoft Software license terms

8.   On the **Which type of installation do you want?** page, select **Custom: Install Microsoft Server Operating System only (advanced)** as shown in the following screenshot:

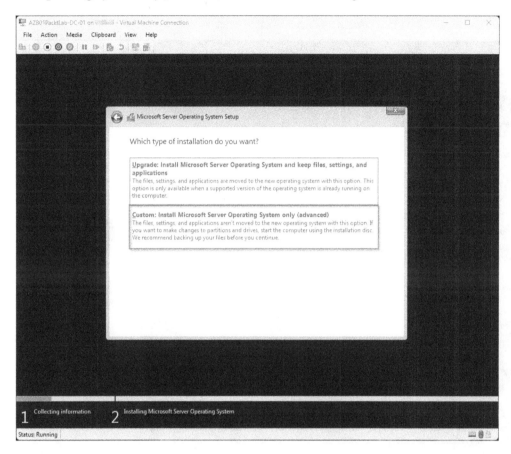

Figure 1.23 – Selecting Custom: Install Microsoft Server Operating System only (advanced)

9.   Select **Next** on the **Where do you want to install the operating system?** page:

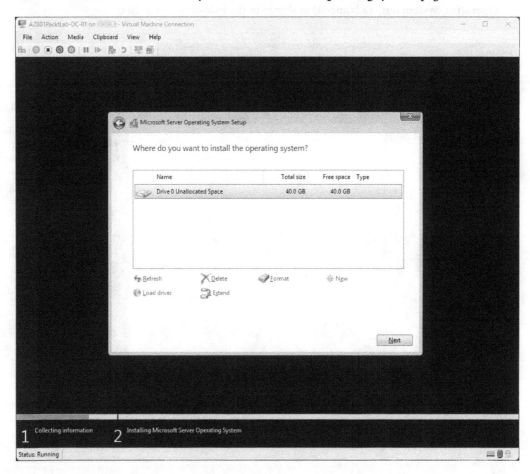

Figure 1.24 – Installing the operating system on the available disk

10. Monitor the **Status** page to confirm that the installation succeeds:

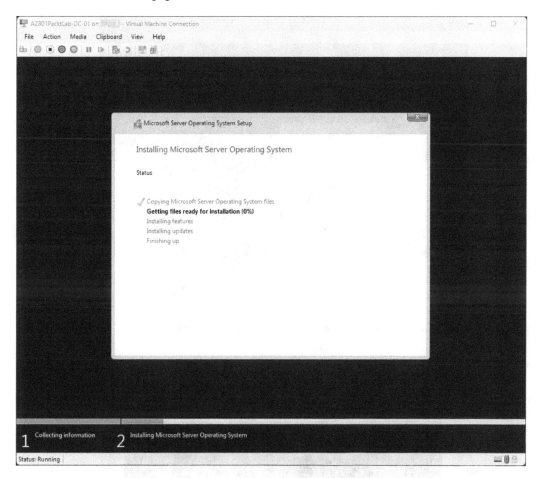

Figure 1.25 – Installing Microsoft Server Operating System

11. After the virtual machine reboots a few times, you will return to the **Customize settings** screen and be prompted for an administrator passcode. We recommend you use `Packtaz801guiderocks` as a passcode for all your lab virtual machines. You may also create your own passcode for these lab machines; just be sure to record the passcode in a password keeper for later use in the lab exercises:

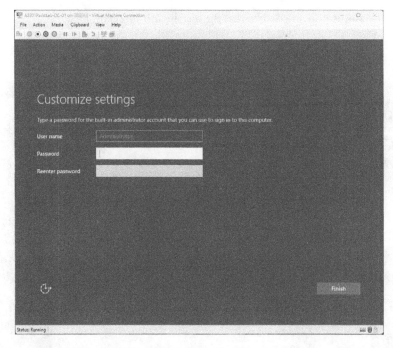

Figure 1.26 – Setting a new password for the administrator account

12. When prompted to adjust the display configuration, choose an appropriate resolution for your virtual machine, and then select **Connect**, as shown in the following screenshot:

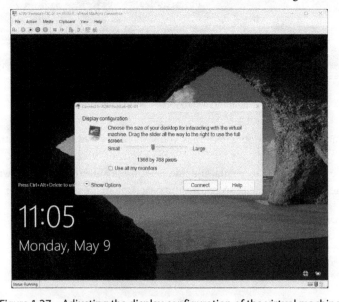

Figure 1.27 – Adjusting the display configuration of the virtual machine

13. Power down the virtual machine.

14. Repeat the preceding steps for each of the four virtual machines within Hyper-V Manager to complete the lab setup.

This concludes the initial Hyper-V lab setup for use throughout the rest of the book and will position you for success throughout all the lab exercises. There will be some additional steps needed throughout the following chapters in this book, and these steps will be detailed in the technical requirements at the beginning of each chapter.

Exam objectives

This is a very broad and deep exam covering the administration, design, and security surrounding on-premises workloads, hybrid workloads and services, and cloud-native applications and infrastructure. This book will cover the specific objectives of the exam guide and outline, while reinforcing hands-on skills in addition to reading and general exercises.

The objectives covered in the AZ-801 exam are listed in the following table:

Overall objective	Weight (in percent or percent range)
Secure the Windows server on-premises and hybrid infrastructures	25 – 30%
Implement and manage the Windows server's high availability	10 – 15%
Implement disaster recovery	10 – 15%
Migrate servers and workloads	20 – 25%
Monitor and troubleshoot Windows server environments	20 – 25%

Table 1.1 – Overall exam objectives

A more complete exam skills outline that covers these objectives in greater detail can be found at https://aka.ms/az-801examguide. Taking and reviewing this comprehensive exam skills outline allows you to review areas of comfort, while also helping to identify and highlight areas of improvement prior to you sitting the exam.

Who should take the AZ-801 exam?

This exam is intended for cloud and data center management administrators and engineers, enterprise architects, Microsoft 365 administrators, network engineers, and anyone seeking to gain additional working knowledge with Windows Server operating systems and managing on-premises, hybrid, and cloud workloads with administrative tools.

The current state of on-premises, hybrid, and cloud workflows (and how they can work together to deliver for you and your business needs)

Every business has a customized set of goals to achieve, no matter the size or business strategy, and they are all vying to create their cloud transformation vision to achieve their initiatives with minimal disruption to the current business model. Building and translating the organization's existing digital estate and goals into a cloud adoption framework takes time and must involve every aspect of the business, including processes, procedures, current management frameworks, upskilling, operational support, and most importantly – your business culture.

In the following section, we will learn about current cloud trends. We will define each of the on-premises, hybrid, and cloud-native infrastructure models, and will discuss key factors in determining an organization's path to the cloud.

Identifying current hybrid cloud trends

In an ever-expanding digital world for businesses, cloud adoption has blossomed over the past few years and is showing no signs of slowing down; new and creative trends in utilization appear almost overnight. Businesses have hundreds or thousands of servers, applications, and digital resources and most organizations are combining their use of both private and public clouds to embrace a multi-cloud approach with this increasing cloud acceleration. Let's define what each of the infrastructure types brings to the business, and debate whether there is a one-size-fits-all approach to cloud transformation and adoption.

Defining on-premises infrastructure

The term on-premises infrastructure refers to the use of the business' own software, hardware, technology, and data that is traditionally set up and running within the four walls of the organization. This approach allows for complete control over the infrastructure, from setup to management, architecture, and security.

Defining hybrid infrastructure

The term hybrid infrastructure refers to the integration of a mixed computing environment, comprising an on-premises data center (or private cloud), private cloud services, and a public cloud. Arguably the most important component of this infrastructure is a resilient connection to both the hybrid cloud's private and public cloud computing environments.

Defining cloud-native infrastructure

The term cloud-native refers to the packaging of application code and dependencies into containers, deploying these applications as microservices or APIs, and managing them by utilizing DevOps practices, processes, and tools for a reliable and secure approach at scale. This typically includes a cloud-provided **Platform-as-a-Service (PaaS)**, **Software-as-a-Service (Saas)**, **Infrastructure-as-a-Service (IaaS)**, **Databases-as-a-Service (DbaaS)**, and even **Container-as-a-Service (CaaS)**.

On-premises versus hybrid versus cloud-native – is there a one size fits all?

This *cloud menu* approach allows businesses to consciously harness the power and agility of each of these computing components and determine when and where it assists them in their transformation while taking advantage of investments in the existing data center architecture. For businesses to be successful, they must successfully assess their current digital estate, not only for existing workloads but for optimizing cost control and management. This must also include what's called the 5Rs approach: rehost, refactor, rearchitect, rebuild, and replace.

While this is not a new digital paradigm shift, interest continues to grow in finding advancements to provide ample operational scale, decrease deployment and management speed, and increase the overall resiliency and efficiency of any business workloads running in the cloud.

No, there is no one-size-fits-all approach for cloud services. Instituting a *Crawl > Walk > Run* approach that's customized to your business needs will help to identify the right overall benefits, migration speed, priorities, and selection of services. This measurable approach will ultimately help you choose the correct approach to select services that meet you and your business where you are in your transformation efforts.

Summary

In this chapter, we covered the details of the AZ-801 Microsoft certification exam and included strategies to review the overall objectives and plan for success. We learned how the exam functions and flows, as well as learning how to prepare for the exam depending on your choice of exam location. We also worked to set up a Microsoft 365 E5 license and advanced Azure Active Directory Premium licensing for our Azure tenant lab. We then established a local Hyper-V lab on a device of our choice and established Windows Server 2022 virtual machines for use throughout the exercises in this book. Finally, we completed a cursory review of the state of the cloud and how these services help your business thrive.

In the next chapter, we will be learning about security for on-premises and hybrid infrastructures. We will start with securing the Windows Server operating system and will investigate a best practice approach to enhance the security posture of your devices utilizing the available Microsoft tools and controls, consistent with the AZ-801 exam objectives.

Part 2: Secure Windows Server On-Premises and Hybrid Infrastructures

This section will primarily focus on how to properly secure Windows Server for both on-premises and hybrid infrastructures, using security baselines, best practices, and best-of-breed security tools to protect data, communications, identity and access management, and analysis and response.

This part of the book comprises the following chapters:

- *Chapter 2, Securing the Windows Server Operating System*
- *Chapter 3, Securing a Hybrid Active Directory (AD) Infrastructure*
- *Chapter 4, Identifying and Remediating Windows Server Security Issues by Using Azure Services*
- *Chapter 5, Securing Windows Server Networking*
- *Chapter 6, Securing Windows Server Storage*

2

Securing the Windows Server Operating System

While the Windows Server operating system is designed to be secure out of the box, additional security features and controls can be implemented to enhance the security posture of your devices and computing environments. In this chapter, we will cover how to configure and manage settings such as exploit protection and SmartScreen, **Windows Defender Application Control** (**WDAC**) and Credential Guard, firewall, and antimalware features such as Microsoft Defender for Endpoint. We will also learn how to implement additional system security configurations using Group Policy and configuration baselines.

In this chapter, we are going to cover the following main topics:

- Technical requirements and lab setup
- Configuring and managing exploit protection
- Configuring and managing Windows Defender Application Control
- Configuring and managing Microsoft Defender for Endpoint
- Configuring and managing Windows Defender Credential Guard
- Configuring Microsoft Defender SmartScreen
- Implementing operating system security by using group policies

Technical requirements and lab setup

To successfully follow along and complete tasks and exercises throughout this chapter and the following chapters in this book, we will need to establish both a Windows Server 2022 domain controller and a Windows 2022 file server. While there are a few ways for administrators to configure Windows

Server 2022 as a domain controller, I believe it is beneficial to utilize PowerShell to complete this configuration simply for ease and speed of deployment. Let's begin with the following set of steps to configure your Windows Server 2022 domain controller.

Installing a new Windows Server 2022 domain controller

In *Chapter 1*, we successfully created a virtual machine named **AZ801PacktLab-DC-01** and installed the Windows Server 2022 operating system onto the virtual machine. We followed our first best practice, which is naming the domain controller appropriately so that it can be properly identified. A second prerequisite for a successful domain controller installation is to set a static IP address on the device. Let's begin the configuration using the following steps as guidance:

1.   Open **Hyper-V Manager** on your device hosting the virtual machines we created in *Chapter 1*.

2.   Locate the **AZ801PacktLab-DC-01** virtual machine in the **Virtual Machines** list as shown in *Figure 2.1*:

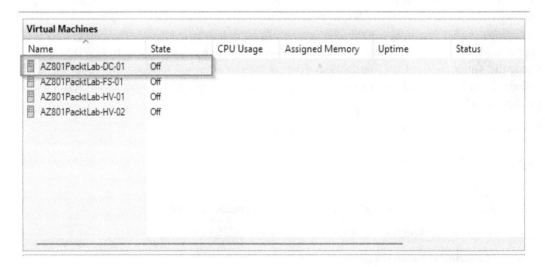

Figure 2.1 – Locating the domain controller in the list of Hyper-V virtual machines

3.   We will first need to enable guest services for this virtual machine to allow us to easily copy and paste clipboard text and configuration components onto the virtual machine from our host device. Let's begin to enable these services by right-clicking on the **AZ801PacktLab-DC-01** virtual machine and selecting **Settings**.

4.   We will then select **Integration Services** on the left and place a checkmark in front of **Guest Services** on the right. Click the **OK** button to complete this configuration task, as shown in *Figure 2.2*:

Figure 2.2 – Enabling guest services inside of our virtual machine configuration

5.   Right-click on the **AZ801PacktLab-DC-01** virtual machine and select **Connect**.

6.   From the **Virtual Machine Connection** window, we can either click the **Start** button in the center of the screen or click the green **Start button** in the taskbar at the top of the screen, as displayed in *Figure 2.3*:

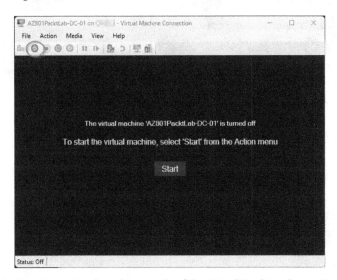

Figure 2.3 – Starting a virtual machine inside of the Virtual Machine Connection console

7. Use the **Action | Ctrl+Alt+Delete** menu options to begin, then log in to **AZ801PacktLab-DC-01** as an administrator using `Packtaz801guiderocks` as the password.

8. In the **Virtual Machine Connection** window, click on the **View** menu option. Then, select **Enhanced Session**. The connection screen will disconnect and then reconnect, allowing you to copy and paste text and files between your host device and the virtual machine.

9. Open the **Start** menu and right-click on **Windows PowerShell**, then select **More | Run as administrator**, as shown in *Figure 2.4*:

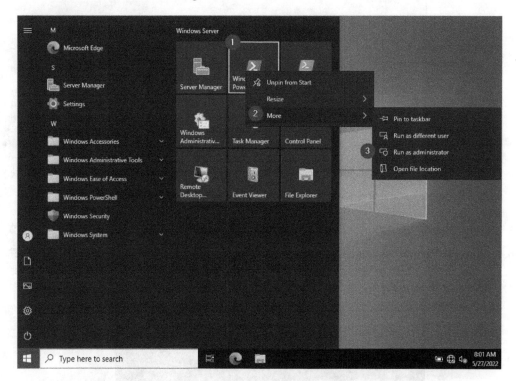

Figure 2.4 – Running Windows PowerShell as an administrator from the Start menu

10. In the open **Windows PowerShell** window, run the following command and press *Enter* to complete the configuration:

```
New-NetIPAddress -IPAddress 10.10.10.1 -DefaultGateway
10.10.10.1 -PrefixLength 24 -InterfaceIndex
(Get-NetAdapter).InterfaceIndex
```

The results should look like those returned in *Figure 2.5*:

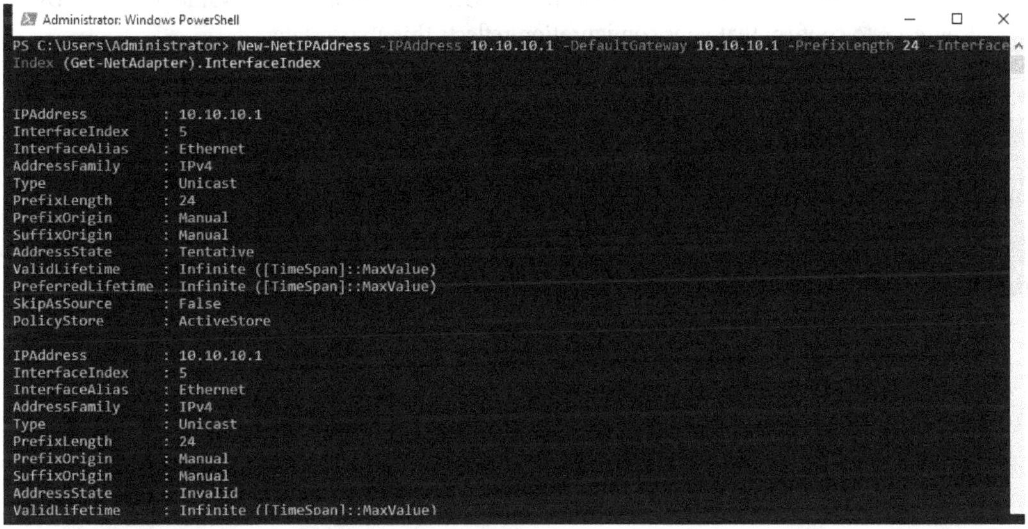

```
Administrator: Windows PowerShell                                                    —   □   ×
PS C:\Users\Administrator> New-NetIPAddress -IPAddress 10.10.10.1 -DefaultGateway 10.10.10.1 -PrefixLength 24 -Interface
Index (Get-NetAdapter).InterfaceIndex


IPAddress            : 10.10.10.1
InterfaceIndex       : 5
InterfaceAlias       : Ethernet
AddressFamily        : IPv4
Type                 : Unicast
PrefixLength         : 24
PrefixOrigin         : Manual
SuffixOrigin         : Manual
AddressState         : Tentative
ValidLifetime        : Infinite ([TimeSpan]::MaxValue)
PreferredLifetime    : Infinite ([TimeSpan]::MaxValue)
SkipAsSource         : False
PolicyStore          : ActiveStore

IPAddress            : 10.10.10.1
InterfaceIndex       : 5
InterfaceAlias       : Ethernet
AddressFamily        : IPv4
Type                 : Unicast
PrefixLength         : 24
PrefixOrigin         : Manual
SuffixOrigin         : Manual
AddressState         : Invalid
ValidLifetime        : Infinite ([TimeSpan]::MaxValue)
```

Figure 2.5 – Establishing a static IP address for our lab domain controller

11. We will now configure the virtual network adapter's DNS settings by running the following command and pressing *Enter* to complete the configuration:

```
Set-DNSClientServerAddress -InterfaceIndex
(Get-NetAdapter).InterfaceIndex -ServerAddresses
10.10.10.1
```

12. The next configuration follows the best practice of setting our domain controller to match our naming convention established in the lab. While still in the open PowerShell window, we will run the following command to rename and reboot our server:

```
Rename-Computer -NewName AZ801Lab-DC-01 -Restart -Force
-PassThru
```

13. Continuing, we will now install the **Active Directory Domain Services (AD DS)** server role to our **Domain Controller** VM. We will run the following command by pressing *Enter*:

```
Install-WindowsFeature -Name AD-Domain-Services
-IncludeManagementTools
```

14. After a few minutes, the AD DS server role will be installed, and you will see a message returned in the PowerShell window stating that the `Success` column shows a value of `True`. Use *Figure 2.6* to confirm that your configuration reflects this `True` state:

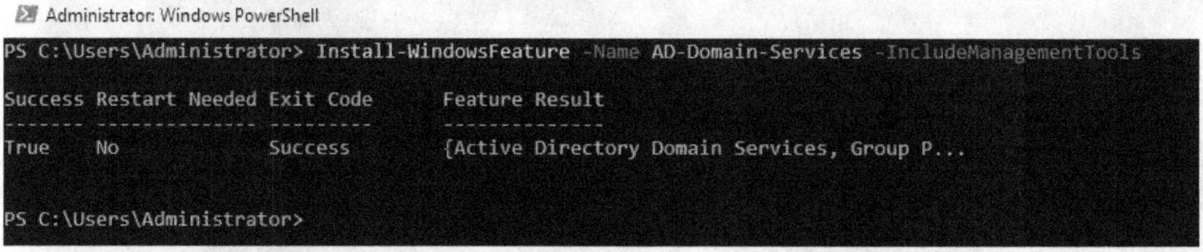

```
Administrator: Windows PowerShell
PS C:\Users\Administrator> Install-WindowsFeature -Name AD-Domain-Services -IncludeManagementTools

Success Restart Needed Exit Code    Feature Result
------- -------------- ---------    --------------
True    No             Success      {Active Directory Domain Services, Group P...


PS C:\Users\Administrator>
```

Figure 2.6 – Installing the Active Directory Domain Services role and Management Tools

15. To configure our virtual machine as a domain controller, we will now need to run `Install-ADDSForest` as shown in the following example. Copy and paste or type the following commands into Windows PowerShell, and press *Enter* to run the following commands:

```
$SafeModeAdmPwdValue = 'P@cktP@ssw0rd'
$SafeModeAdmPwd = ConvertTo-SecureString -AsPlainText
$SafeModeAdmPwdValue -Force


Install-ADDSForest -CreateDNSDelegation:$False
-DatabasePath "c:\Windows\NTDS" -DomainMode
'WinThreshold' -DomainName "AD.az801.com"
-DomainNetbiosName "AZ801" -ForestMode 'WinThreshold'
-InstallDNS:$true -LogPath "C:\Windows\NTDS"
-NoRebootOnCompletion:$false -Sysvolpath "C:\Windows\
SYSVOL" -Force:$true -SafeModeAdministratorPassword
$SafeModeAdmPwd
```

> **Note on forest functional and domain functional levels for Active Directory**
>
> Functional levels not only determine which Windows operating systems you can support on the domain controllers in your environments but also help determine the availability of certain features within **AD DS** reflective of the operating system version. For example, Windows Server 2008 R2 forest functional level brought us the AD recycle bin, while Windows Server 2012 R2 domain functional levels brought us authentication policies and authentication policy silos.
>
> The following table details the domain and forest functional levels, their AD schema versions, and the string value of each of the functional levels. Note that there have been no new domain or forest functional levels since the release of Windows Server 2016. For additional details on the domain and forest functional levels, I recommend visiting `https://docs.microsoft.com/windows-server/identity/ad-ds/active-directory-functional-levels`.

Functional level	Schema version	String
Windows Server 2003	30	`Win2003`
Windows Server 2008	44	Win2008
Windows Server 2008 R2	47	Win2008R2
Windows Server 2012	56	Win2012
Windows Server 2012 R2	69	Win2012R2
Windows Server 2016	87	WinThreshold
Windows Server 2019	88	WinThreshold
Windows Server 2022	88	WinThreshold

Note on the safe mode administrator password

The safe mode administrator password, also referred to as the safe mode password, is used in **Directory Services Restore Mode** (**DSRM**). This credential should be considered as tier 0, or super sensitive and secret within any organization, and needs to be protected as such. I highly recommend storing this in an established **privileged access management** (**PAM**) system, such as **Azure Key Vault** (**AKV**). While this book briefly touches on AKV, I recommend a quick review of what AKV has to offer by visiting this URL to get started: `https://aka.ms/AzureKeyVault`.

The DSRM functionality and the use of this credential for disaster recovery purposes will be discussed in *Troubleshooting Active Directory* in *Chapter 21*.

16. You will notice a few yellow warnings appear on the screen indicating default security settings for **Windows Server 2022**. These can be safely ignored. After a brief pause, the newly promoted domain controller will automatically reboot to complete the configuration, as shown in *Figure 2.7*:

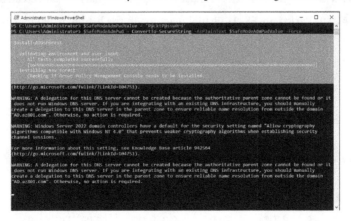

Figure 2.7 – Installing our AD forest onto the domain controller

17. Use the **Action | Ctrl+Alt+Delete** menu options, then log in to the **AZ801PacktLab-DC-01** as the administrator using `Packtaz801guiderocks` as the password.

18. Once again, open the **Start** menu and then right-click on **Windows PowerShell**. Then, select **More | Run as administrator**.

19. We will finally utilize the following PowerShell script to automatically create our necessary **Organizational Units (OUs)** and AD computer accounts for the lab:

```
New-ADOrganizationalUnit "File Servers"
New-ADOrganizationalUnit "Hyper-V Servers"
New-ADComputer -Name "AZ801Lab-FS-01"
-SAMAccountName "AZ801Lab-FS-01" -Path "OU=File
Servers,DC=AD,DC=AZ801,DC=com"
New-ADComputer -Name "AZ801Lab-HV-01"
-SAMAccountName "AZ801Lab-HV-01" -Path "OU=Hyper-V
Servers,DC=AD,DC=AZ801,DC=com"
New-ADComputer -Name "AZ801Lab-HV-02"
-SAMAccountName "AZ801Lab-HV-02" -Path "OU=Hyper-V
Servers,DC=AD,DC=AZ801,DC=com"
```

We have completed the necessary steps to configure our domain controller and prepare our lab for the next steps in this chapter and exercises in this book. In the next section, we will follow the steps to establish our Windows Server 2022 file server for use throughout this book.

Installing a new Windows Server 2022 file server

We will now begin the following steps for configuring a new Windows 2022 file server for our AZ-801 lab environment to help with future labs and exercises throughout this book:

1. Open **Hyper-V Manager** on the device hosting the virtual machines we created in *Chapter 1*.

2. Locate the **AZ801PacktLab-FS-01** virtual machine in the list of virtual machines, as shown in *Figure 2.8*:

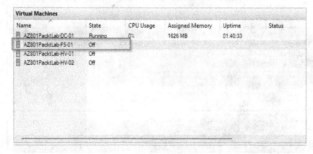

Figure 2.8 – Locating the domain controller in the list of Hyper-V virtual machines

3. We will first need to enable guest services for this virtual machine to allow us to easily copy and paste clipboard text and configuration components onto the virtual machine from our host device. Let's begin to enable these services by right-clicking on the **AZ801PacktLab-FS-01** virtual machine and selecting **Settings**.

4. We will then select **Integration Services** on the left and place a checkmark in front of **Guest services** on the right. Then, click the **OK** button to complete this configuration task as shown in *Figure 2.9*:

Figure 2.9 – Enabling guest services inside of our virtual machine configuration

5. Right-click on the **AZ801PacktLab-FS-01** virtual machine and select **Connect**.

6. From the **Virtual Machine Connection** window, we can either click the **Start** button in the center of the screen or click the green **Start** button in the taskbar at the top of the screen, as displayed in *Figure 2.10*:

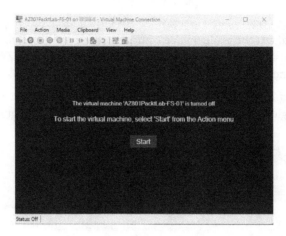

Figure 2.10 – Starting a virtual machine inside of the Virtual Machine Connection console

7. Use the **Action | Ctrl+Alt+Delete** menu options to begin, then log in to **AZ801PacktLab-FS-01** as the administrator using `Packtaz801guiderocks` as the password.

8. In **Virtual Machine Connection**, select the **View** menu option, and select **Enhanced Session**. The connection screen will disconnect and then reconnect, allowing you to copy and paste text and files between your host device and the virtual machine.

9. Right-click on the **Start** menu and then select **Windows PowerShell (Admin)** as shown in *Figure 2.11*:

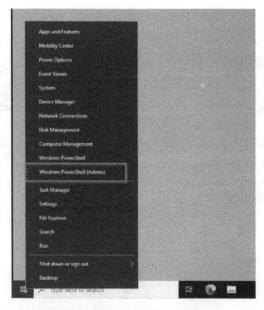

Figure 2.11 – Running Windows PowerShell as an admin from the Start menu

10. In the open **Windows PowerShell** window, run the following command and press *Enter* to complete the configuration:

```
New-NetIPAddress -IPAddress 10.10.10.2 -DefaultGateway
10.10.10.1 -PrefixLength 24 -InterfaceIndex
(Get-NetAdapter).InterfaceIndex
```

11. We will now configure the virtual network adapter's DNS settings by running the following command and pressing *Enter* to complete the configuration:

```
Set-DNSClientServerAddress -InterfaceIndex
(Get-NetAdapter).InterfaceIndex -ServerAddresses
10.10.10.1
```

12. Continuing with our best practices, we will match our naming convention established in the lab by renaming our file server. While still in the open PowerShell window, we will run the following command to rename and reboot our server:

```
Rename-Computer -NewName AZ801Lab-FS-01 -Restart -Force
-PassThru
```

13. Log back in to the device using *Step 6*. Right-click on the **Start** menu and then select **Windows PowerShell (Admin)**.

14. Next, we will add our computer to the **AD.az801.com** domain using the following Windows PowerShell command:

```
Add-Computer -ComputerName AZ801Lab-FS-01 -DomainName
AD.az801.com -Credential AZ801\Administrator -Restart
```

15. You will be prompted to enter the credential for the **AZ801\Administrator** account, which is `Packtaz801guiderocks`. Click the **OK** button to continue and have the server automatically join the domain and reboot.

16. If **Server Manager** is not already open on the `AZ801PacktLab-FS-01` virtual machine, select the **Start** menu and then select **Server Manager** from the tiles on the menu. Once **Server Manager** opens, you will receive a window that says **Try managing servers with Windows Admin Center**. This is a fantastic (and *free*) administrative tool that we will install and utilize in an upcoming chapter, so we can select the checkbox for **Don't show this message again** and close the window.

17. Select **Manage | Add Roles and Features** from the **Server Manager** window, then click the **Next** button to continue.

18. On the **Select installation type** screen, click the **Next** button.

19. On the **Select destination server** screen, click the **Next** button.

20. On the **Select roles** screen, we will select the following roles, as shown in *Figure 2.12*:

* Under the **File and Storage Services** section, we will select the following:

  * **File Server**

  * **iSCSI Target Server**

  * **iSCSI Target Storage Provider (VDS and VSS hardware providers)**

  * **Server for NFS**:

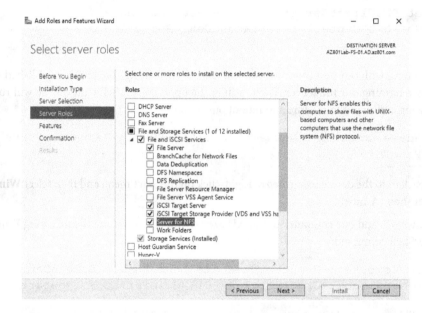

Figure 2.12 – Selecting server roles using Server Manager

21. On the **Select Features** screen, we will select the following features. You will see **Add features that are required** for both **NFS** and **Storage Migration Service** – simply click the **Add Features** button to continue:

* **Containers**

* **Storage Migration Service** (which will also select **Storage Migration Service Proxy**)

* **Storage Replica**

On the confirmation page, place a checkbox in the **Restart the destination server automatically if required** field and click the **Install** button to complete the configuration.

We have completed the necessary steps to configure our file server and prepare our lab for the next steps in this chapter and exercises in this book. In the next section, we will learn about the configuration and management of exploit protection from Microsoft Defender.

Configuring and managing exploit protection

This section will begin by defining exploit protection, explaining how the features work collectively, and discussing what this set of features brings to the Windows Server operating system (as well as Windows 10/11). We will review all the available exploit protection mitigation options and learn how to evaluate and enable these mitigations using the configuration and management tools available to us.

What is exploit protection?

Exploit protection from Microsoft Defender for Endpoint provides IT pros with a feature-rich set of advanced protections that can be applied to applications that have been compiled and distributed to devices, or at an overall operating system level. Exploit protection was designed as an on-by-default security system for all devices based on the original Microsoft tool called **Enhanced Mitigation Experience Toolkit** (EMET). This exploit protection feature contains mitigations that help prevent intrusion by malicious executables and code, reducing attack vectors, and blocking common and known patterns in malware attacks.

How does exploit protection work?

At a very high level, exploit protection works to intelligently block many popular attacks and exploitative behavior such as memory exploits or corruption, having malicious code injected into running processes, and even randomization of memory address space to name a few. The mitigations are applied per application, and each application configured will have its own set of mitigation options in the registry (set in HKEY_LOCAL_MACHINE \ Software \ Microsoft \ Windows NT \ CurrentVersion \ Image File Execution Options \ {ApplicationImageFileName} \ MitigationOptions). The configuration changes take effect only after you restart the program, and remain effective until an update to the configuration, where you restart the program again.

> **Note on removing or reverting exploit protection mitigations from devices**
>
> The **Image File Execution Options** area only accounts for a specific filename or path and do not consider application version numbers, architecture differences, code signing certificates, and other differences. In addition, if exploit protection mitigations are enforced using an XML configuration file via PowerShell, Group Policy, Mobile Device Management, or Microsoft Endpoint Configuration Manager and you later want to exempt a device from this policy, these settings will not be automatically removed. To correct this configuration approach, you will need to utilize the default XML file to reset the exploit protection settings to a well-known security baseline, as detailed in the following article:
>
> https://docs.microsoft.com/microsoft-365/security/defender-endpoint/exploit-protection-reference?view=o365-worldwide

Exploit protection mitigations

The following table identifies the native mitigations within exploit protection. While you will not need to know every single one of these in depth to pass the exam, it will be incredibly helpful to know of the existence of these various mitigations and what each brings to the table for overall operating system security:

Mitigation	Description
Arbitrary Code Guard (ACG)	This feature protects against malicious attackers by loading selected code into memory through a known safety vulnerability, and ultimately executing that code against the device.
Block low-integrity images	This feature prevents applications from loading files that have been downloaded from the internet and are considered untrusted.
Block remote images	This feature prevents applications from loading images that are sourced from a remote device, such as a file share or UNC path on the network.
Block untrusted fonts	This feature helps mitigate the risk of a flaw or vulnerability with font parsing, trusting, and loading only fonts that have been installed on the device.
Code integrity guard	This feature simply blocks any binaries from loading that are not digitally signed by Microsoft. This also includes approved drivers signed by Windows Hardware Quality Labs signatures.
Control Flow Guard (CFG)	This feature, while aptly using my personal initials, helps to mitigate attacker use of memory corruption vulnerabilities at runtime by protecting calls to functions. This requires applications to be compiled for full CFG support.
Data Execution Prevention (DEP)	This feature prevents an attacker from injecting malicious code into a running process and then executing the code on a device.
Disable extension points	This feature disables the reach of an application that seeks to load persistent code at process initialization or elevate privileges to run malicious code.

Disable Win32k system calls	This feature prevents calls into the `win32k.sys` component and is designed to protect processes that are dedicated non-user interfaces attempting to convert themselves into a GUI thread.
Do not allow child processes	This feature simply prevents any application from creating new child applications.
Export Address Filtering (EAF)	This feature helps mitigate the risk of potentially malicious code by reviewing the address table of loaded modules to determine additional APIs available for an attack vector.
Force randomization for images (mandatory ASLR)	This feature helps mitigate the risk of an attacker using known memory layouts of the device to execute code already running in the process memory. For best protection, this should be combined with randomized memory allocations, as discussed in the following table rows.
Hardware-enforced stack protection	This feature helps to verify the return addresses of function calls at runtime via the CPU.
Import Address Filtering (IAF)	This feature helps reduce the risk of an attacker changing the flow of an application to take control, intercept, or block calls to APIs of interest.
Randomize memory allocations (bottom-Up ASLR)	This feature makes it incredibly difficult for an attacker to identify or guess an address space in memory and needs to be combined with Mandatory ASLR for full functionality.
Simulate Execution (SimExec)	This feature is only available for 32-bit applications only and helps validate that calls to APIs return to the originating and legitimate function caller.
Validate API invocation (CallerCheck)	This feature inspects calls to sensitive APIs and ensures that they come from a valid and expected caller.
Validate Exception Chains (SEHOP)	This feature ensures structured exception handling and validates throughout the handling chain to stay within known and expected stack boundaries and exception handling.
Validate handle usage	This feature helps by raising an exception on any identified invalid handle references in the running code.

Validate heap integrity	This feature immediately terminates a process when memory heap corruption is identified.
Validate image dependency integrity	This feature enforces code signing for dependencies loaded within the Windows operating system.
Validate stack integrity (StackPivot)	This feature validates that there are no calls returned to an illicit stack stored in heap memory, ensuring that the application returns to the correct stack based on pointer validation.

We have finished our review of the available exploit protection mitigations and will now learn about utilizing a best practices model for exploit protection validation called Audit mode.

Using exploit protection Audit mode

Utilizing a feature inside of exploit protection called Audit mode, you can evaluate and test out certain mitigation configurations in advance of enabling this protection in a production environment. Audit mode gives a what-if approach to help you evaluate a defined configuration, or set of configurations, indicating how the configuration will affect your business applications, or will properly protect your devices and environment from malicious attacks or events.

Let's begin by learning about the various ways we can review the available exploit protection settings, and configure, manage, and audit the settings configured.

Enabling exploit protection using the Windows Security app

Utilizing the built-in Windows Security app, let's learn how we can identify configurations and learn how to apply and audit the settings using the following steps:

1. After connecting to the **AZ801PacktLab-FS-01** virtual machine, select the **Start** menu and search for or select **Windows Security** (this will be a shield icon) as shown in *Figure 2.13* to launch the **Windows Security** application:

Figure 2.13 – The Windows Security app icon

2. Inside the **Windows Security** app, select **App & browser control** as shown in *Figure 2.14*:

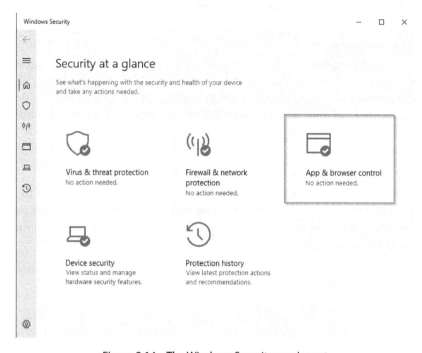

Figure 2.14 – The Windows Security app layout

3. Next, we will select **Exploit protection settings** as shown in *Figure 2.15*:

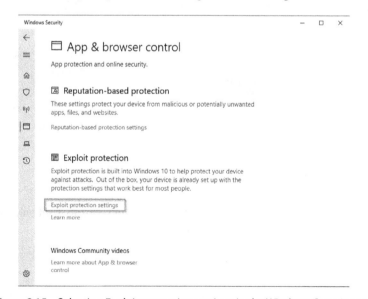

Figure 2.15 – Selecting Exploit protection settings in the Windows Security app

4.   You will notice that there are two sections: system settings and program settings. Let's select **Program settings**, then **Add program to customize**, and then **Add by program name**.

System settings section details

Under the **System settings** section, the configured settings apply to applications that do not have individual configuration settings in the **Program settings** section. There are three available options for these mitigation configurations: **On by default**, **Off by default**, and **Use default**. When changing some of these settings, you may need to either supply the administrator credentials to make the change and/or restart the device to proceed.

---

5.   Let's choose everyone's favorite application and type it in: `notepad.exe`.

6.   In the **Program settings: notepad.exe** screen, scroll down and review each of the options. Once completed, scroll to select the **Override system settings** checkbox for **Do not allow child processes**. Then, select the slider to change it to **On** and select the checkbox for **Audit only**. Click the **Apply** button to finish as shown in *Figure 2.16*:

Figure 2.16 – Setting the Do not allow child processes configuration to Audit only mode

7. To review all program setting configurations, select **Export settings** as shown in *Figure 2.17* and save the file to a known location for review in the next step:

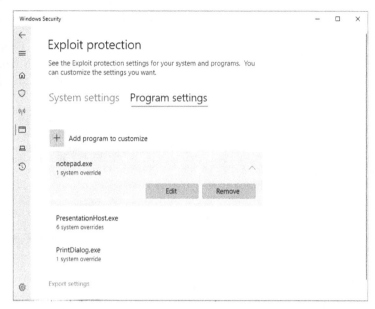

Figure 2.17 – Export your exploit protection settings in the Windows Security app

8. Locate the **Settings** file you just exported, then open the file with notepad.exe for a review of the configured settings. When done with your review, simply close the Notepad application.

9. To review any audited events, let's begin by opening **Event Viewer** from the **Start** menu.

10. Select **Applications and Services Log | Microsoft | Windows | Security-Mitigations | Kernel Mode** to begin a review of the captured events. Take note of the event data stating **would have been blocked**, indicating that this setting is in Audit mode and gives additional details on remediation or additional validation.

11. We can now close both **Event Viewer** and the **Windows Security** app, as we have completed our review of configuration and management steps using the Windows Security app.

This completes our review of establishing exploit protection settings on a Windows device using the Windows Security app.

Enabling exploit protection using Microsoft Endpoint Manager

For many organizations in a cloud-first or hybrid management approach, Microsoft Endpoint Manager is the configuration and management tool of choice. Let's walk through the general steps of how to configure exploit protection within Microsoft Endpoint Manager:

1. Open any browser and visit `https://endpoint.microsoft.com`. When prompted to sign in, be sure to utilize the credentials we created for use with Microsoft Azure and Microsoft 365 in *Chapter 1*.

2. Select **Endpoint Security | Attack Surface Reduction**, as shown in *Figure 2.18*:

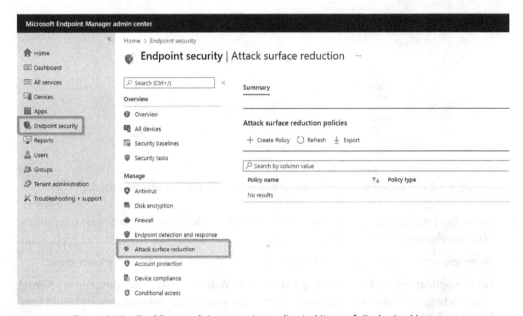

Figure 2.18 – Enabling exploit protection policy in Microsoft Endpoint Manager

3. Select **Create Policy** and then set **Platform** to **Windows 10 and later**.

4. Select **Exploit Protection** from the **Profile** drop-down menu, as shown in *Figure 2.19*, then click the **Create** button:

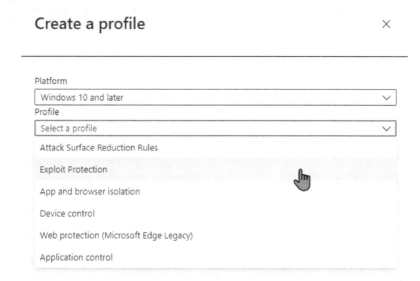

Figure 2.19 – Creating an Attack Surface Reduction profile in Microsoft Endpoint Manager

5.   For the **Name** and **Description** fields, supply `AZ801Lab-ExploitProtectionDemo` as the value and click the **Next** button.

6.   Download the following file to your device for use in the next steps: `https://demo.wd.microsoft.com/Content/ProcessMitigation.xml`.

7.   Change the slider for **Exploit Protection Settings** to **Configured**, then for the **Select a file** prompt, locate the `ProcessMitigation.xml` file you just downloaded. Review the imported XML data to see the imported settings.

8.   Before selecting the **Next** button, I want to draw attention to the Windows Defender Security Center configuration. For this exercise, we will select **Not configured**. But in your production environments, there is a high likelihood that you will be selecting **(Enable) Local users cannot make changes in the exploit protection settings area** to ensure there is no configuration skew. Review *Figure 2.20* for these additional configuration settings:

Figure 2.20 – Reviewing Windows Defender Security Center optional settings

9. For this exercise, we will not be applying assignments and scope tags to the profile and can simply select the **Next** button to continue our configuration. In a production environment with real users and devices, it is recommended that you select appropriate users and groups for **Assignments** to complete the initial pilot validation before proceeding with an enterprise-wide deployment.

10. On the **Review + create** screen, click the **Create** button.

11. Review the newly created policy as shown in *Figure 2.21*:

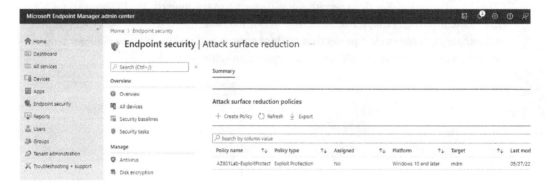

Figure 2.21 – Viewing the Attack surface reduction policy created in Microsoft Endpoint Manager

We have now completed our review and configuration steps within Microsoft Endpoint Manager and will proceed with learning how to complete the same configuration steps within Group Policy.

Enabling exploit protection using Group Policy

Many organizations continue to utilize Group Policy as their tried-and-true approach to configuration management and application of security policy. Let's dig into how exploit protection is configured and managed using Group Policy:

1. Navigating to our **AZ801PacktLab-DC-01** virtual machine, let's start a virtual machine connection and log in using our administrator credentials to begin.

2. Locate the `ProcessMitigation.xml` file we previously downloaded on our local device, right-click it, and select **Copy**.

3. Navigate back to the **AZ801PacktLab-DC-01** virtual machine, create a new folder for `C:\LabFiles`, and paste the copied `ProcessMitigation.xml` file into this new folder.

4. Open the **Start** menu and select **Windows Administrative Tools** | **Group Policy Management**.

5. Expand the forest: **AD.az801.com** | **Domains** | **AD.az801.com** | **Group Policy Objects**.

6. Right-click on **Group Policy Objects** and select **New** from the menu.

7. For the name of the new GPO, enter `AZ801Lab-ExploitProtectionDemo` and select **OK**.

8. Right-click on the newly created GPO and select **Edit** from the menu as shown in *Figure 2.22*:

Figure 2.22 – Editing our new Group Policy Object in AD

9. Navigate **Group Policy Management Editor** by going to **Computer Configuration** | **Administrative Templates** | **Windows Components** | **Windows Defender Exploit Guard** | **Exploit Protection**.

10. Right-click on **Use a common set of exploit protection settings**.

11. Select the **Enabled** radio button and then enter `C:\LabFiles\ProcessMitigation.xml` into the **Type the location** field.

12. Select the **OK** button and then close **Group Policy Management Editor**.

13. Navigate to the **File Servers** OU and right-click, selecting **Link an existing GPO**.

14. Select the **AZ801Lab-ExploitProtectionDemo** Group Policy Object from the list and click the **OK** button to apply.

15. At this point, you can evaluate the changes made directly to the **AZ801PacktLab-FS-01** virtual machine by rebooting the VM. Once the machine is back up and logged in, launch applications and validate that the Windows Security app and the event logs located at **Applications and Services Log | Microsoft | Windows | Security-Mitigations | Kernel Mode** are recording audited events.

16. When finished validating, return to **AZ801PacktLab-DC-01** with the **Group Policy Management** console still open, navigate to the **File Servers** OU, right-click on the lone GPO in the list, and select **Delete** from the menu to remove the GPO link.

We have now completed our review and configuration steps within Group Policy and will proceed with learning how to complete the same configuration steps within Windows PowerShell.

Enabling exploit protection using PowerShell

Windows PowerShell is a tool that almost every administrator uses to manage devices. Let's learn how Windows PowerShell helps us achieve the same application-level mitigations:

1. Navigating to our **AZ801PacktLab-FS-01** virtual machine, let's start a virtual machine connection and log in using our administrator credentials to begin.

2. Right-click on the **Start** menu and select **Windows PowerShell (Admin)**.

3. To start, we can enter `Get-ProcessMitigation` to see all currently configured settings for exploit protection. Review the results returned, and then enter `cls` to clear the screen after your review.

4. To enable a per-application mitigation setting, we can issue the following command to enable mitigations for **Validate heap integrity**: `Set-ProcessMitigation -Name C:\Windows\System32\notepad.exe -Enable TerminateOnError`.

5. To disable a per-application mitigation setting, we can issue the following command as an example: `Set-ProcessMitigation -Name C:\Windows\System32\notepad.exe -Disable TerminateOnError`.

6. To enable a system mitigation setting, we can issue the following command to enable mitigations for **Validate heap integrity**: `Set-ProcessMitigation -System -Enable TerminateOnError`.

7. To disable the system mitigation setting, run the following command as an example: `Set-ProcessMitigation -System -Disable TerminateOnError`.

8.   This completes our PowerShell review. For additional details on all of the available PowerShell options for exploit protection, refer to `https://docs.microsoft.com/microsoft-365/security/defender-endpoint/enable-exploit-protection?view=o365-worldwide#powershell`.

In this section, we learned about exploit protection and how it can be used to plan for, validate, troubleshoot, configure, and manage mitigation settings across a wide range of Microsoft management tools. For additional reading on a deeper exploit protection reference, refer to `https://docs.microsoft.com/microsoft-365/security/defender-endpoint/exploit-protection-reference?view=o365-worldwide`.

Coming up next, we will be working with WDAC configuration and management.

Configuring and managing WDAC

In a nutshell, WDAC works to combat untrusted software, malware, and other vulnerabilities running on your devices. In a *trust what you know and know what you trust* approach, this means that only known and approved code will be able to run on your devices. This approach is super useful in high-security departments or government agencies, as you are equipped with an incredible amount of control over what can and cannot run.

For the most part, Windows Store apps, signed drivers, Windows operating system base components, Configuration Manager clients and software deployments, and updates that target standard Windows components are all marked as allowed and can run. The key differentiator with WDAC is that this feature makes use of the Microsoft Intelligent Security Graph, allowing you to trust applications that have been vetted by Microsoft and reducing your overall WDAC policy management.

We can utilize several deployment methods to deploy WDAC policies and configuration to endpoints and some of those are Microsoft Intune, Microsoft Endpoint Manager, Microsoft Endpoint Configuration Manager, Script, and Group Policy. For this section, we will primarily focus on utilizing Microsoft Endpoint Manager to configure and evaluate the policies in audit mode, just as we learned previously with exploit protection. Let's begin learning how to complete these configurations in Microsoft Endpoint Manager!

Enabling WDAC using Microsoft Endpoint Manager

In the following steps, we will learn how to enable WDAC using Microsoft Endpoint Manager:

1.   Open any browser and visit `https://endpoint.microsoft.com`. When prompted to sign in, be sure to utilize the credentials we created for use with Microsoft Azure and Microsoft 365 in *Chapter 1*.

2.   Select **Endpoint Security | Attack Surface Reduction**.

3.   Select **Create Policy** and then set **Platform** to **Windows 10 and later**.

4. Select **Application control** from the **Profile type** drop-down menu.

5. For the **Name** and **Description** fields, supply `AZ801Lab-ApplicationControlDemo` as the value and select the **Next** button.

6. For AppLocker's application control, select **Audit Components**, **Store Apps**, and **Smartlocker** from the drop-down menu as shown in *Figure 2.23*. Leave the remaining defaults and click the **Next** button to continue:

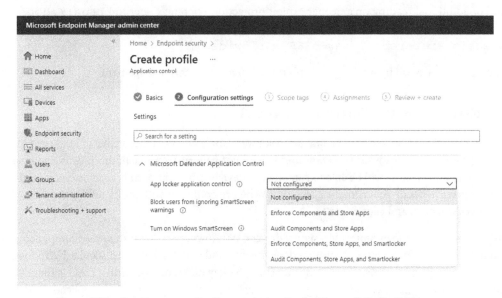

Figure 2.23 – Creating an application control policy in Microsoft Endpoint Manager

7. For this exercise, we will not be applying **Assignments** and **Scope Tags** to the profile and can simply click the **Next** button to continue our configuration. In a production environment with real users and devices, it is recommended that you select appropriate users and groups for **Assignments** to complete initial pilot validation before proceeding with an enterprise-wide deployment.

8. When running a WDAC policy in audit mode, any activity that would have been denied is logged in the **Applications and Services Logs** | **Microsoft** | **Windows** | **CodeIntegrity** | **Operational** event log. Any script and MSI events are also logged in the **Applications and Services Logs** | **Microsoft** | **Windows** | **AppLocker** | **MSI and Script** event log.

9. We have now completed our review and configuration steps within Microsoft Endpoint Manager and will proceed on to learning how to complete the same configuration steps within AD Group Policy.

> **WDAC example policies**
>
> As a best practice, it's best to always start with an existing or basic policy, and then incorporate your own customizations or configurations as you go to build your custom policy. To that end, you will be able to navigate to `C:\Windows\schemas\CodeIntegrity\ExamplePolicies` and review the example policies as developed and provided by Microsoft. For more details on these base policies, be sure to review `https://docs.microsoft.com/windows/security/threat-protection/windows-defender-application-control/example-wdac-base-policies`.

In this section, we learned about WDAC and how it can be used to not only audit and discover running applications in your environment, but also to identify additional scripts, binaries, and other components that might be missing from your Application Control policy. We also learned that Application Control can help mitigate vulnerabilities by restricting unknown or untrusted applications, while also allowing approved and known applications to function as they normally would.

Next up, we will tackle the configuration and management of Microsoft Defender for Endpoint.

Configuring and managing Microsoft Defender for Endpoint

Microsoft Defender for Endpoint is designed to be an **endpoint detection and response tool** (**EDR**) that helps businesses monitor and detect, further investigate, and respond to attacks on their devices and networks. This one topic could easily take up a full chapter on its own – we will cover the major topics of system requirements, onboarding, and offboarding processes to fulfill the AZ801 exam objectives. Let's start by reviewing the system and software requirements.

Licensing, software, and hardware requirements for Microsoft Defender for Endpoint

The basic requirements for Microsoft Defender for Endpoint are as follows:

- The client's computer needs to be managed by Microsoft Intune or running the Configuration Manager client
- A current subscription to Microsoft Defender for Endpoint (Plan 1, Plan 2, or Microsoft Defender for Endpoint Server)
- A supported operating system (covers Windows 8.1 through Windows Server 2022)
- An administrator user assigned the Endpoint Protection Manager security role
- Internet connectivity, either directly or via a proxy service
- The diagnostic data service needs to be enabled (diagnostic log level needs no specific value)

Let's now discuss how to achieve onboarding of devices into the Microsoft Defender for Endpoint service in the Microsoft cloud.

Onboarding devices to Microsoft Defender for Endpoint

There are two types of devices to consider and will require differing approaches for onboarding: up-level operating systems and down-level operating systems.

Down-level operating systems include Windows 8.1, Windows Server 2012 R2, and Windows Server 2016. For these systems, they will require a workspace key and a workspace ID for successful onboarding. Recent changes to the onboarding process for down-level operating systems are now easier using the unified solution package (and more importantly, have removed the requirement for the **Microsoft Monitoring Agent** (**MMA**) on devices).

Up-level operating systems include all operating systems of Windows Server 2019 or later and will only require the onboarding configuration file for successful onboarding.

While we have several tools available for onboarding Windows, macOS, Linux Server, iOS, and Android, we will focus primarily on onboarding via Microsoft Endpoint Configuration Manager.

Let's begin stepping through the general configuration steps for the onboarding process:

1. Open any browser and navigate to `https://security.microsoft.com`. When prompted to sign in, be sure to utilize the credentials we created for use with Microsoft Azure and Microsoft 365 in *Chapter 1*.

2. On the **Microsoft 365 Defender** portal, select **Settings | Endpoints | Onboarding**, as shown in *Figure 2.24*:

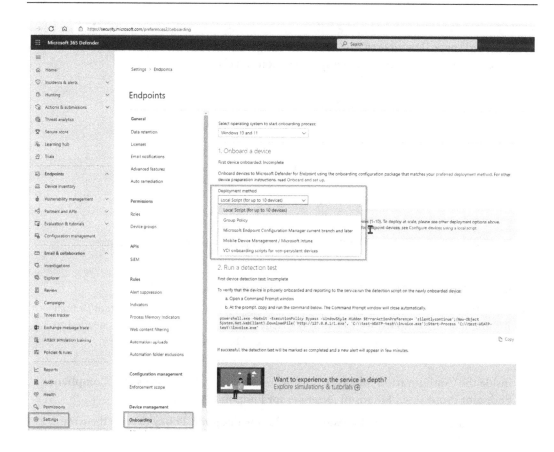

Figure 2.24 – Accessing the onboarding options for Microsoft Defender for Endpoint

3. For the operating system selection, let's select **Windows 10 and 11**.

4. For the deployment method, we will select **Microsoft Endpoint Configuration Manager current branch and later**.

5. We will then select the **Download onboarding package** button and save this sensitive zip package in our C:\AZ801PacktLab\Scripts folder.

6. Now select **Windows 7 SP1 and 8.1** from the operating system list to make **Workspace key** and **Workspace ID** visible to you, as shown in *Figure 2.25*. Note that these values will be the same regardless of which operating system option you choose:

Figure 2.25 – Viewing the sensitive workspace ID and workspace key for your tenant

---

Demonstration only of Microsoft Endpoint Configuration Manager

Knowing that we do not have a Microsoft Endpoint Configuration Manager Current Branch environment available to us, we will simply discuss and demonstrate the steps needed to complete this configuration in Configuration Manager 2107 and newer.

---

7. Inside the **Configuration Manager** console, we will navigate to **Assets and Compliance | Endpoint Protection | Microsoft Defender for Endpoint Policies**.

8. We will then select **Create Microsoft Defender for Endpoint Policy** to open the wizard, as shown in *Figure 2.26*:

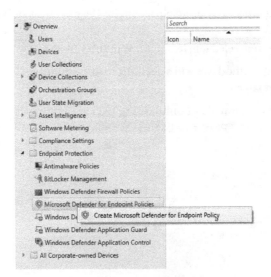

Figure 2.26 – Create Microsoft Defender for Endpoint Policy

9. We will give a valid name and description of this new policy, select **Onboarding**, and then select the **Next** button to continue as shown in *Figure 2.27*:

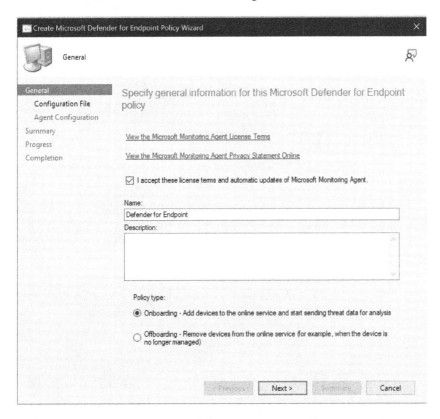

Figure 2.27 – Selecting an onboarding policy type for Microsoft Defender for Endpoint

10. We will then be prompted to browse the onboarding configuration file we downloaded previously.

11. Next, we need to supply the correct values of the workspace key and ID and click the **Next** button, as shown in *Figure 2.28*:

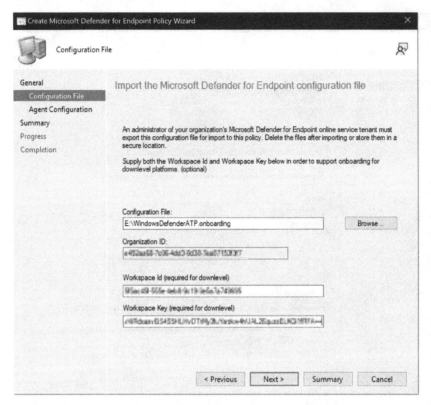

Figure 2.28 – Uploading the onboarding configuration file and supplying a workspace ID and key

12. On the **Agent Configuration** screen, set **Specify which file samples are shared for analysis by the Microsoft Defender for Endpoint online service:** to **All file types** and click the **Next** button.

13. Review the summary page and select the **Finish** button to complete the onboarding configuration in Configuration Manager.

14. At this point, the policy is now ready for deployment to targeted clients.

15. We could also monitor the status by visiting the **Monitoring | Security | Microsoft Defender for Endpoint Status** section to validate both the onboarding status and the agent health dashboard.

This completes the steps for onboarding devices to Microsoft Defender for Endpoint with Microsoft Endpoint Configuration Manager.

Offboarding devices from Microsoft Defender for Endpoint

Let's begin stepping through the general configuration steps for the onboarding process:

1. Open any browser and navigate to `https://security.microsoft.com`. When prompted to sign in, be sure to utilize the credentials we created for use with Microsoft Azure and Microsoft 365 in *Chapter 1*.

2. On the Microsoft 365 Defender portal, select **Settings | Endpoints | Offboarding**.

3. For the operating system selection, let's select **Windows 10 and 11**.

4. For the deployment method, we will select **Microsoft Endpoint Configuration Manager current branch and later**.

5. We will then select the **Download package** button and save this sensitive zip package in our `C:\AZ801PacktLab\Scripts` folder. Note that you will receive another prompt saying **Are you sure you want to download the device offboarding package?** that will expire in 30 days. Click the **Download** button.

> **Demonstration only of Microsoft Endpoint Configuration Manager**
>
> Knowing that we do not have a Microsoft Endpoint Configuration Manager environment available to us, we will simply discuss and demonstrate the steps needed to complete this configuration.

6. Inside the **Configuration Manager** console, we need to navigate to **Assets and Compliance | Endpoint Protection | Microsoft Defender for Endpoint Policies**.

7. We will then select **Create Microsoft Defender for Endpoint Policy** to open the wizard.

8. We will give a valid name and description of this new policy, then select **Offboarding**, and then click the **Next** button to continue.

9. We will then be prompted to browse the offboarding configuration file we downloaded previously.

10. Review the summary page and select the **Finish** button to complete the onboarding configuration in **Configuration Manager**.

11. At this point, the policy is now ready for deployment to targeted clients.

We have now completed the section on configuring and managing the basics of Microsoft Defender for Endpoint, including a review of the requirements and steps for successful onboarding and offboarding of devices. In the next section, we will be learning about Windows Defender Credential Guard, including how to configure, evaluate, and manage the Credential Guard security configurations.

Configuring and managing Windows Defender Credential Guard

A lot of businesses are still using numerous security measures to enforce password complexity, recommend password best practices and hygiene, and even require password changes or resets on a scheduled basis. Let's be honest – the world of security and password management has changed drastically, and passwords are now the weakest link in the security chain.

One of the ways we can protect and isolate these frequently used secrets is to utilize a virtualization-based security model called Windows Defender Credential Guard, which is backed by hardware security to deter targeted and credential theft attacks in the wild. This model blocks unauthorized access to the secrets so that only the Windows-privileged subsystem can gain access to them. For instance, when Credential Guard is enabled on a Windows system, the `lsass.exe` process is wrapped with Credential Guard and runs the `Lsalso.exe` process with the added protections as shown in *Figure 2.29*:

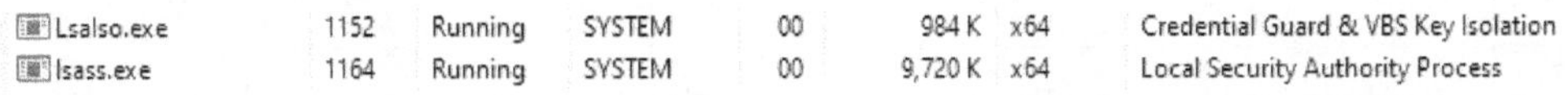

| Lsalso.exe | 1152 | Running | SYSTEM | 00 | 984 K | x64 | Credential Guard & VBS Key Isolation |
| lsass.exe | 1164 | Running | SYSTEM | 00 | 9,720 K | x64 | Local Security Authority Process |

Figure 2.29 – Credential Guard protection lsass.exe with the Lsalso.exe guarded process

Let's learn together how this feature can be enabled for compatible devices.

Requirements for enabling Windows Defender Credential Guard

To provide initial protections against credential theft and the use of derived credentials, Windows Defender Credential Guard uses the following:

- Secure boot (required)
- Virtualization-based security support (required)
- **Trusted Platform Module (TPM)**, with versions 1.2 or 2.0 supported (preferred)
- UEFI lock, to prevent disabling in the registry (preferred)
- Additional requirements for virtualization-based security:
    - 64-bit CPU
    - CPU virtualization extensions
    - Windows Hypervisor (no requirement for Hyper-V features to be installed, however)

There are also application requirements that may potentially break or have certain authentication capabilities blocked. This level of depth is beyond the scope of the AZ-801 exam objectives, but you can refer to `https://docs.microsoft.com/windows/security/identity-protection/credential-guard/credential-guard-requirements` for additional details.

Now that we have gone through an overview of the hardware, software, and application requirements, let's learn how to enable Windows Defender Credential Guard.

Enabling Windows Defender Credential Guard using the hardware readiness tool

One of the easiest ways to enable Windows Defender Credential Guard is by using the Windows Defender Credential Guard hardware readiness tool:

1. To begin, let's download the tool from `https://aka.ms/DGCGTool`.

2. Utilize the **Copy** button to copy the PowerShell code to your clipboard.

3. Open **Notepad** on your device, then paste the copied code into the new Notepad file.

4. Navigate to `C:\AZ801PacktLab`, and create a new folder named `Scripts`.

5. Save the file with a filename of `DG_Readiness_Tool.ps1` and set **Save as type** to **All Files** (*.*), ensuring that the file has been saved to `C:\AZ801PacktLab\Scripts`.

6. Open Windows PowerShell as an administrator and enter the following command to determine if your device is ready to run Credential Guard, as shown in *Figure 2.30*: `C:\AZ801PacktLab\Scripts\DG_Readiness_Tool.ps1 -Ready`

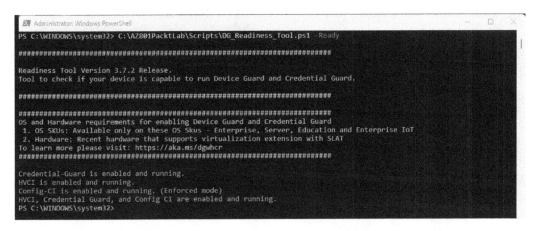

Figure 2.30 – Running the Credential Guard hardware readiness tool in Ready mode

7. For example, if you were ready to configure a device to enable Windows Defender Credential Guard, you could issue the following command in PowerShell: `DG_Readiness_Tool.ps1 -Enable -AutoReboot`.

8. At this point, our review of using the hardware readiness tool is complete, and you can close out of any open PowerShell or Windows Explorer windows.

Now, let's look at how the same configuration can be completed using Group Policy to configure Windows Defender Credential Guard.

Enabling Windows Defender Credential Guard using Group Policy

Group Policy can be used to enable Windows Defender Credential Guard, and in doing so, will also enable any necessary virtualization-based security features if they have not yet been enabled on the devices:

1. Navigating to our **AZ801PacktLab-DC-01** virtual machine, let's start a virtual machine connection and log in using our administrator credentials to begin.

2. Open the **Start** menu and select **Windows Administrative Tools | Group Policy Management**.

3. Expand the forest: **AD.az801.com | Domains | AD.az801.com | Group Policy Objects**.

4. In **Group Policy Objects**, open the `AZ801Lab-ExploitProtectionDemo` GPO for this exercise.

5. From within **Group Policy Management Console**, navigate to **Computer Configuration | Administrative Templates | System | Device Guard**.

6. Double-click the setting for **Turn On Virtualization Based Security**, and then click the **Enabled** radio button.

7. In the **Select Platform Security Level** setting, choose **Secure Boot and DMA Protection**.

8. In the **Credential Guard Configuration** box, click **Enabled with UEFI lock**.

9. If we were to deploy these changes to devices, we could select the **OK** button at this point and link this GPO to a set of pilot devices for evaluation first. However, these steps only show where the configuration settings can be administered, so choose the **Cancel** button to destroy our changes and exit Group Policy Management.

We have now completed this section on what Windows Defender Credential Guard is and how to configure and manage the features on devices. We will now cover the configuration of SmartScreen in the following section.

Configuring Microsoft Defender SmartScreen

Microsoft Defender SmartScreen is a feature that primarily protects against internet websites, downloads, and applications that could engage in sophisticated phishing or malware attacks or blocking of unwanted applications. It does not protect against malicious files that are stored on-premises or within network shares.

Microsoft Defender SmartScreen works by validating the reputation, digital signature, and additional checks done by a Microsoft cloud service. If the file has not been run before or has no prior reputation, the user may be warned that the file has no history and asks whether the app should run or be blocked. Utilizing enterprise-grade controls, you can orchestrate how this feature works in your environment.

Configuring Microsoft Defender SmartScreen on an individual device

To demonstrate where Microsoft Defender SmartScreen settings are managed within the user interface, let's complete the following steps to review a configuration on an individual device for initial validation:

1. On any Windows device, open the **Start** menu and search for or select **Windows Security** (this will be a shield icon).

2. Once the Windows Security app opens, select **App & Browser control** from the left menu, then select **Reputation-based protection** settings as shown in *Figure 2.31* and *Figure 2.32*:

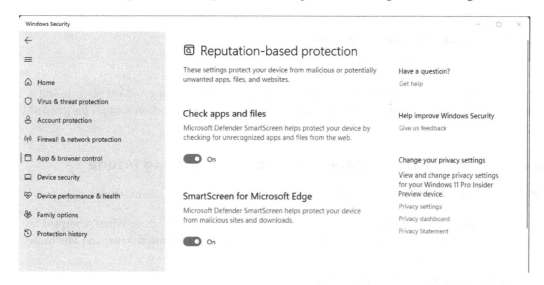

Figure 2.31 – Reputation-based protection options for SmartScreen, part 1

The details in *Figure 2.32* show additional settings surrounding potentially unwanted app blocking and SmartScreen for Microsoft Store apps:

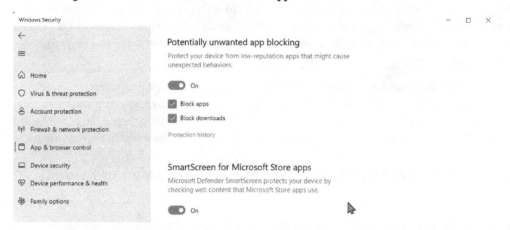

Figure 2.32 – Reputation-based protection options for SmartScreen, part 2

3. Notice that there are currently four sections pertaining to SmartScreen that can be independently managed:

    *   Check apps and files

    *   Microsoft Edge protections

    *   Potentially unwanted app blocking

    *   SmartScreen for Microsoft Store apps

At this point in the walk-through, we can close the Windows Security app and get ready to learn how to create a configuration profile in Intune to manage these configuration settings for a broad collection of devices.

Configuring Microsoft Defender SmartScreen using Intune

We will now quickly review how to create and manage a configuration profile in Intune:

1. Open any browser and visit `https://endpoint.microsoft.com`. When prompted to sign in, be sure to utilize the credentials we created for use with Microsoft Azure and Microsoft 365 in *Chapter 1*.

2. Select **Devices | Configuration Profiles**.

3. Select **Create profile** and then set **Platform** to **Windows 10 and later**.

4. Select **Settings catalog** from the **Profile type** drop-down menu, then select the **Create** button.

5. For the **Name** and **Description** fields, supply `AZ801Lab-SmartScreenDemo` as the value and select the **Next** button.

6. On the **Configuration settings** screen, select the + **Add settings** button.

7. On the **Settings picker** blade, enter `smartscreen` as the search term and select the **Search** button. You should get a list of settings returned, grouped by category, as shown in *Figure 2.33*:

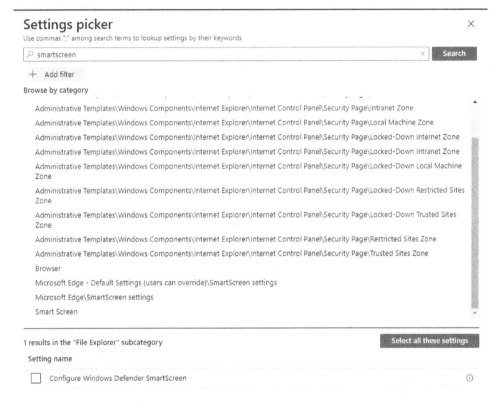

Figure 2.33 – Reviewing available configuration items from Settings picker

8. Scroll down to **Microsoft Edge\SmartScreen settings** and select **category** to show all the available settings. Click the **Select all these settings** button to display all the SmartScreen settings available within Microsoft Edge and review the list paying attention to the small information button by each setting.

9. Once you have reviewed some or all the settings, click the **X** in the upper right corner of the **Settings** and **Create profile** pages to discard the changes, as we have no cloud-managed devices at this point to apply this configuration profile to.

10. At this point, we have completed all the review steps and can sign out of Intune.

Next, we will review some additional topics surrounding Microsoft Defender SmartScreen management that will help you in your learning path for the exam objectives.

Additional notes on Microsoft Defender SmartScreen

Users that have identified a malicious website are empowered to report an unsafe site directly to Microsoft from within Microsoft Edge. To do this, select **…|Help and Feedback|Report unsafe site**.

Files can be manually submitted for review to either remove an inadvertent block or block a file you believe contains undetected malware. To complete this manual review step, you can visit `https://www.microsoft.com/wdsi/filesubmission/` for your file submission and select the proper analysis options.

For additional details on the Intune, Group Policy, and Mobile Device Management options available currently for configuration, please refer to `https://docs.microsoft.com/windows/security/threat-protection/microsoft-defender-smartscreen/microsoft-defender-smartscreen-available-settings`.

In this section, we learned what Microsoft Defender SmartScreen brings to the device management and security table, including covering an overview of where the settings can be managed both on an individual device as well as a corporate-owned and managed device. We also learned a few additional tips that will help solidify our exam objectives review. Next, we will review the available options for implementing operating system security by using Group Policy in combination with security baselines.

Implementing operating system security by using Group Policy

As we learned throughout this chapter, Windows Server was designed to be secure out of the box, with a myriad of features and by-default configurations. However, there are many organizations that are required to have more granular control over their configuration, management, and security configurations.

Microsoft has, over the years, provided security baselines that help to give a deeper understanding of and insight into guidance on security-related settings beyond the default configuration. There are thousands of configuration items and policy settings to go through, and it takes time and effort to review each one of them.

The security baselines can be downloaded from the following URL: `https://docs.microsoft.com/windows/security/threat-protection/windows-security-configuration-framework/security-compliance-toolkit-10`.

They are considered a part of the Security Compliance Toolkit, which helps to provide administrators with additional tools and recommendations on how to manage the security baselines. For example, many of the security baselines are Group Policy Object backups that can be imported into a test or production environment for review and verification.

In addition, Microsoft has also released a **Mobile Device Management** (**MDM**) security baseline that can be used to complete the same Group Policy management approach inside of Intune and Microsoft Endpoint Manager.

Migrating Group Policy settings to MDM

Note that Microsoft had created a tool called the **MDM Migration Analysis Tool** (**MMAT**) that generated a report based on the current Group Policy Object set for a user and listed the appropriate MDM policy settings to configure.

This has since been replaced with a public preview of Group Policy analytics in Microsoft Endpoint Manager that not only analyzes your on-premises GPOs, but identifies supported and deprecated settings, and can also migrate your settings into a catalog policy that can be reused to deploy to devices. To read more about this feature, refer to `https://docs.microsoft.com/mem/intune/configuration/group-policy-analytics`.

In this section, we learned about some of the incredible tools that are available to administrators that can be used to incorporate additional security compliance based on individual business and industry requirements. We also learned that there are newer tools helping to advance the migration of legacy configurations, where appropriate or required.

Summary

In this chapter, we expanded our lab setup to include an AD Domain Controller, a Windows Server 2022 File Server, and reviewed both default security configurations as well as identified some best practices. We then learned about some of the more advanced attack mitigation features such as exploit protection, WDAC, Microsoft Defender for Endpoint, Windows Defender Credential Guard, SmartScreen, and the implementation of additional security protections using security baselines within Group Policy.

In the next chapter, we will be learning how to secure a hybrid AD infrastructure. This will include the configuration of password policies, additional hardening of domain controllers, additional hybrid features for the protection of passwords and identities, and the administration of protected users and administrative groups.

3

Securing a Hybrid Active Directory (AD) Infrastructure

Arguably one of the most important tasks is to ensure that your hybrid identity infrastructure is properly secured using feature-rich tools, rigid controls with consistent configurations, and solid processes and procedures to ensure a high level of security.

In this chapter, we will learn how to apply appropriate layers of security to protect **Active Directory** (**AD**) domain controllers against attack, while allowing for continued productivity and secure workloads. We will also learn how to manage protected users and the delegation of privileges and administrators, secure the administrative workflows and authentications of and to domain controllers, and successfully implement and manage Microsoft Defender for Identity. In this chapter, we will cover the following topics:

- Technical requirements and lab setup

- Managing password policies

- Configuring and managing account and user security

- Securing domain controllers

- Managing AD built-in groups and delegation

- Implementing and managing Microsoft Defender for Identity

Technical requirements and lab setup

To successfully follow along and complete tasks and exercises throughout this chapter and the following chapters in this book, we will need to establish a custom domain name for use with Azure AD, set up Azure AD Connect to synchronize our on-premises lab AD with Azure AD, and finally establish **Self-Service Password Reset** (**SSPR**). Let's begin with the following set of steps to configure these lab components in our Azure Active Directory tenant and our on-premises Hyper-V lab.

Establishing a custom domain name

When we established our tenant subscription within Azure back in *Chapter 1*, there is a default domain registered to the tenant that reflects the naming convention of `tenantname.onmicrosoft.com`. This **Fully Qualified Domain Name (FQDN)** is unique throughout all of Azure AD, and this tenant FQDN becomes a permanent and public part of your Azure AD tenant. As part of your organization's brand and naming conventions, it may be desired to utilize an already-in-place or new custom domain that must be registered and ultimately verified within Azure AD.

Adding and verifying an existing or new custom domain as the primary domain

The process for incorporating an existing or a new custom domain into Azure AD is relatively the same. However, we need to cover some details and additional steps if you are purchasing a new custom domain for use within your tenant.

Custom domains that are intended for public use for both email and website traffic, among other public services, must first be purchased from a public domain registrar. For the purposes of this book and incorporated labs, while Microsoft offers the ability to purchase a domain, we recommend utilizing a public domain registrar such as GoDaddy.com or NameCheap.com to establish an affordable and unique-to-you custom domain name.

The following steps will now focus on the configuration of your existing or newly purchased custom domain within Azure AD:

1. Visit `https://portal.azure.com`, utilizing your Global Administrator account created in *Chapter 1*.

2. Select **Azure Active Directory** from the list of Azure services, or simply search for **Azure Active Directory**.

3. Select **Custom domain names** under the **Manage** section, then select **+ Add custom domain** to continue, as shown in *Figure 3.1*. Notice that you will see your default `tenantname.onmicrosoft.com` listed here as one of the custom domain names:

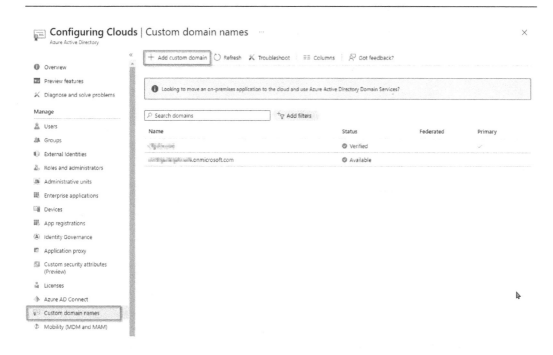

Figure 3.1 – Configuring custom domain names and adding a new custom domain to your tenant

4. When prompted for a custom domain name, enter your desired custom domain name that already exists or that you recently purchased, as shown in *Figure 3.2*, and click **Add domain** to continue:

Figure 3.2 – Entering your custom domain name into the configuration

5. After clicking **Add domain**, you will now be presented with a screen that contains **TXT** information with instructions to visit your domain registrar and add the information provided, as shown in *Figure 3.3*. For additional assistance with this and step-by-step instructions to complete verification for varying public domain registrars, refer to https://docs.microsoft.com/azure/active-directory/fundamentals/add-custom-domain.

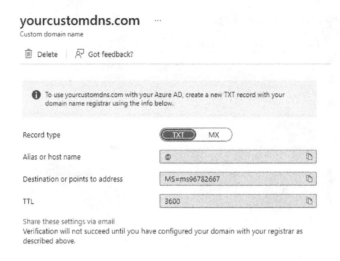

Figure 3.3 – Custom domain name verification information

6. Once the settings have been properly configured at the domain registrar, select the **Verify** button to complete the verification. If there are still errors there will be a red warning, such as the one shown in *Figure 3.4*, allowing you to select **Click here for more information** to complete additional in-portal troubleshooting:

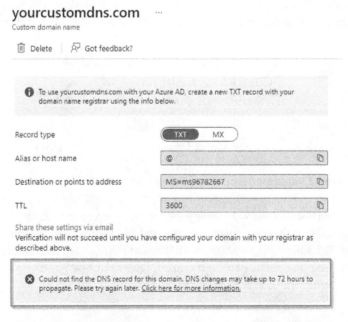

Figure 3.4 – Additional custom domain name troubleshooting link

7. Once the verification step has been successfully completed, you will see a screen listing all your custom domain names as part of your tenant as shown in *Figure 3.5*. Select your recently added custom domain name, select **Make Primary**, and then click **Yes** to set this new custom domain to the primary custom domain name:

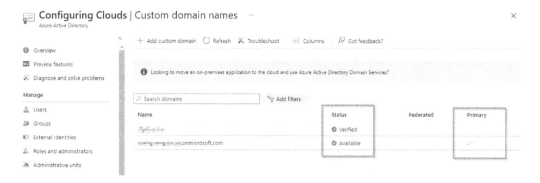

Figure 3.5 – Reviewing newly added custom domain name in Azure AD

8. Confirm that your custom domain name is reflecting that it is now the primary custom domain name.

You have now completed the steps necessary to add and verify a new custom domain name in your Azure AD tenant. In the next section, we will discuss and learn how to establish the synchronization of our on-premises AD with Azure AD.

Synchronizing on-premises AD with Azure AD

In a hybrid world, most organizations already have an existing on-premises identity infrastructure that they are managing, specifically managing *domain controllers* and **Active Directory Domain Services** (**AD DS**). As IT pros, we are tasked with the challenge of synchronizing these identities (whether users or groups) to Azure AD. The **Azure AD Connect** tool helps to simplify **Identity and Access Management** (**IAM**) by providing a consistent and resilient way to manage these hybrid identities across platforms while having updates reflected on-premises and in Azure AD almost instantaneously.

Terms such as *hybrid identity* mean that an organization has users and groups utilizing resources available both on-premises and within cloud platforms, such as **Software as a Service** (**SaaS**), **Platform as aService** (**PaaS**), and **Infrastructure as a Service** (**IaaS**), to name a few. Azure AD helps to further manage cloud identities, whether synchronized or cloud-native, utilizing industry-standard open source authentication and authorization exchange such as OAuth 2.0, **Security Assertion Markup Language** (**SAML**), and OpenID Connect protocols. To provide access to these cloud applications, it is required to have the identity available within Azure AD, thus requiring the identities to be synchronized to the Azure AD tenant in some fashion.

While we could go into great depth on all the features, patterns, and practices within Azure AD Connect, to focus on the AZ-801 exam objectives we will discuss the following four options for configuring Azure AD Connect for synchronization within your organizations and your lab:

- **Password Hash Synchronization (PHS)**

  - This is the default option when selecting the express setup within **Azure AD Connect**. This authentication topology persists in both an on-premises and a cloud identity for users, allowing users to use the same authentication credentials for both on-premises and cloud applications even when one of these methods become unavailable for authentication. The synchronized hash is both encrypted and salted for credential protection and is a great option for organizations quickly migrating services to cloud-native infrastructure models. However, for organizations with mature authorization and complex authentication models, there are better choices.

- **Seamless Single Sign-On (SSSO)**

  - This authentication topology option allows users already signed in on a corporate device that's connected to the corporate network (laptop, desktop, or virtual desktop) to sign users easily and automatically into cloud-based applications. This option does not need any additional on-premises components but does, however, create a new AD computer account on-premises named AZUREADSSOACC. This should be placed in a unique OU and protected just as you would a domain controller computer account, ensuring that it cannot be deleted or tampered with.

- **Pass-Through Authentication (PTA)**

  - This authentication topology requires an on-premises PTA agent to be installed and managed on a minimum of three servers running in your environment, and also must not exceed 40 authentication agents per tenant. These servers should be hardened just like domain controllers (which we will cover later in this chapter) as a **Control Plane** asset. This option does require on-premises AD connectivity for the users and devices to complete successful authentication, which is vastly different from the PHS method previously discussed, where either on-premises AD or Azure AD can be used for authentication.

- **Active Directory Federation Services (AD FS)**

  - This authentication topology is typically seen in organizations that have incredibly complex AD infrastructures with multiple forests and/or domains with additional security requirements surrounding third-party **Multi-Factor Authentication (MFA)**, certificate authentication, or smart card authentication. This is arguably the most complex (and expensive) topology of all the topologies discussed and requires the most additional infrastructure to deliver a resilient authentication platform. This option is also referred to as a federated identity trust model.

Now that we have completed an overview of what each of the Azure AD Connect authentication topologies brings to the organization and hybrid identity management, let's start with the initial steps for installing and configuring Azure AD Connect to our Azure AD tenant within a hybrid identity infrastructure:

> **Azure AD Connect requirements review and best practices**
>
> While our lab machines meet the requirements for a successful Azure AD Connect installation, you may want to review the full installation prerequisites listed at the following URL: `https://docs.microsoft.com/azure/active-directory/hybrid/how-to-connect-install-prerequisites`.
>
> It is also not recommended to install the Azure AD Connect on a domain controller, but instead, on another member server within your infrastructure.
>
> Please also note that we have another tool to help simplify and properly automate the management of AD and on-premises users – **Azure AD Connect Cloud Sync**. While this may not be on the AZ-801 exam at this point, I fully expect that it will appear in the future and `https://aka.ms/cloudsync` will certainly help you to differentiate between the two sync tools.

1. Open **Hyper-V Manager** on the device hosting the virtual machines we created in *Chapter 1*.

2. Locate the **AZ801PacktLab-FS-01** virtual machine in the list of virtual machines.

3. Right-click on the **AZ801PacktLab-FS-01** virtual machine and select **Settings…**.

4. Select **Add Hardware** under the **Hardware** section, then select **Network Adapter** and click the **Add** button, as shown in *Figure 3.6*:

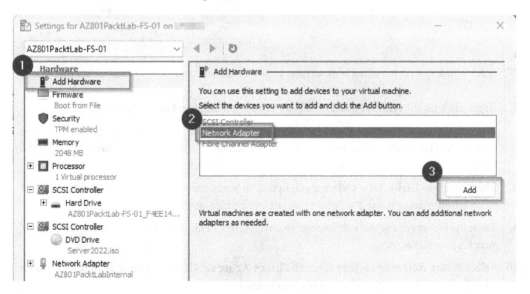

Figure 3.6 – Adding a new network adapter to our lab file server

5. After clicking the **Add** button, open the **Virtual Switch** drop-down menu and select AZ801PacktLabExternal from the list, as shown in *Figure 3.7*. Then, click the **OK** button to confirm the changes and continue:

Figure 3.7 – Adding an external network adapter to allow internet connectivity in our virtual machine

6. Right-click on the **AZ801PacktLab-FS-01** virtual machine and click **Connect**.

7. From the **Virtual Machine Connection** window and if not already started/running, we can either select the **Start** button in the center of the screen or click the green **Start button** in the taskbar at the top of the screen.

8. Use the **Action|Ctrl + Alt + Delete** menu option to begin, then log in to **AZ801PacktLab-FS-01** as an administrator using Packtaz801guiderocks as the password.

9. From the file server, use the web browser to navigate to and log in to Azure AD using https://portal.azure.com.

10. Select **Azure Active Directory** from the list of Azure services, or simply search for Azure Active Directory and select it to continue.

11. In the **Azure Active Directory** blade, select the **Azure AD Connect** option under the **Manage** section, as shown in *Figure 3.8*:

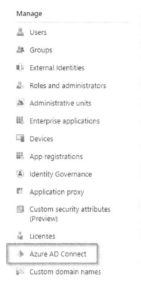

Figure 3.8 – Selecting Azure AD Connect from the Azure AD|Manage section

12. Once the **Azure AD Connect** blade loads, locate **Download Azure AD Connect** under the **Azure AD Connect Sync** section, as shown in *Figure 3.9*:

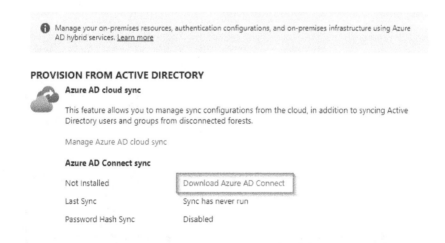

Figure 3.9 – Locating the Azure AD Connect download link within the Azure AD portal

13. Click the **Download** button on the rendered **Microsoft Azure Active Directory Connect** download page, as displayed in *Figure 3.10*:

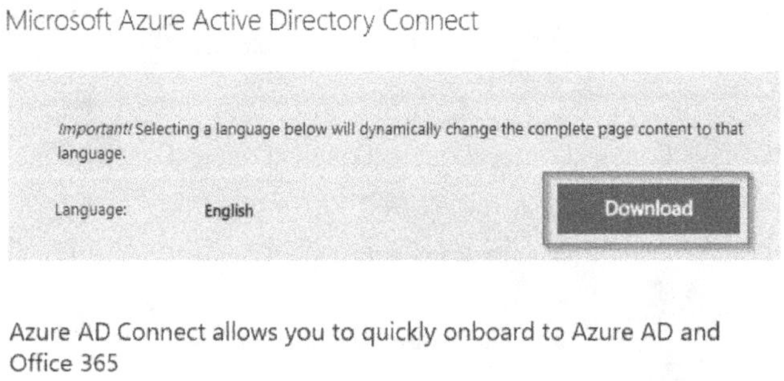

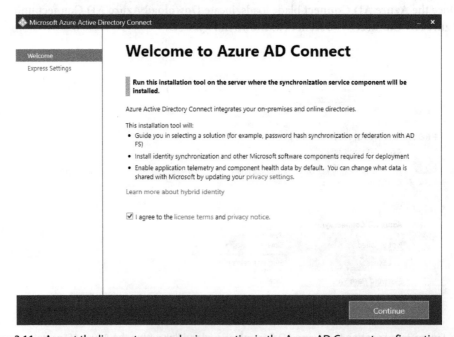

Microsoft Azure Active Directory Connect

Important! Selecting a language below will dynamically change the complete page content to that language.

Language: English   Download

Azure AD Connect allows you to quickly onboard to Azure AD and Office 365

Figure 3.10 – Click the Download button on the Microsoft Azure Active Directory Connect page

14. After the download is completed, navigate to the Downloads folder, and install the application. A few initial installation screens will quickly display and once the installation has been completed, you will receive a **Welcome to Azure AD Connect** screen to begin the configuration.

15. Select the checkbox for **I agree to the license terms and privacy notice** then click the **Continue** button for the configuration, as shown in *Figure 3.11*:

Figure 3.11 – Accept the license terms and privacy notice in the Azure AD Connect configuration wizard

16. Click **Use express settings** on the **Express Settings** screen as shown in *Figure 3.12*:

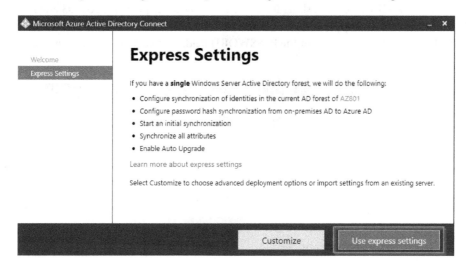

Figure 3.12 – Reviewing the express settings for Azure AD Connect

17. Enter the Global Administrator or hybrid identity administrator credentials for your Azure AD account, as shown in *Figure 3.13*. If you have taken any steps to enable multi-factor authentication on this Global Administrator account, you will need to acknowledge and fulfill the request to continue the setup:

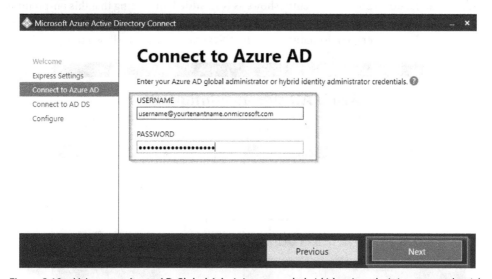

Figure 3.13 – Using your Azure AD Global Administrator or hybrid identity administrator credentials

18. After the Azure AD credentials have been successfully verified, you will be presented with a **Connect to AD DS** screen asking for credentials to the on-premises AD DS environment we have in our lab. Supply AZ801\administrator for the **USERNAME** field and Packtaz801guiderocks for the **PASSWORD** field, then click **Next** to continue, as shown in *Figure 3.14*:

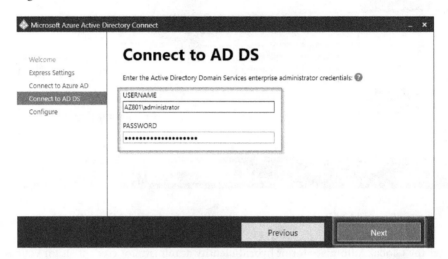

Figure 3.14 – Enter the AD DS administrator credentials

19. On the **Azure AD sign-in configuration** screen, review the UPN suffix entry or entries and notice that the ad.az801.com entry shows as **Not Added**, indicating that this on-premises domain has not been registered in Azure AD. Select the checkbox for **Continue without matching all UPN suffixes to verified domains**, then click **Next** to continue, as shown in *Figure 3.15*:

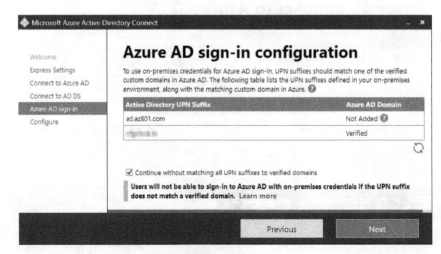

Figure 3.15 – Azure AD sign-in configuration

20. The configuration will now be verified once we click **Install**, and a summary of the changes will be displayed as shown in *Figure 3.16*. Leave the defaults on this page to immediately synchronize any users after configuration completion and click the **Install** button to continue:

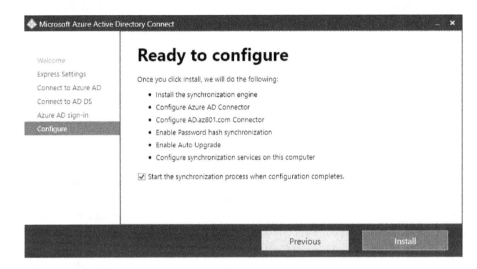

Figure 3.16 – Ready to configure Azure AD Connect

21. We have now completed all the initial steps to install and configure Azure AD Connect and the health agent and can review our successful configuration details as per *Figure 3.17*:

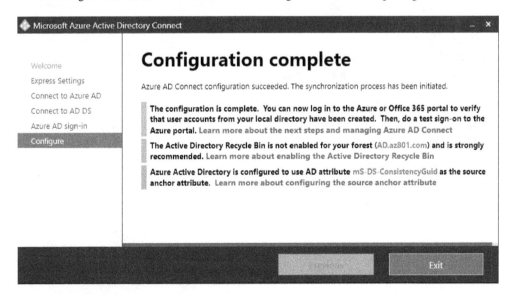

Figure 3.17 – Review of the Azure AD Connect completed configuration screen

> **Azure AD Connect sync health and troubleshooting**
>
> Note that as part of this installation, the Azure AD Connect health agent also gets installed and configured to assist in the health, monitoring, and troubleshooting of the synchronization agent by default. The **Azure AD Connect Health** dashboard, including a blade for **Troubleshooting + Support**, can be accessed by visiting `https://portal.azure.com/#blade/ Microsoft_Azure_ADHybridHealth/AadHealthMenuBlade/SyncServicesList`.

Next, we will continue our lab configuration by enabling self-service password reset for a few soon-to-be-created on-premises user accounts, reviewing permission grants, policy recommendations and best practices, and overall secure configuration.

Configuring self-service password reset (SSPR)

Continuing the focus on securing a hybrid AD infrastructure, one of the focal points of any organization regarding security and productivity is the availability of a good password reset tool or utility. Typically, this process walks a user through a self-service application or portal, where they need to confirm their identity using some factor, be it a generated code, a security key, an SMS text, a phone call, or a series of challenge and response questions. Getting this correct in an organization not only makes an impact on security benefits but overall user and IT administrator productivity, as well as providing a huge financial benefit with a user self-service approach.

Azure AD brings a **Self-Service Password Reset** (**SSPR**) approach to the table, reducing the need for administrator or help desk support while allowing the user to reset, unblock themselves, and return to work in a short amount of time for hybrid accounts. In a relatively short amount of time, this configuration can be established and validated by enabling the self-service reset for a small group of Azure AD users, using defined authentication methods and an array of registration options based on the organization's requirements. Let's begin a walk-through of this configuration using our lab environments to get some hands-on experience:

1. Open **Hyper-V Manager** on the device hosting the virtual machines.

2. Right-click on the **AZ801PacktLab-DC-01** virtual machine and click **Connect**.

3. From the **Virtual Machine Connection** window, we can either click the **Start** button in the center of the screen or click the green **Start button** in the taskbar at the top of the screen.

4. Use the **Action|Ctrl + Alt + Delete** menu options to begin, then log in to **AZ801PacktLab-DC-01** as an administrator using `Packtaz801guiderocks` as the password.

5. Using the **Start** menu, search for or select **Active Directory Users and Computers**.

6. Expand **Active Directory Users and Computers|AD.az801.com|Users**.

7. Right-click on **Users**, then click **New|User** as shown in *Figure 3.18*:

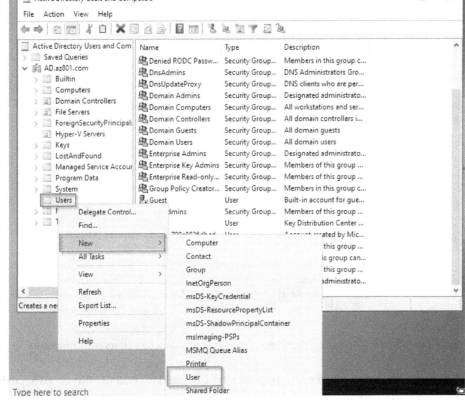

Figure 3.18 – Create a new test user

8. Create a new user using the following details:

- **First name**: Test

- **Last name**: User1

- **Full name**: Test User1

- **User logon name**: testuser1

- **Password** (and confirmation): Packtaz801guiderocks

9. Click **Finish** to complete the setup and repeat *step 8*, replacing the number 1 with 2 to create testuser2.

10. While remaining in **Active Directory Users and Computers**, select the **View** menu and ensure that you have **Advanced Features** selected, as shown in *Figure 3.19*, to begin the necessary steps for enabling SSPR writeback to AD:

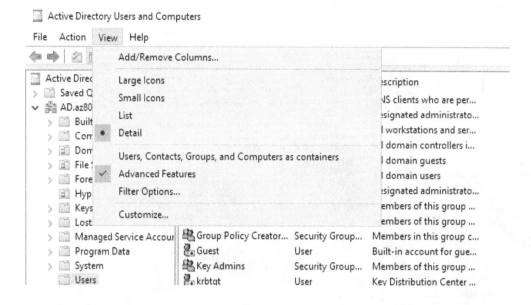

Figure 3.19 – Showing Advanced Features options in Active Directory Users and Computers

11. Go through **AD.az801.com|Properties|Security|Advanced** to open the **Advanced Security Settings for AD** screen.

12. Click the **Select a principal** link, then type MSOL to find our Azure AD Connect account and select **OK**.

13. In the **Applies to** drop-down list, select **Descendant User objects** and then select the following properties and permissions from the open screen:

   - Reset password (under permissions)

   - Write lockoutTime (under properties)

   - Write pwdLastSet (under properties)

14. Click the **OK** button three times to accept and apply the changes.

15. Right-click on the **AZ801PacktLab-FS-01** virtual machine and click **Connect**.

16. From the **Virtual Machine Connection** window, we can either click the **Start** button in the center of the screen or click the green **Start button** in the taskbar at the top of the screen.

17. Use the **Action | Ctrl + Alt + Delete** menu options to begin, then log in to **AZ801PacktLab-FS-01** as an administrator using `Packtaz801guiderocks` as the password.

18. Using the **Start** menu, find and select **Azure AD Connect configuration wizard**.

19. On the **Welcome** screen, click **Configure** to continue when prompted.

20. On the **Additional tasks** screen, click **Customize synchronization options** and then click **Next**.

21. On the **Connect to Azure AD** screen, use your Global Administrator credential for your Azure tenant and click **Next** to continue.

22. On the **Connect your directories** and **Domain and OU filtering** screens, click **Next**.

23. On the **Optional features** screen, click **Password writeback** as shown in *Figure 3.20* and click **Next** to continue:

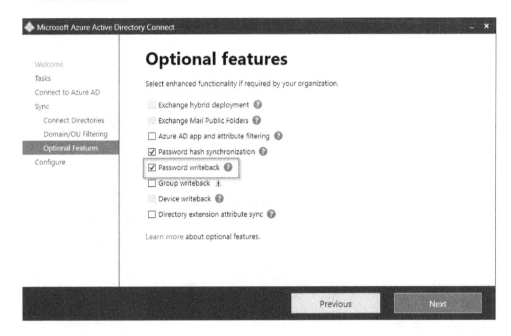

Figure 3.20 – Selecting Password writeback as an optional feature in Azure AD Connect

24. On the **Ready to configure** screen, select **Configure** and wait for the **Configuration complete** screen to appear. Click **OK** to close the configuration editor.

25. By this point in the lab, the two test accounts we created should have synchronized to Azure AD successfully, so we will continue our last steps for enabling password writeback for SSPR by visiting `https://portal.azure.com` as Global Administrator in a web browser.

26. As a best practice, let's first create a new Azure AD group containing the two users we created by searching for and selecting **Azure Active Directory** from the portal, then selecting **Groups** from the blade. Select **New group** to begin the process.

27. Create the new **Security** group using the name, description, and members specified (by searching for a user and selecting both **testuser1** and **testuser2** to add as members), as shown in *Figure 3.21*. Click **Create** to continue with the creation of the new group:

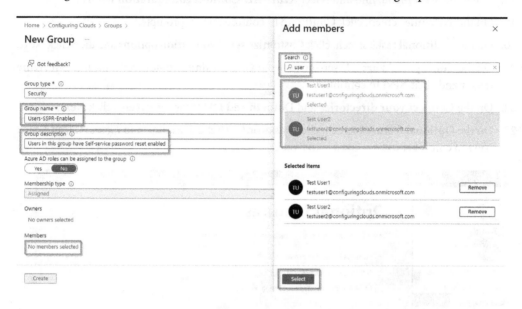

Figure 3.21 – Creating a new Users-SSPR-Enabled group with two test users as members

28. Return to **Azure Active Directory** and select **Password reset** from the blade, then select **Properties**. For **Self-service password reset enabled**, choose **Selected**, then select **Users-SSPR-Enabled** as shown in *Figure 3.22*. Click the **Save** button to commit the changes:

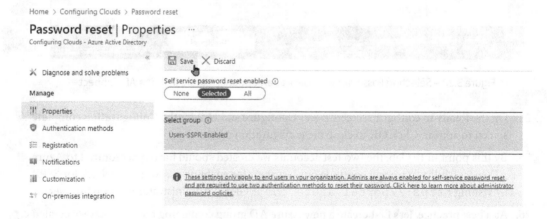

Figure 3.22 – Enabling SSPR for a selected group

29. Select **Authentication methods** from the blade, change **Number of methods required to reset** to 2 to achieve best practice configuration, and place a checkmark in **Mobile App notification** and **Mobile app code**. Click **Save** to commit the changes.

30. Finally, click **On-premises integration** and place a checkmark on **Write back passwords to your on-premises directory** to begin the final setup step. You should see the page reflect *Figure 3.23*, indicating that the on-premises client is running:

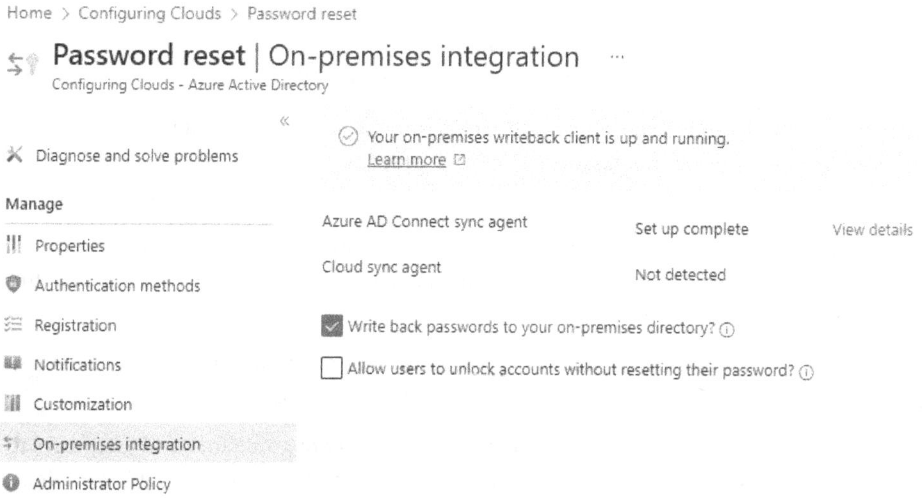

Home > Configuring Clouds > Password reset

⟵ **Password reset** | On-premises integration   ⋯
   Configuring Clouds - Azure Active Directory

《

✕ Diagnose and solve problems

Manage

⏸ Properties

🛡 Authentication methods

≋ Registration

🖼 Notifications

▥ Customization

⤴ On-premises integration

ⓘ Administrator Policy

✓ Your on-premises writeback client is up and running.
   Learn more ☑

Azure AD Connect sync agent   Set up complete   View details

Cloud sync agent   Not detected

✓ Write back passwords to your on-premises directory? ⓘ

☐ Allow users to unlock accounts without resetting their password? ⓘ

Figure 3.23 – Completing the password writeback setup and configuration on-premises

In this section, we learned about custom domain names regarding acquisition, application, and validation in a hybrid identity infrastructure. We discussed the various hybrid identity approaches with a comparison of what each brings to the organization. We then worked to establish Azure AD Connect and an Azure AD Connect health agent with Azure AD in a best-practices configuration (avoiding installation on a domain controller). We then established SSPR and created a few test accounts to validate SSPR later in this chapter. The next section will discuss managing password policies across a hybrid infrastructure, including validation of SSPR and additional password protection features.

Configuring and managing password policies

The protection of identities in general has become the foremost security focus for companies, and with that comes a myriad of choices to help protect the identities to decrease or mitigate risks. For managing identities in a hybrid fashion, we have many options in both on-premises and in Azure AD to help guide advanced protection and avoid security breaches. In this section, we will discuss the available default AD password policies and advanced password policies and review banned password lists, otherwise known as Azure AD password protection.

The configuration location and various types of default password policy options in AD encompass complexity rules, length, age, expiration rules, and encrypted storage and are detailed as follows:

- On your domain controller within the Group Policy Management Console, all policy settings are located under `Computer Configuration\Windows Settings\Security Settings\Account Policies\Password Policy`.

- **Enforce password history** simply helps to deter password reuse by setting a specific number of unique new passwords that must be used for an account before an old password can be reused. The best practice setting is **24 passwords remembered** in combination with a **Minimum password age** value higher than 1 day and a **Maximum password age** value of at least 90 days.

- **Maximum password age** is a setting that determines the time in days that a password can be used before needing to reset the password. The best practice setting is at least 90 days, with a maximum of 999 days.

- **Minimum password age** is a setting that determines the time in days that a password can be used before needing to reset the password. The best practice setting is 1 day, with a countermeasure of 2 days and a maximum of 998 days.

- **Minimum password length** is a setting that determines the minimum number of characters that can make up a password. The best practice setting is a value of 14.

- **Password must meet complexity requirements** enables a set of guidelines and categories to establish a strong password. When this setting is used (the best practice setting is **Enabled**), the password may not contain the user's display name or the **samAccountName** value, and the password must contain characters from three of the following categories:

  - Lowercase letters

  - Uppercase letters

  - Base 10 digits (0-9)

  - Non-alphanumeric characters

  - Any Unicode characters

- **Store passwords using reversible encryption** is a setting that allows for compatibility with applications that utilize certain protocols that require the user's password for authentication, allowing the password to be decrypted. The best practice setting is **Disabled**, and this setting is not required when creating a PSO as discussed next.

While AD has only one password policy that is applied to all user accounts on-premises (defined in the default domain policy), we also have advanced policies called **Fine-Grained Password Policies** (**FGPPs**) that allow for even stricter control on privileged accounts within AD or Azure AD DS. These controls allow for setting and managing multiple password, lockout, and restriction policies based on the type of account being protected by the policy. Only a domain administrator or delegated administrator can set these policies, and the policies can only be applied to a security group or an OU shadow group. Let's discuss how these FGPPs are created and stored in AD DS.

The Password Settings Container (**PSC**) is a default container as part of the AD schema and is located under the system domain container. This object stores **Password Settings Objects** (**PSOs**) for the domain, and can only be viewed when advanced features are enabled in **Active Directory Users and Computers**.

Password Settings Objects (**PSOs**) contain all the settings that can be configured in the default domain policy (minus the Kerberos settings) and contain the following additional attributes:

- Account lockout settings:

  - **Account lockout duration**

  - **Account lockout threshold**

  - **Reset account lockout after**

- Additional attributes:

  - **PSO link** identifies the group or users to whom the password settings object directly applies to

  - **Precedence** is a number value that is used to prioritize multiple PSOs

A PSO must contain all nine password settings attributes. Policy creation along with settings management can be completed using AD Administrative Center, so let's walk through how this is completed:

1. While connected to the **AZ801PacktLab-DC-01** virtual machine, from the **Start** menu, select **Windows Administrative Tools**, then **Active Directory Administrative Center**.

2. Select the **AD (local)** domain, then double-click the **System** container.

3. Locate **Password Settings Container** and right-click to select **New|Password Settings**, as shown in *Figure 3.24*:

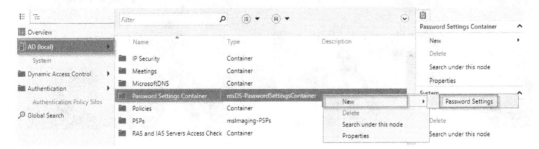

Figure 3.24 – Creating a new password settings object

4.   Configure your password settings object as shown in *Figure 3.25* and click **OK** to apply the best practices configuration:

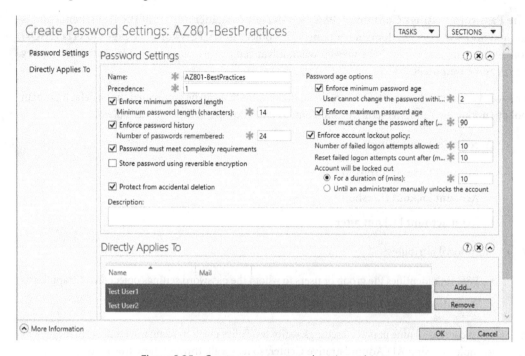

Figure 3.25 – Create a new password settings object

5.   This new PSO can also be managed and viewed using PowerShell, as shown in *Figure 3.26*:

```
PS C:\Users\Administrator> Get-ADFineGrainedPasswordPolicy -filter *

AppliesTo                    : {CN=Test User2,CN=Users,DC=AD,DC=az801,DC=com, CN=Test
                               User1,CN=Users,DC=AD,DC=az801,DC=com}
ComplexityEnabled            : True
DistinguishedName            : CN=AZ801-BestPractices,CN=Password Settings Container,CN=System,DC=AD,DC=az801,DC=com
LockoutDuration              : 00:10:00
LockoutObservationWindow     : 00:10:00
LockoutThreshold             : 10
MaxPasswordAge               : 90.00:00:00
MinPasswordAge               : 2.00:00:00
MinPasswordLength            : 14
Name                         : AZ801-BestPractices
ObjectClass                  : msDS-PasswordSettings
ObjectGUID                   : a52e3303-8592-4c0d-b68a-a9b95829bced
PasswordHistoryCount         : 24
Precedence                   : 1
ReversibleEncryptionEnabled  : False
```

Figure 3.26 – Reviewing the new PSO using Get-ADFineGrainedPasswordPolicy

While FGPPs cover on-premises and Azure AD DS, Azure AD Password Protection provides another layer of security for all users. Now that we have reviewed FGPPs, let's discuss what custom banned passwords for Azure AD Password Protection provide for hybrid AD security.

Enabling password block lists

Let's be honest, we all have moments of balancing convenience and security and tend to create passwords that indicate our favorite food, college or college mascot, special date, location, or (gasp) the name or nickname of a well-known person or organization. With the advent of Azure AD's custom banned password list, we not only rely on Microsoft's global list of curated bad passwords but can also apply custom terms, organization, and internal buzzwords to adhere to organizational requirements and adopt an added layer of security protection for the users and the organization.

Let's discuss the steps necessary to enable a blocked password list in Azure AD, as well as additional options for the protection of on-premises AD environments:

1. Visit `https://portal.azure.com` as Global Administrator in a web browser.

2. Search for or select **Azure AD Password Protection**.

3. On the **Password Protection** screen, select **Yes** for **Enforce custom list**. Then, add any custom terms you would like to validate with and change the mode from **Audit** to **Enforced**, then select **Save** as shown in *Figure 3.27*. Note that this may take a few hours to be enabled by the tenant:

Figure 3.27 – Configuring a custom banned password list in Azure AD

4. Sign out of the Azure portal and then sign in as `testuser1@{yourtenantname}.onmicrosoft.com` to `https://aka.ms/ssprsetup` to complete registration for an SSPR.

5. Once you have completed setting up at least two factors (recommending **Microsoft Authenticator** and **SMS**), select **Done** to continue.

6. Now, navigate to `https://aka.ms/sspr` to complete a password reset using one of the custom blocked terms. You should be presented with an error message, such as the one shown in *Figure 3.28*:

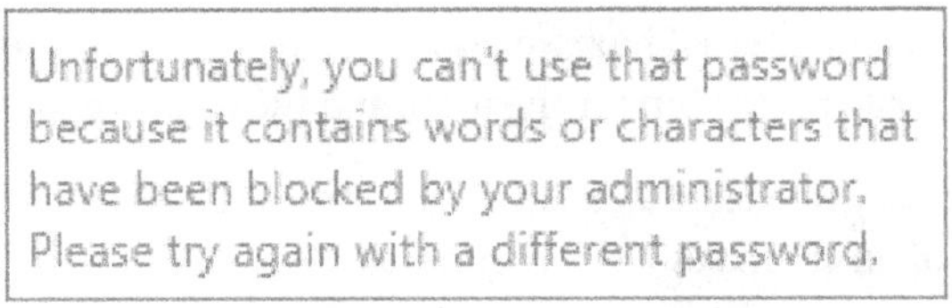

> Unfortunately, you can't use that password because it contains words or characters that have been blocked by your administrator. Please try again with a different password.

Figure 3.28 – Warning from use of a custom banned password

At this point, we have completed the custom blocked password demo and will now discuss key details surrounding similar password protection on-premises.

For a hybrid environment with AD also running on-premises, we can deploy an Azure AD Password Protection proxy and DC agent. There are myriad details surrounding the planning and deployment of Azure AD Password Protection on-premises, with a great read at `https://docs.microsoft.com/azure/active-directory/authentication/howto-password-ban-bad-on-premises-deploy`. For the purposes of this exam guide, the following are key details surrounding Azure AD Password Protection:

- Password change events do not occur on read-only domain controllers, so the **Azure AD Password Protection DC agent** software does not need to be installed (it's also not supported).

- Any AD domain must be fully running **DFSR** for **sysvol** replication, and not **FRS** (additional migration details from **FSR** to **DFSR** can be found at this URL: `https://docs.microsoft.com/en-us/windows-server/storage/dfs-replication/migrate-sysvol-to-dfsr`).

- A general recommendation of two Azure AD Password Protection proxy servers in an environment is sufficient for a high-availability approach.

- All machines running the agent and proxy services must have at least **.NET Framework 4.7.2** and **Universal C Runtime** installed.

- **Key Distribution Service** must be enabled on all domain controllers.

- There must be network line-of-sight connectivity between at least one domain controller and the member server hosting the proxy service and must allow RPC endpoint mapper port 135 and the RPC server port on the proxy service.

- The proxy service must be able to communicate outbound via TLS 1.2 to known network endpoints of https://login.microsoftonline.com and https://enterpriseregistration.windows.net for Azure service connectivity.

- Member servers hosting the proxy service must allow domain controllers the ability to log onto the proxy service via **Access this computer from the network** policy assignment located at **Security Settings | Local Policies | User Rights Assignment**.

- It is strongly recommended to deploy in audit mode first to generate password results in an event log for further reviewing insecure passwords, existing policies, and procedures, and establishing user experience documentation and training prior to enabling an enforcement mode.

- Do not install the **Azure AD Password Protection Proxy** and **Azure AD Application Proxy** agents on the same machine as they will be in conflict with each other.

Now that we have learned about Azure AD Password Protection for both on-premises and in the cloud to help users achieve better password management and hygiene, let's move toward managing an account and user security.

Configuring and managing account and user security

One of the most important aspects of managing a hybrid identity model is that the user identity lifecycle is fully managed from account creation, through daily usage, to eventual account deletion. This holds true for both standard user accounts as well as system or service accounts. In this section, we will discuss the protection of accounts, users, credentials, and overall account security.

Managing protected users

Within every AD environment, there exists a default collection of highly privileged groups and accounts that are secured with an automatic service that enforces template permission policies on the group and accounts (via the **AdminSDHolder** object in AD), returning the object's permissions back to defaults rather quickly. As they are well-known objects, they have this permission and service persistence no matter where the object is stored in AD. Such objects in AD are considered protected accounts and protected groups, and are listed as follows:

- Account operators
- Administrator
- Administrators
- Backup operators

- Domain admins
- Domain controllers
- Enterprise admins
- Krbtgt
- Print operators
- Read-only domain controllers
- Replicator
- Schema admins
- Server operators

The best practices for managing these accounts and groups are as follows:

- Institute a least-privileges model that incorporates both a new group and new low-privilege management accounts to give just enough access permission set to trusted administrators. Do not use a standard account as a member of these groups.

- Ensure there are restrictions on the management accounts, controlling where they can be used and what they can be used for. It is recommended to utilize administrative workstations that help to achieve this privileged management.

- Limit the use of privileged groups and accounts with no standing membership in the privileged groups to help mitigate pass-the-hash and other credential theft risks.

In addition, there is a security group that is designed to be proactively secure and restrictive by default, which helps limit the availability of weak encryption algorithms, outdated authentication protocols, and sensitive system accounts. This group is named the **Protected Users** group, and while it does not add 100% protection to the group members, it does add a strong layer of protection for many organizations. Let's quickly discuss what the protected user's group provides and some caveats to point out:

- Caching of the users' credentials will be denied (such as **NTLM**, **CredSSP**, or a cached verifier during unlock or sign-in). Moreover, the Kerberos delegation will be blocked as the accounts are considered sensitive.

- Membership of the protected user group does not change the **AdminCount** attribute, a value that is frequently looked for to indicate membership of other protected groups, such as enterprise admins or domain admins.

- Do not add accounts for computers and services as authentication will fail.

- Do not add **Managed Service Accounts (MSAs)** or **group Managed Service Accounts (gMSAs)** as they utilize Kerberos constrained delegation and will break.

- This must be thoroughly tested with a few accounts before placing all management accounts into the protected user group.

Continuing with managing accounts in an AD or hybrid model, we will next discuss how to configure account security using **Local Administrator Password Solution (LAPS)**.

Configuring account security

To address years of debate over the age-old question of "Should I rename and/or disable the local administrator account on my devices?", Microsoft created a tool called LAPS to help mitigate against pass-the-hash attacks for both Windows workstations and Windows Servers. Originally part of **Securing Lateral Access Movement** (yes, it was once called **SLAM**), this LAPS component ensures that the local administrator passcode is securely rotated and stored securely within AD with delegated **Access Control Lists (ACLs)**. This ensures that every device has a different local administrator password and only approved administrators can retrieve the stored passwords for the devices (when needed).

While we will not walk through the entire setup and configuration of LAPS in an environment, it is important to note that this is a GPO-based **client-side extension (CSE)** that fires each time a GPO attempts to update (whether in the foreground or as a background GPO refresh). This CSE then checks to determine whether the password has expired, generates a new password if it has, reports and stores the password and the new expiration date and time into AD, and then finally changes the password on the local administrator account.

The following two supporting links identify how to download, plan for, and configure LAPS, and how Microsoft Defender for Identity helps to enforce this mitigation by assessing devices that have not updated their local administrator account passwords:

- `https://www.microsoft.com/download/details.aspx?id=46899`
- `https://docs.microsoft.com/defender-for-identity/cas-isp-laps`

In addition to LAPS, there are also extensive controls that can be established for the built-in administrator accounts in AD. At a high level, we understand that this account is sensitive and as such should require additional security for interactive logon, should not be delegated, and should have additional restrictions set on both user rights and permissions for the administrator account. For full in-depth recommendations on how to secure the built-in administrator account, see `https://docs.microsoft.com/windows-server/identity/ad-ds/plan/security-best-practices/appendix-d--securing-built-in-administrator-accounts-in-active-directory` for a step-by-step walk-through of the configurations.

Continuing with managing accounts in an AD or hybrid model, we will next discuss how to manage account security on a **Read-Only Domain Controller (RODC)**.

Managing account security on a Read-Only Domain Controller (RODC)

An RODC's purpose in an AD environment is to provide domain services to a branch office that lacks good physical security (or where the storage of domain credentials is prohibited), lacks local expertise for management support, or experiences poor network bandwidth.

Management of an RODC is somewhat flexible and can be delegated to a group or user to help administer the branch controller, noting that this delegation does not allow the delegated administrator access to any other domain controller in the domain. The RODC only requires inbound replication and prevents any accidental or unnecessary information from being written to the branch controller through filtered attribute sets to help designate what can and cannot be replicated (including confidential attributes such as BitLocker recovery keys and PKI master keys).

When deploying an RODC for the first time, you must configure a **Password Replication Policy** (**PRP**) that is based on one of the three following models:

- **No accounts cached** ensures that there are no passwords replicated to the RODC, except for the computer account of the RODC and its own **krbtgt** account.

- **Most accounts cached** includes a significant section of the user accounts and ultimately permits offline operation.

- **Few accounts cached** is restrictive and requires administrators to explicitly identify a set of user and computer accounts that can be cached and replicated.

PRP administration options for the RODC (and necessary administrative credentials or roles) are as follows:

Administrative task	Administrative role or credential needed
Configure the PRP for an RODC	Must be a member of the domain admins group
View the current credentials cached on an RODC	Viewable by any domain user
Prepopulate the password cache for an RODC	Must be a member of the domain admins group
Review the accounts that attempted authentication to an RODC	Viewable by any domain user
Reset the currently cached credentials on an RODC (if compromised)	Must be a member of the domain admins group

Table 3.2 – PRP administrative tasks and the role or credential needed

Now that we have reviewed how to manage account security on an RODC, let's move on to securing domain controllers in a hybrid AD infrastructure.

Securing domain controllers

Anyone in the world of security knows that a knowledge base was published by Microsoft well over 20 years ago discussing the 10 immutable laws of security, covering years of experience to help guide organizations in establishing solid security principles (this is still available at `https://docs.microsoft.com/previous-versions//cc722488(v=technet.10)?redirectedfrom=MSDN`). Though there was a version 2.0 published at one point, the original set of laws and specifically lucky #7 stuck with me all these years: *The most secure network is a well-administered one*. This couldn't be truer even to this current day – a solid set of documentation, policies, and procedures ensure proper governance and compliance with responsibilities and operating procedures. Failure to adhere to this set of governing policies could lead to vulnerabilities, misconfigurations, outdated applications, and missing patches that can be mitigated and even more easily exploited by an attacker. With that said, let's begin the discussion on some tactics and best practices to properly secure domain controllers in a hybrid identity model.

Hardening domain controllers

Domain controllers must be treated with protection and security as part of the privileged access plane as part of the enterprise access model, with additional reading available at `https://docs.microsoft.com/security/compass/privileged-access-access-model`. As such, they must be properly managed by trained IT staff, requiring a high level of control, secure installation, and configuration to protect these devices against compromise. Let's discuss some of the best practice settings and recommendations on how to harden domain controllers in your organization.

For data center domain controllers, you must ensure solid physical security practices are in play for both physical and virtual domain controller workloads:

- Physical domain controllers should be securely installed onto physical hosts that are in disparate server racks from other member servers. These servers should utilize BitLocker drive encryption on locally attached storage and have a **Trusted Platform Module** (**TPM**) to mitigate vulnerabilities.

- Virtual domain controllers should also be securely installed onto separate physical hosts away from other server workloads. These virtual domain controllers should also utilize non-shared storage to prevent inadvertent access to the DC virtual machine files. If the virtual DCs are to be run on shared workload hosts, it is recommended to utilize Hyper-V and Microsoft's Shielded VM technology to incorporate a more comprehensive security solution for the protection of your virtual DC workloads. For more details on Shielded VM technology, visit `https://docs.microsoft.com/windows-server/security/guarded-fabric-shielded-vm/guarded-fabric-and-shielded-vms`.

- Remote or branch location domain controllers should be protected as detailed previously for physical and virtual workloads, ensuring that the domain controller is installed on a host in a locked room or secured area. If these physical security recommendations cannot be met, it is advised to consider deploying an RODC in the branch location(s). The planning and deployment considerations are located at `https://docs.microsoft.com/previous-versions/windows/it-pro/windows-server-2008-r2-and-2008/cc771744(v=ws.10)`.

- Utilize the Security Compliance Toolkit, as discussed in the *Implementing operating system security by using Group Policy* section of *Chapter 2*, paying close attention to the MSFT Windows Server 2022 policies for the `Domain Controller Virtualization Based Security` and `Domain Controller` templates, depending on the type of domain controller being protected. This includes enabling proper audit policy settings in Group Policy.

- Patch management and vulnerability scanning for domain controllers should be handled separately from other member server patching configurations and routines. This ensures that if or when there is a compromise on your management system(s), the compromise is isolated and does not affect the privileged access tier. This also assists in stricter management of patches and reduces unnecessary applications on the domain controllers.

- Minimize or avoid additional software or server role installation on domain controllers, ensuring that only the necessary management tools or applications are installed and a minimalistic set of roles/features are configured (e.g., DNS and DHCP).

- Use the latest Windows Server operating systems that are supported within your organization whenever possible to incorporate any new secure by default settings within the operating system, while staying within available support models and updated security and hotfix availability.

- Have a solid backup and recovery plan that includes an incident response policy and plan, procedures for incident handling and reporting, prioritization of servers and recovery mode options, and scheduled training and walk-throughs of the disaster recovery plan with the regular validation of object-level backup and restoration efforts.

- Blocking internet access and preventing web browsing from domain controllers. Domain controllers should have this traffic strictly controlled if needed.

- Additional perimeter firewall controls to prevent outbound connections from the domain controllers to the internet, while still allowing inter-site traffic. A great resource for the AD DS firewall recommended configuration can be found at `https://docs.microsoft.com/troubleshoot/windows-server/identity/config-firewall-for-ad-domains-and-trusts`.

- Use the latest **Active Directory Federation Services** (**AD FS**) and Azure security features such as smart lockout and IP lockout, attack simulations, MFA authentication, Azure AD Connect Health, and both banned and custom banned passwords to thwart password spraying, phishing, and account compromise.

Having covered a great set of best practices for hardening domain controllers, let's now learn about how best to restrict access to domain controllers to ensure secure identity management for your organization.

Restricting access to domain controllers

As a general best practice, only **secure administrative workstations** (**SAWs**) should be used to remotely manage domain controllers in an organization. This *allow only authorized users on authorized computers* approach requires a mutually trusted host that is just as secure as the domain controller, strong authentication and MFA or SmartCard for access, and should abide by physical security for these administrative workstations (sometimes called *jump servers*). Like the recommendations made in the *Hardening domain controllers* section in this chapter, SAWs should follow the same rigors and practices to ensure utmost security, utilizing RDP restrictions and tools such as AppLocker and Windows Defender Application Control.

More details on SAWs can be found at the following URL:

```
https://docs.microsoft.com/windows-server/identity/ad-ds/plan/
security-best-practices/implementing-secure-administrative-hosts
```

For our final discussion in this section, we will review authentication policies and authentication policy silos and what these features bring to identity management.

Configuring authentication policies silos

In addition to protecting users, authentication policies were introduced to provide additional configurations for restricting accounts at the authentication and ticket-granting exchanges of Kerberos authentication to help mitigate credential exposure and theft. This authentication policy can then be applied to either a user, computer, or MSA/gMSA. These restrictions allow for custom lifetimes of tokens and the restriction of a user account to specific devices, and they can also help to sequester workloads that are unique to an organization by defining an object relationship within an authentication policy silo.

To get a basic idea of how authentication policies and authentication policy silos are created, you can follow along with the images that follow in your **AZ801PacktLab-DC-01** virtual machine using **Active Directory Administrative Center**, as shown in *Figure 3.29*:

Figure 3.29 – Using Active Directory Admin Center to review the
creation of authentication policies and silos

In *Figure 3.30*, you will notice that once you assign a display name to your authentication policy, you have two options: **Only audit policy restrictions** and **Enforce policy restrictions**. The best practice here is to always start by auditing your configuration so that you can review logging and make an assessment on how best to proceed with configuration changes. Notice that you can assign multiple accounts, and if the policy was already assigned to a silo, it would appear under **Assigned Silos**:

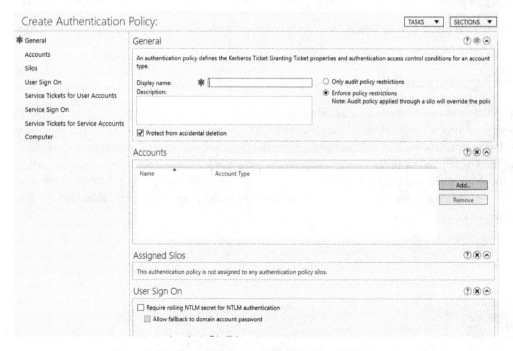

Figure 3.30 – The initial Create Authentication Policy screen

Moving on through the authentication policy screen as shown in *Figure 3.31* and *Figure 3.32*, you will see additional control conditions for handling **NTLM authentication**, **Ticket Granting Service**, **Service Tickets for User Accounts**, and **Service Sign On**. These settings allow for rule-based and compound authentication and Kerberos armoring configurations using dynamic access control as the basis:

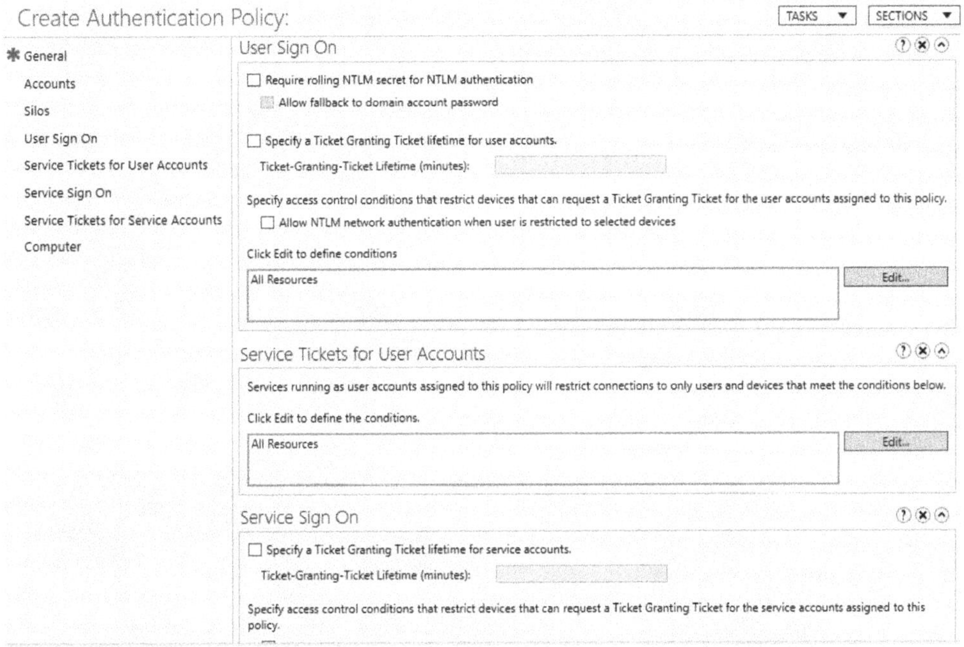

Figure 3.31 – Create Authentication Policy, continued

Note that you can specify conditions for issuing service tickets for user accounts in addition to specifying conditions for specific computers as part of the authentication policy, as shown in *Figure 3.32*:

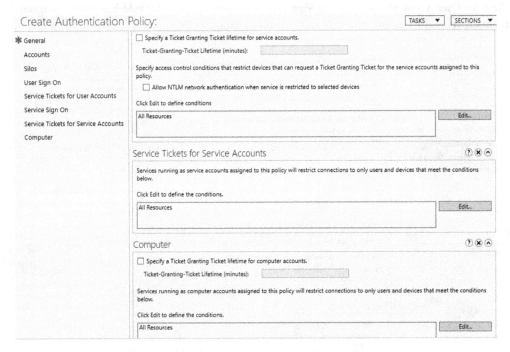

Figure 3.32 – Create Authentication Policy, continued

As you can see from these screenshots, this is quite an in-depth topic and a deeper step-by-step review of this advanced topic is located at the following URL: `https://docs.microsoft.com/windows-server/identity/ad-ds/manage/how-to-configure-protected-accounts`.

Managing AD built-in groups and delegation

Over the course of the past few chapters and discussions, we have learned that several built-in accounts and groups are automatically created during the installation of AD, and a handful of additional built-in groups are installed as well when adding different services to AD, such as Microsoft Exchange. These users and groups that are created are in two specific containers:

- **Builtin** – objects in this container have a domain local scope
- **Users** – objects in this container can have either a domain local, global, or universal scope

A great resource that covers more in-depth details on security groups, group scope, and the default security groups can be found at this URL: `https://docs.microsoft.com/previous-versions/windows/it-pro/windows-server-2012-R2-and-2012/dn579255(v=ws.11)`

Managing AD built-in administrative groups

In the *Managing protected users and groups* section earlier in this chapter, we identified all the built-in AD users and groups and alluded to AD as having a model that follows the least privilege model for rights and permission assignments. Additional privileges based on role requirements can then be delegated to additional administrators so that they can perform necessary tasks for their roles and responsibilities.

For the purposes and scope of this AZ-801 exam guide, we will focus on the four main built-in groups:

- The **Enterprise Admins** group appears at the forest root domain and allows members full control over every domain in the forest. Members of this group are automatically added to the Administrators group on every forest domain controller. This group should only be used during initial forest creation or when setting up a new outbound forest trust.

- The **Domain Admins** group has full control in a domain and adding accounts as members should be used sparingly. A proper delegation model should be established to delineate roles and permissions as necessary.

- The **Administrators** group is the most powerful of these four and has full control over the domain. By default, Domain Admins, Enterprise Admins, and the Administrator account are members of this group.

- The **Schema Admins** group allows members to modify the AD schema and has few rights and administrative permissions beyond modification of the schema.

To complete an in-depth review of all of the built-in administrative groups and the recommended user rights assignments for each, please review the following URL: `https://docs.microsoft.com/windows-server/identity/ad-ds/plan/security-best-practices/appendix-b--privileged-accounts-and-groups-in-active-directory`.

To end this section, we will learn how to manage the delegation of privileges in an AD DS environment.

Managing AD delegation of privileges

To begin, **Delegation of Control Wizard** is accessed through the **Active Directory Users and Computers** tool. This wizard is typically used to select groups or users to whom you would like to delegate certain tasks, permissions, and responsibilities within your domain. This tool is primarily utilized so that privileged accounts are granted just enough administration permissions to complete the task without adding to a built-in group or additional protected group.

Please feel free to follow along by using your **AZ801PacktLab-DC-01** virtual machine to complete the steps that follow:

1. From the **Start** menu, open **Active Directory Users and Computers** and then expand the root domain AD.az801.com.

2. Locate and right-click on the **Users** container, then select **Delegate Control…**, as shown in *Figure 3.33*:

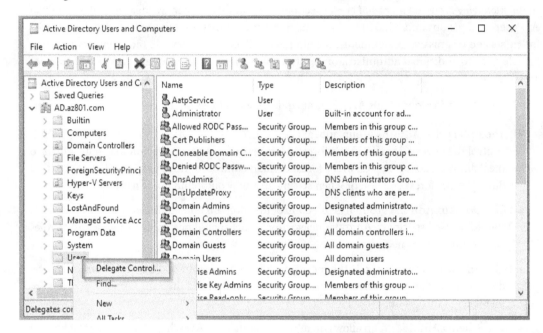

Figure 3.33 – Starting Delegation of Control Wizard

3. Click **Next** on the **Welcome to the Delegation of Control Wizard** screen, then on the **Users or Groups** screen, click the **Add** button and search for testuser2. Then, select and add the user, then click **Next** to continue.

4. On **Tasks to Delegate**, feel free to scroll through the list to see all available common tasks, select **Reset user passwords and force password change at next logon**, then click **Next** to continue, as shown in *Figure 3.34*. Note that you can also create a custom task to delegate to users and groups:

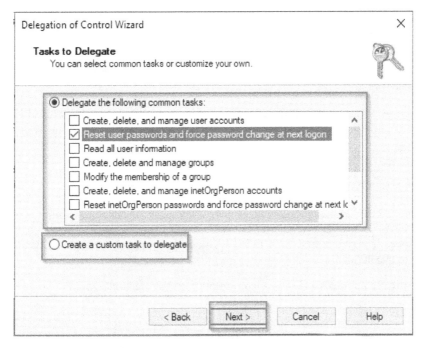

Figure 3.34 – Delegate the reset-user-passwords privilege to testuser2

5. Click **Finish** on the final screen to complete this exercise.

This completes the section on how to manage the delegation of privileges in an AD DS environment. Next, we will learn about getting started with implementing and managing Microsoft Defender for Identity.

Implementing and managing Microsoft Defender for Identity

As a best practice, Microsoft recommends utilizing the power of the cloud to protect all identities, whether on-premises or in the cloud, and the tool of choice is Microsoft Defender for Identity (formerly known as **Advanced Threat Protection (ATP)**). As we learned previously with Azure AD Password Protection, Microsoft Defender for Identity relies on a cloud configuration and agents/sensors installed on all identity management servers in your on-premises environment. This includes all domain controllers, ADFS servers, and any additional devices identified as part of the Access Management plane, and all one-way traffic can be configured and controlled through a proxy service that allows traffic to specific endpoints.

Utilizing Microsoft Defender for Identity provides security professionals the ability to quickly detect and respond to attacks in a hybrid infrastructure, utilizing best-in-class user activity monitoring and analytics based on machine learning and AI capabilities; identify and prevent suspicious user activities; protect all user identities no matter where they are stored; and provide a clear and consistent incident that is actionable from the start.

Let's walk through an overview of the implementation of the Microsoft Defender for Identity setup and installation and configuration of the agent on-premises, as well as getting a view into the Microsoft Defender for Identity configuration and management portal.

Microsoft Defender for Identity prerequisites

The prerequisites for the installation and management of Microsoft Defender for Identity are as follows:

- The **Npcap** driver must first be installed (without the loopback settings) on the device you intend to install the Microsoft Defender for Identity sensor

- **Microsoft .NET Framework 4.7** or later installed on the machine

- A **Microsoft Defender for Identity** instance with trial or full licensing

- A Global Administrator account or security administrator on the tenant you are using to access and manage the **Identity** section within the **Microsoft 365 Defender** portal

- An AD services account configured with specific privileges on-premises to be used for connection to Active Directory on-premises environment

- The access key from your Defender for Identity instance as well as the downloaded package containing the **Defender for Identity agent** installation

Installation and management process overview

There are two separate approaches and portal experiences currently, so we will walk through a high-level setup using the new Microsoft 365 Defender experience:

1. After going to https://security.microsoft.com, then **Microsoft 365 Defender,** and establishing your tenant for **Microsoft Defender for Identity**, we will continue by visiting the **Settings** and **Identities** blades, as shown in *Figure 3.35*:

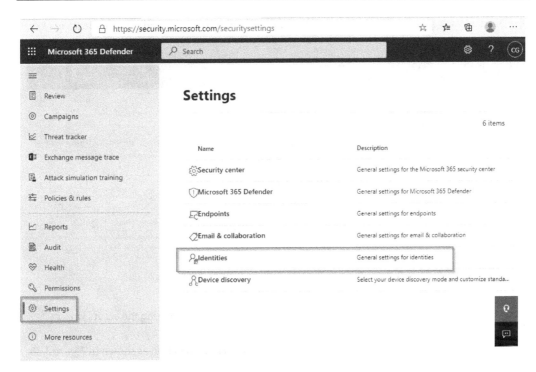

Figure 3.35 – Initial setup of Microsoft Defender for Identity via Microsoft 365 Defender

2. Ensure the prerequisites are met in terms of the domain controller, AD FS server, or other identity management server and that the appropriate AD services account has been properly established, as shown in *Figure 3.36*:

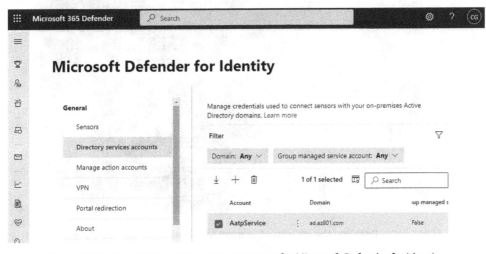

Figure 3.36 – Establishing AD services accounts for Microsoft Defender for Identity

3. Next, we will download the **Microsoft Defender for Identity** agent and capture the access key, as shown in *Figure 3.37*:

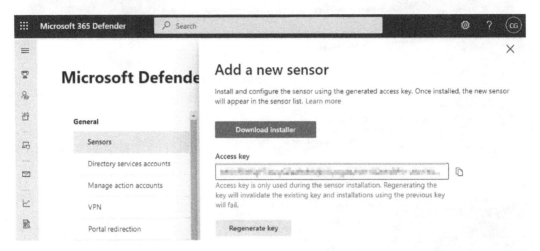

Figure 3.37 – Adding a new sensor, downloading the agent, and capturing the access key

4. At this point, we can begin the Defender for Identity agent installation as shown in *Figure 3.38*:

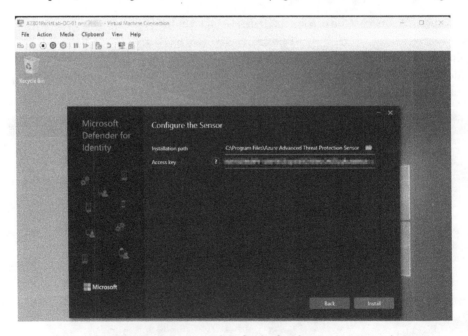

Figure 3.38 – Installing the Microsoft Defender for Identity agent on a device

5. Upon reviewing Microsoft 365 Defender at this point, we can see a healthy agent status and initial Microsoft Defender for Identity protections already being managed, as shown in *Figure 3.39*:

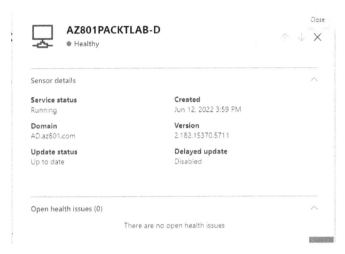

Figure 3.39 – Healthy Defender for Identity agent status in the Microsoft 365 Defender portal

One additional key component of Microsoft Defender for Identity is that the syslog alerts can be sent to any SIEM using the generalized **Common Event Format** (**CEF**) for security alerts. This allows for a wide variety of third-party tools to assist in managing alerts within your organization, and more details can be found at `https://docs.microsoft.com/defender-for-identity/cef-format-sa`.

This concludes the section on Microsoft Defender for Identity configuration and management, where we learned how Microsoft Defender for Identity can be designed, installed, and managed for an organization.

Summary

In this chapter, we learned how to secure a hybrid AD infrastructure. This included configuration of password policies, Azure AD Connect Sync, and SSPR with custom banned password lists, additional hardening of domain controllers, additional hybrid features for the protection of passwords and identities, and the administration of protected users and administrative groups. We also discussed implementing and managing Microsoft Defender for Identity.

In the next chapter, we will be learning how to identify and remediate Windows Server security issues by using Azure services, learning how to monitor on-premises servers and Azure IaaS virtual machines with Microsoft Sentinel, and utilizing the power of Microsoft Defender for Cloud to identify and remediate security issues for hybrid servers.

4

Identifying and Remediating Windows Server Security Issues Using Azure Services

We will continue building upon our security knowledge and experience from the previous chapter by building and reviewing additional depth-in-defense approaches to assist in monitoring and responding to performance and security. We will learn how to successfully monitor virtual machines running both on-premises and in Azure using **Azure Arc**, **Azure Monitor**, and **Microsoft Sentinel**, allowing for telemetry and metrics insights, analysis, and response. We will also cover how to onboard devices into **Microsoft Defender for Cloud** (**MDC**) so that we can proactively identify and remediate security issues wherever the virtual machine may be running within the infrastructure. In this chapter, we will cover the following topics:

- Technical requirements and lab setup

- Monitoring on-premises servers and Azure IaaS VMs using Azure Arc, Azure Monitor, and Microsoft Sentinel

- Identifying and remediating security issues for on-premises servers and Azure-hosted VMs with MDC

Technical requirements and lab setup

To successfully follow along and complete the tasks and exercises throughout this chapter and the following chapters of this book, we will need to establish **Microsoft Sentinel** (formerly **Azure Sentinel**) within the Microsoft Azure tenant we created in *Chapter 1, Exam Overview and the Current State of Cloud Workloads*. If you have not done so yet, we highly recommend you review the *Technical requirements and lab setup* section of *Chapter 1, Exam Overview and the Current State of Cloud Workloads*, so that you are prepared to follow along with the examples in this chapter.

To begin, let's gain some understanding surrounding the history, requirements, design, and architecture for integrating on-premises security monitoring with Azure Services, including integration with **Microsoft Defender for Cloud**.

Introduction to hybrid security using Microsoft Sentinel and Microsoft Defender for Cloud

Traditionally, there have been multiple offerings from Microsoft granting IT professionals insights into configuration, monitoring, change management, and auditing in general. For organizations using System Center offerings, **System Center Operations Manager** (**SCOM**) was top of mind, giving insights out of the box and delivering on pre-configured management packs that helped users manage the varying server roles and configurations used in most organizations.

Windows Event Forwarding (**WEF**) was also introduced (much earlier mind you, with the release of Windows Vista) to not only read administrative events of the organization's choosing, but also to send them to a **Windows Event Collector** (**WEC**) server. While this approach incurs no cost and requires no agent installations, the architecture and design of WEF do raise scalability concerns for organizations with large amounts of endpoints as both storage and weaker query abilities become prohibitive to quick insights of events collected. Note that this WEF approach continues to be a very powerful and recommended way of quickly gathering both baseline and suspect events for intrusion detection. The optional reading surrounding this approach can be found at `https://docs.microsoft.com/windows/security/threat-protection/use-windows-event-forwarding-to-assist-in-intrusion-detection`.

In 2015, Microsoft introduced a new service offering called **Operations Management Suite** (**OMS**) that incorporated a collection of cloud-based services to assist with insights, automation, security, and protecting your on-premises environment in a single pane within Azure. The major services included at that time were **Log Analytics**, **Azure Backup**, **Azure Site Recovery**, and **Azure Automation**. These were used to extend on-premises controls and insights into a management-as-a-service approach in the cloud with an array of preconfigured solutions and the ability to customize data collection.

Why does this history lesson matter, you ask? Fast forward to now, where the service names and management approaches may have changed, but the same principles still apply to hybrid architectures and scalability for quickly correlating and inspecting events:

- A virtual machine(s), physical machine(s), **Infrastructure-as-a-Service** (**IaaS**) virtual machine(s), or even **Internet of Things** (**IoT**) devices that have an agent installed (the **Azure Monitor Agent**, or **AMA**)

- An **Azure Log Analytics** workspace created for use with **Microsoft Sentinel** and **Azure Monitor**, collecting metrics and logs from a myriad of sources

- An extension into **Azure Stack**, where integrated systems are in data centers of your choosing, remote offices, or other edge locations and provide system insights in a familiar Azure security logging and auditing approach

- **Azure Arc** and the **Azure Arc Connected Machine agent** allow you to extend a cohesive management approach to on-premises, multi-cloud, and edge resources with supported operating systems such as Ubuntu, CentOS, and Linux and management of Kubernetes clusters running anywhere and on many supported distributions

- A network and firewall architecture that allows you to proxy traffic from a private data center on-premises into a cloud management infrastructure

- **Microsoft Defender for Cloud** allows for a holistic security management platform that protects multi-cloud and hybrid workloads where security events can utilize the speed of the cloud and **artificial intelligence** (**AI**) and **machine learning** (**ML**) components to investigate, detect, and respond to security events in near-real time

> **Azure Arc pricing and management**
>
> Azure Arc offers a considerable amount of services as part of the core control plane at no cost to customers, simply as an extension of their Azure services. Additional Azure services and management controls are at an additional cost.
>
> The free control plane services include Azure Resource Graph indexing and searching, resource inventory and organization through familiar and consistent Azure resource groups and tagging abilities, familiar security and access controls via Azure **resource-based access control** (**RBAC**) and subscription management, and template-based environments and automation, including available extensions and extension management.

Now that we have had a brief history lesson and an overview of how Azure services can be utilized to create a single pane of glass management approach to on-premises, multi-cloud, and edge computing resources, let's learn how to establish a hybrid security model using Microsoft's recommended tools for telemetry and security monitoring.

Establishing a Data Collection Rule for Azure Monitor Agent

To collect the necessary performance data and events from machines using Azure Monitor Agent, let's walk through how to configure **Data Collection Rules** (**DCRs**) for data being sent to Azure Monitor and associate those DCRs with virtual machines:

1. To begin, visit `https://portal.azure.com` while utilizing the Global Administrator account you created in *Chapter 1, Exam Overview and the Current State of Cloud Workloads.*

2. Select **Resource groups** from the list of Azure services or simply search for `Resource groups` and select it to continue.

3. Select **+ Create** to begin creating a new resource group, selecting the **Subscription** option you created in *Chapter 1, Exam Overview and the Current State of Cloud Workloads*. Use `rg-log-Sentinel` for the **Resource Group** name and select the **Region** option you are geographically closest to.

4. Select the **Review + create** button to continue creating the new resource group.

5. Select **Monitor** from the list of Azure services or simply search for `Monitor` and select it to continue.

6. Select **Data Collection Rules** under the **Settings** section, then select **+ Create** to continue creating a new **Data Collection Rule**.

7. Set **Rule Name** to `AZ801-Lab_Rules`, select the **Subscription** option you created in *Chapter 1, Exam Overview and the Current State of Cloud Workloads*, set **Resource Group** to **rg-log-Sentinel**, select the **Region** option you are geographically closest to, and set **Platform Type** to **Windows**. Click the **Next : Resources >** button to continue, as shown in *Figure 4.1*:

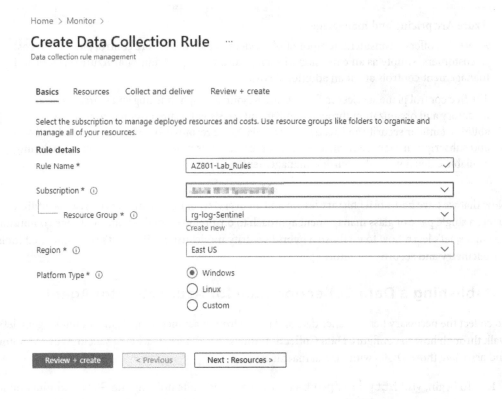

Figure 4.1 – Creating a Data Collection Rule in Azure Monitor

8. Select **Collect and deliver**, then **+ Add data source**. For **Data source type**, select **Performance counters** from the drop-down list and select **Add data source**.

9. Select + **Add data source** again, this time setting **Data source type** to **Windows event logs**. Configure the **Event logs** settings as shown in *Figure 4.2* and select **Add data source** when done:

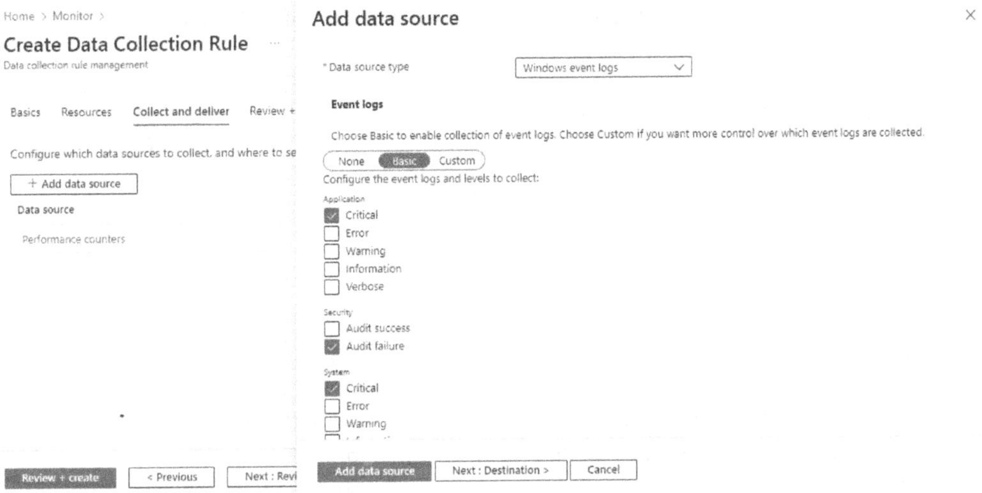

Figure 4.2 – Adding performance counters and Windows event logs data sources

10. Select **Next : Review + create** > to review the **Data Collection Rule** deployment and select **Create** to complete this task.

With that, you have established a DCR for Azure Monitor. Next, we will walk through how to set up and install the Azure Arc Connected Machine Agent for connecting to and collecting data for use within Azure Monitor and Microsoft Sentinel.

Installing the Azure Arc Connected Machine Agent

Azure Arc is one of the newer tools in the single pane of glass management tools offering, allowing customers to manage and govern hybrid resources across disparate data centers, resources running within Azure, and resources running in multi-cloud deployments. This centralized approach gives a unified feel to applying **Azure Policy** or **Template Specs** to resources, unified tags, monitoring, update management, security, and even integration with **Windows Admin Center** (which we will cover in more depth in *Chapter 8, Managing Failover Clustering*). The following Infrastructure, data, and application services components can be managed and abstracted via Azure Arc:

- Azure Arc virtual machines
- Azure Stack HCI
- Kubernetes clusters
- Servers

- SQL Servers

- VMware vCenters

- **System Center Virtual Machine Manager (SCVMM)** management Servers

- PostgreSQL Hyperscale data services

- SQL Managed instances data services

- API Management

- App Services

- Event Grid topics

- Functions

- Logic apps

In addition, familiar tools such as the Azure portal, the Azure CLI, and the Azure SDK can be used to manage these Azure Arc-connected devices. To read and learn more about Azure Arc and Arc-enabled Servers, go to `https://docs.microsoft.com/azure/azure-arc/servers/overview`.

Let's begin with the setup and walkthrough steps regarding Azure Arc and the Connected Machine Agent to get a feel for what Azure Arc helps deliver in terms of security monitoring, insights, remediation, and compliance:

1. To begin, visit `https://portal.azure.com` while utilizing the Global Administrator account you created in *Chapter 1, Exam Overview and the Current State of Cloud Workloads*.

2. Select **Resource groups** from the list of Azure services or simply search for `Resource groups` and select it to continue.

3. Select **+ Create** to begin creating a new resource group, selecting the **Subscription** option you created in *Chapter 1, Exam Overview and the Current State of Cloud Workloads*. Use `rg-monitor-Arc` for the **Resource Group** name and select the **Region** option you are geographically closest to.

4. Select the **Review + create** button to continue creating the new resource group.

5. Select **Servers – Azure Arc** from the list of Azure services or simply search for `Servers – Azure Arc` and select it to continue.

6. Under the **Infrastructure** section, select **Servers** and then select **+ Add**, as shown in *Figure 4.3*:

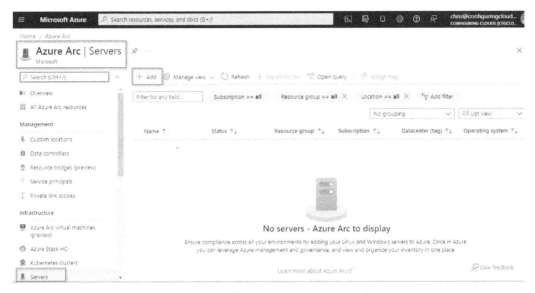

Figure 4.3 – Adding a new server to Azure Arc

7. Select **Add a single server** and then select **Generate script** to begin, as shown in *Figure 4.4*:

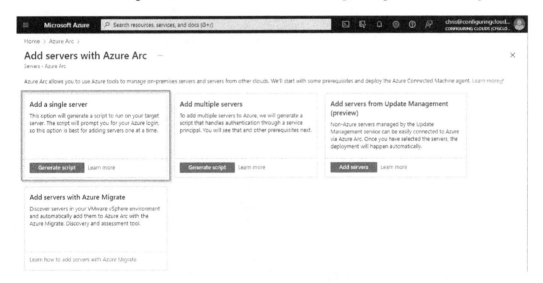

Figure 4.4 – Adding a single server to Azure Arc using the Generate script option

8. A screen will appear where you can review the prerequisites to ensure your VM has appropriate access to Azure services, proper network connectivity, local Administrator permissions on the source server, and an existing resource group in Azure. Select **Next** to continue after your review is complete.

9. For our lab setup, select the **Subscription** option you created in *Chapter 1, Exam Overview and the Current State of Cloud Workloads*, and set **Resource group** to **rg-monitor-Arc**. Then, select the **Region** option you are geographically closest to, **Windows** under **Operating System**, and finally **Public endpoint** under **Connectivity method**. Select **Next**, as shown in *Figure 4.5*:

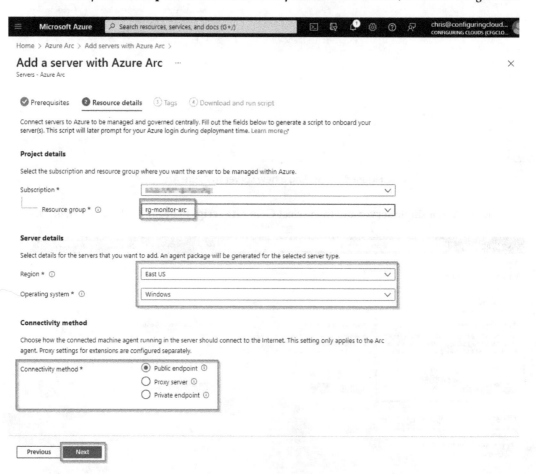

Figure 4.5 – Adding a server with Azure Arc

Azure Private Link (private endpoint) recommendation

Recently, **Azure Private Link** has become available for securely connecting **Azure Arc**-connected machines or services using a private endpoint, sending all traffic over a site-to-site VPN or Azure ExpressRoute connection without requiring the use of public networks.

While the deeper details of this feature may be outside the scope of the exam objectives, Azure Private Link greatly increases the security of your Azure Arc connections and is highly recommended. For additional details, be sure to review the following URL: `https://docs.microsoft.com/azure/azure-arc/servers/private-link-security`.

10. Fill out the **Physical location tags** area as best you can regarding the location of your virtual machine and be sure to add a custom tag under **Custom tags** with **Name** set to `Project` and **Value** set to `AZ-801-Arc`. Then, select **Next**, as shown in *Figure 4.6*:

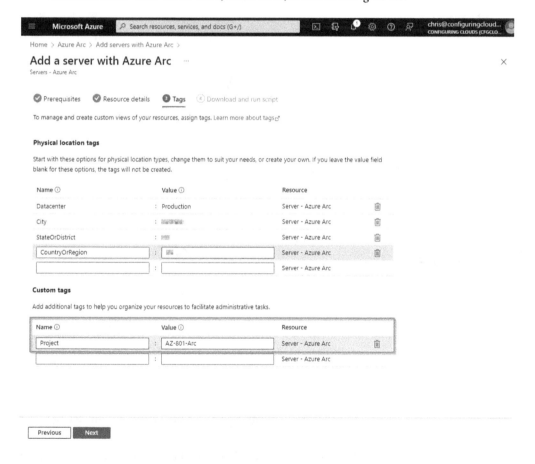

Figure 4.6 – Adding tags to the Azure Arc server

11. On the **Download and run script** page, select **Register** to establish Azure Arc in your tenant. Next, you can either **Download** or **Copy** the script from the screen. Then, you will need to follow the instructions in **Step 3 – Open a PowerShell console to run the script** against the virtual machine you want to onboard, as shown in *Figure 4.7*:

Figure 4.7 – Registering your subscription for Azure Arc and downloading the onboarding script

12. At this point, you will also want to open **Hyper-V Manager** on your device that's hosting the virtual machines.

13. Right-click on the **AZ801PacktLab-DC-01** virtual machine and select **Start**. Complete the same steps for **AZ801PacktLab-FS-01** and ensure that both virtual machines are running.

14. Right-click on the **AZ801PacktLab-FS-01** virtual machine and select **Connect**.

15. Use the **Action | Ctrl+Alt+Delete** menu option to begin, then log in to **AZ801PacktLab-FS-01** as an Administrator while using `Packtaz801guiderocks` as the password.

16. Copy the script to your **AZ801PacktLab-FS-01** virtual machine and run the script in an administrative PowerShell console. You will be prompted to visit `https://microsoft.com/devicelogin` to enter a code and verify the Azure Connected Machine Agent. Then, select **Continue** to complete the sign-in and application setup on the device. Once the script has been completed, you should see results on the screen similar to those shown in *Figure 4.8*:

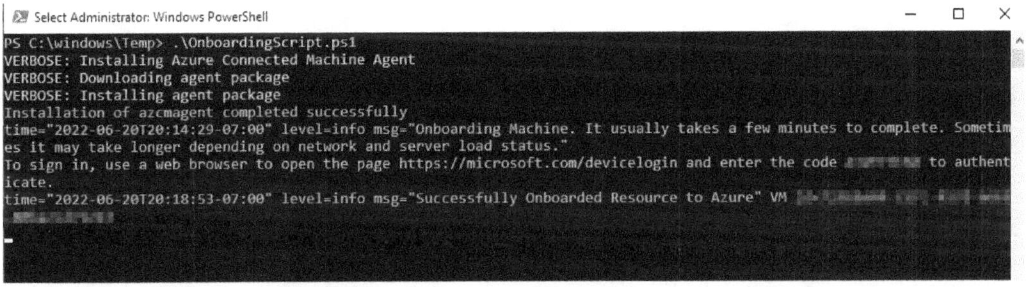

Figure 4.8 – Running the Azure Connected Machine agent onboarding script

17. While on the **AZ801PacktLab-FS-01** virtual machine, make a copy of `calc.exe` and place it on the desktop of the virtual machine. Rename `calc.exe` to `ASC_AlertTest_662jfi039N.exe` and save it.

18. At this point, you will want to complete alert validation within Defender for Cloud Alerts by forcibly raising an alert for a known alert simulation test within Microsoft Defender. To complete these alert trigger steps, simply open an administrative Command Prompt and enter the following commands:

```
Cd desktop
ASC_AlertTest_662jfi039N.exe -validation
ASC_AlertTest_662jfi039N.exe -foo
```

19. Return to the `https://portal.azure.com/#view/Microsoft_Azure_HybridCompute/AzureArcCenterBlade/~/servers` page in the Azure portal and ensure that you can see your lab VM in **Azure Arc** with a **Status** of **Connected**, as shown in *Figure 4.9*:

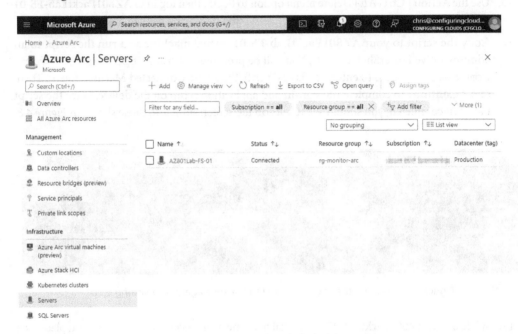

Figure 4.9 – Reviewing your newly onboarded Azure Arc server

20. Select **Monitor** from the list of Azure services or simply search for `Monitor` and select it to continue.

21. Under **Settings**, select **Data Collection Rules**, then select **AZ801-Lab_Rules** from the list.

22. Under **Configuration**, select **Resources**, then select **+ Add** to continue.

23. Select the scope of `rg-monitor-arc` (to include any virtual machines you add to this resource group, including your **AZ801Lab-FS-01** device) and select **Apply** to complete the necessary changes and begin installing the extension on the remote virtual machine.

In this section, we learned how to set up and install the Azure Arc Connected Machine Agent for connecting to and collecting data for use within Azure Monitor and Microsoft Sentinel. In the next section, we will establish the necessary Log Analytics workspaces for use with Azure Monitor and Microsoft Sentinel and learn how to monitor both on-premises and Azure **Infrastructure-as-a-Service** (**IaaS**) virtual machines.

Monitoring on-premises servers and Azure IaaS VMs using Microsoft Sentinel

Microsoft Sentinel is a cloud-based **Security Information and Event Management** (**SIEM**) and **Security Orchestration, Automation, and Response** (**SOAR**) solution that provides threat intelligence and security analytics at scale. Arguably one of my favorite tools to work with, this single-pane collection of solutions, dashboards, playbooks, notebooks, hunting, analytics, and, most importantly, data source connectors make this tool unbelievably powerful for threat response and security analytics in your organizations.

Highly customizable with custom alerts, immersive dashboard design, community components available, and built off the **Kusto Query Language** (**KQL**) to help with data queries and manipulation, Microsoft Sentinel brings an arsenal of tools and great community support to help everyone achieve their desired threat detection and response goals. It's only right to share the following great KQL resources so that we can all learn how to use this incredibly powerful language that is used across all of Microsoft Azure:

- **KQL overview**: `https://aka.ms/KQL`

- **KQL in Microsoft Sentinel**: `https://docs.microsoft.com/azure/sentinel/kusto-overview`

- **Microsoft Sentinel Ninja Training**: `https://techcommunity.microsoft.com/t5/microsoft-sentinel-blog/become-a-microsoft-sentinel-ninja-the-complete-level-400/ba-p/1246310`

Following up on what we've learned so far in this chapter by feeding data into Azure using connectors and agents, we will continue establishing a **Log Analytics** workspace as the base for our Microsoft Sentinel deployment. It is highly recommended to have a single, dedicated workspace for Microsoft Sentinel, so we will follow that best practice in our walkthroughs.

Log Analytics region information

It is important to note that for you to send data into Log Analytics and Microsoft Sentinel, the data collection rule *must* be created in the same geographic region as the Log Analytics workspace! Recommended reading on Microsoft Sentinel and Log Analytics workspace best practices and template deployments can be found at `https://docs.microsoft.com/azure/sentinel/best-practices-workspace-architecture` and `https://docs.microsoft.com/azure/azure-monitor/logs/resource-manager-workspace`, respectively.

Let's begin by onboarding Microsoft Sentinel in the Azure tenant we created in *Chapter 1, Exam Overview and the Current State of Cloud Workloads*:

1. To begin, visit `https://portal.azure.com` while utilizing the Global Administrator account you created in *Chapter 1, Exam Overview and the Current State of Cloud Workloads*.

2. Select **Microsoft Sentinel** from the list of Azure services or simply search for `Microsoft Sentinel` and select it to continue.

3. To set up a new workspace, select **+ Create** from the menu, then **Create a new workspace**.

4. For our walkthrough, select the **Subscription** option you created in *Chapter 1, Exam Overview and the Current State of Cloud Workloads*. Then, under **Resource group**, select **Create new**, setting **Name** to `rg-log-Sentinel`.

5. Set **Instance name** to `logAZ801Sentinel`, then select the **Region** option you are geographically closest to and select **Review + Create**.

6. Once the **Microsoft Sentinel** workspace has been completed, select `logAZ801Sentinel` from the list. Then, under **Configuration**, select **Data Connectors** to continue.

7. First, let's search for `Azure`. From the list of connectors returned, select **Azure Active Directory**, then **Open Connector page**, as shown in *Figure 4.10*:

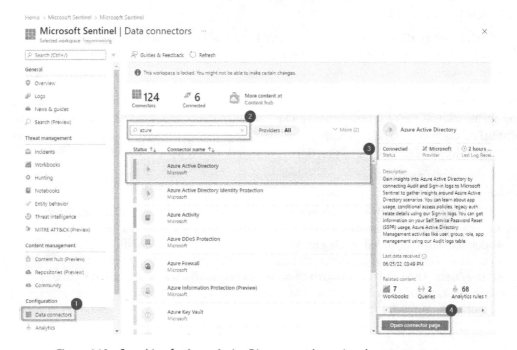

Figure 4.10 – Searching for Azure Active Directory and opening the connector page

8. Under the **Configuration** section, select **Connect** for the **Azure Active Directory** connector to send alerts to the Microsoft Sentinel workspace. In addition, under the **Create incidents – Recommended!** section, select **Enabled** to automatically create incidents from all generated alerts in this service.

9. Select the **X** button in the top right-hand corner to close the connector blade, then search for `Security` to return all the security-related connectors. Select **Windows Security Events via AMA** from the list, and then select **Open Connector Page**.

10. Notice that, on the left, you should be able to see events that have already been sent from your connected virtual machine, as shown in *Figure 4.11*:

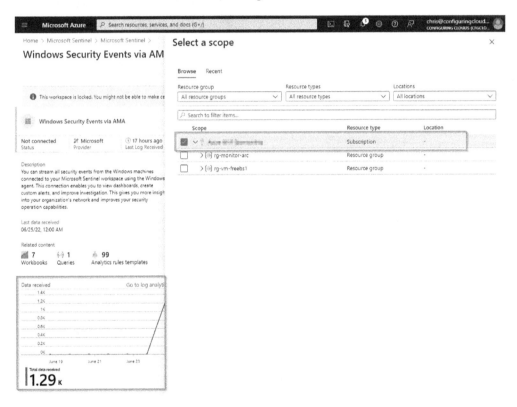

Figure 4.11 – Windows Security Events via AMA and selecting a scope

11. Under **Configuration**, select **+ Create data collection rule** to continue. For the **Rule Name** option, enter `AZ801-WinSecurityAMA`, keeping the default settings for the remaining **Subscription** and **Resource Group** selections.

12. For the **Resources** and **Select a scope** pages, select your subscription from the list to automatically add any Azure Arc-connected machines from on-premises or other environments, as well as any Azure Virtual Machines running in IaaS. Select the **Next : Collect >** button to continue.

13. Accept the default of **All Security Events** under **Select which events to stream**, then select **Next : Review + Create** > to continue.

14. Finish by selecting the **Create** button. You will be returned to the **Windows Security Events via AMA** page; you may need to select **Refresh** under the **Configuration** section to see your newly created data collection rule.

15. Under **Threat Management**, select **Workbooks** and then search for `Security Alerts`. Select **Save** and then select **View saved workbook** to review the **Security Alerts** dashboard for your environment, as shown in *Figure 4.12*:

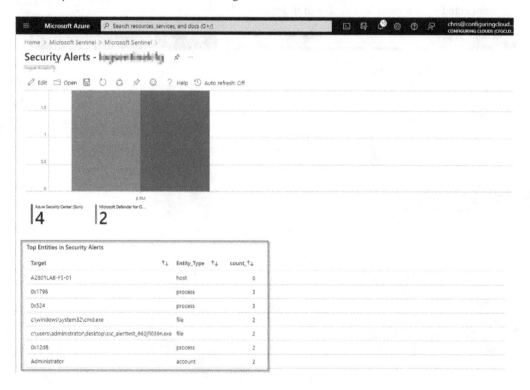

Figure 4.12 – Reviewing the Security Alerts dashboard within Microsoft Sentinel

16. Once you've finished reviewing the security alerts, select the **X** button in the top right-hand corner to close the **Security Alerts** dashboard blade.

In this section, we learned how to onboard Microsoft Sentinel by establishing a new Log Analytics workspace dedicated to Microsoft Sentinel, creating the Microsoft Sentinel workspace for collecting and investigating security alerts, establishing the necessary data connectors for the configuration, and reviewing the workbooks to successfully review and investigate security events within Microsoft Sentinel.

We will now continue to our final section, where we will learn how to identify and remediate security issues for on-premises and Azure IaaS within a Microsoft Defender for Cloud environment.

Identifying and remediating security issues for on-premises servers and Azure-hosted VMs with MDC

Microsoft Defender for Cloud (**MDC**) brings advanced defense-in-depth approaches to all your Windows and Linux, machines no matter where they are running (Azure, AWS, GCP, or your on-premises environments), combining best-in-class security features for workloads of any type and size.

The advanced protection capabilities within MDC include, but are not limited to, the following:

- **Just-in-Time** (**JIT**) virtual machine access to lock down inbound access to your virtual machines
- An integrated **Microsoft Defender for Endpoint** license giving comprehensive **endpoint protection and response** (**EDR**) capabilities for your endpoints
- Vulnerability assessment and management features for devices
- Adaptive application controls to reduce the overall device attack surfaces
- File integrity or change monitoring for indications of an attack based on registry and file/OS changes
- Fileless attack detection with insights into process metadata
- **Adaptive network hardening** (**ANH**) to help improve the overall network security posture by providing recommendations for hardening and monitoring known trusted configurations and baseline activities
- Docker host hardening for unmanaged containers hosted on Linux VMs or other machines running Docker containers

As we learned in the *Installing the Azure Arc Connected Machine Agent* section, the preferred route for onboarding multi-cloud and hybrid devices into MDC is to utilize **Azure Arc** as the control plane (note that this integration of MDC with Azure Arc does incur additional costs, which can be reviewed at https://azure.microsoft.com/en-us/pricing/details/azure-arc/).

Knowing that we have already onboarded an Azure Arc server into our tenant, let's review how this server can be managed within MDC:

1. To begin, visit https://portal.azure.com while utilizing the Global Administrator account you created in *Chapter 1, Exam Overview and the Current State of Cloud Workloads*.

2. Select **Microsoft Defender for Cloud** from the list of Azure services or simply search for Microsoft Defender for Cloud and select it to continue.

3. As this is your first time visiting this page, you will be presented with a **Getting Started** page, asking you to **Upgrade** to a **Standard** plan. Select the **Enable** button at the bottom of the screen, as shown in *Figure 4.13*:

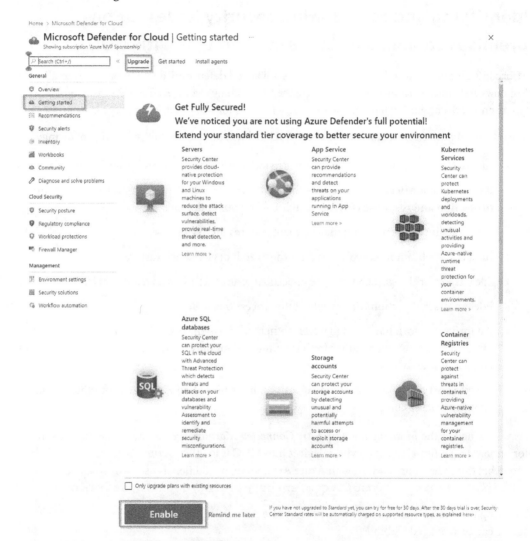

Figure 4.13 – Initializing Microsoft Defender for Cloud and enabling the Standard plan

4. You will then be directed to the **Get Started** tab. Under **Add non-Azure servers**, select the **Configure** button to continue.

5. On the **Onboard servers to Defender for Cloud** page, select **logAZ801Sentinel** from the list. You may also need to select the **Upgrade** button to expand coverage for a 30-day free trial on the VMs and servers in your tenant.

6. Return to **Microsoft Defender for Cloud** and select **Inventory** from the **General** section.

7. Review the resources that can be protected and review any unhealthy resources. Note that you can also select multiple filters on this screen to focus on certain resources, such as **Virtual Machines** and **Azure Arc Servers**, as shown in *Figure 4.14*. This page is super valuable for viewing the security posture across your entire infrastructure and resource estate and will also provide recommendations on how to improve overall resource security:

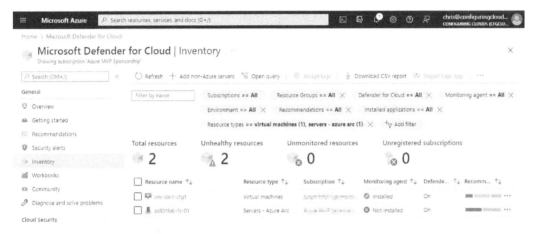

Figure 4.14 – Reviewing the Asset Inventory section in Microsoft Defender for Cloud

8. From this screen, select **az801lab-fs-01** from the list to initially view the overall **Resource health** and recommendations for this resource. Select **az801lab-fs-01** from this page, then select **Insights** from the **Monitoring** section, and finally select **Performance** to get an overall real-time view of virtual machine performance. Finally, select **Map** to view the current connections to and from the machine, as well as details about log events, alerts, and changes, as shown in *Figure 4.15*:

Figure 4.15 – Reviewing Monitoring Insights for a virtual machine in Microsoft Defender for Cloud

9. Return to **Microsoft Defender for Cloud** and select **Security alerts** under the **General** section. You will see the two alerts that were triggered in the previous lab activity in this chapter, indicating that a test malicious event happened in an on-premises server and alerted Microsoft Defender for Cloud for additional support, as shown in *Figure 4.16*:

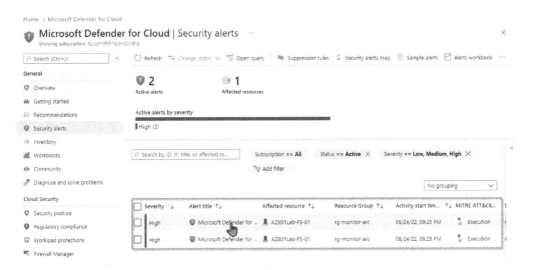

Figure 4.16 – Reviewing the security alerts that have been raised
from a hybrid device in Microsoft Defender for Cloud

10. Select **Recommendations** under the **General** section, and review the recommendations being made by **Microsoft Defender for Cloud** based on best practices and industry-standard controls, as shown in *Figure 4.17*:

Figure 4.17 – Reviewing the recommendations made by Microsoft Defender for Cloud

11. When your review of the **Recommendations** area is completed, select **Regulatory Compliance** to review an overall assessment of your compliance with known security benchmarks. This page continually updates over time to keep you in the know on compliance drift across your environment. It also provides a quick and easy reporting engine to give you audit reports for both your tenant as well as Microsoft's overall compliance and auditing. These options can be reviewed, as shown in *Figure 4.18*:

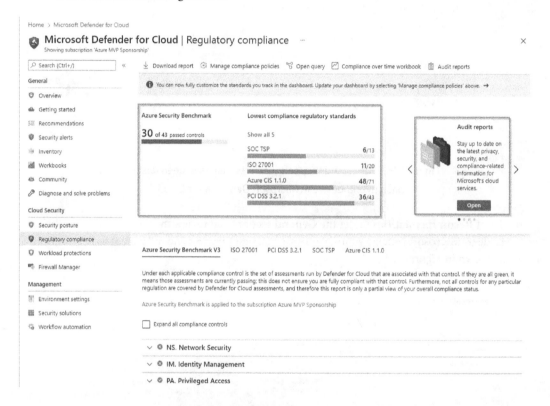

Figure 4.18 – Reviewing the Regulatory compliance section within Microsoft Defender for Cloud

12. Feel free to examine the **Security Alerts**, **Workbooks**, **Security Posture**, and **Workload Protections** sections within **Microsoft Defender for Cloud** and close out of the Azure portal once your review is completed.

In this section, we learned how devices are onboarded, managed, and responded to within a **Microsoft Defender for Cloud** instance. We learned about the various advanced protection capabilities that **Microsoft Defender for Cloud** brings to the organization, and how quickly and easily this can be incorporated into an existing hybrid or cloud environment. We also learned how valuable each of the sections within **Microsoft Defender for Cloud** is and how they can be used to audit, administer, and protect your organization from advanced threat activity.

Summary

In this chapter, we learned how to identify and remediate Windows Server security issues by using Azure Services. We did so by building and reviewing additional depth-in-defense approaches to assist in monitoring and responding to performance and security. We learned how to successfully monitor virtual machines running both on-premises and in Azure using Azure Arc, Azure Monitor, and Microsoft Sentinel, allowing for telemetry and metrics insights, analysis, and response. We also covered how to onboard devices into MDC so that we can proactively identify and remediate security issues wherever the virtual machine may be running within the infrastructure.

In the next chapter, we will learn how to secure Windows Server networking on workloads, adding security to the communications layer for shared workloads. We will also learn how to configure and manage **Microsoft Defender Firewall**, how to plan for and configure domain isolation, and how to implement connection security and authentication request rules for your **Windows Servers** in a hybrid environment.

5

Secure Windows Server Networking

Next up, we will dive into establishing secure networking for our Windows Server workloads to add security to the communications layer for shared workloads. This chapter will cover how to configure and manage **Windows Defender Firewall**, how to successfully plan for and configure domain isolation, and how to implement connection security and authentication request rules for your Windows servers.

By the end of this chapter, we will have learned how to configure and manage **Windows Defender Firewall**, how to plan for and configure domain isolation, and how to implement connection security and authentication request rules for your Windows servers in a hybrid environment.

In this chapter, we will cover the following topics:

- Technical requirements and lab setup
- Managing **Windows Defender Firewall**
- Implementing domain isolation
- Implementing connection security rules

Technical requirements and lab setup

To successfully follow along and complete the tasks and exercises throughout this chapter and in the following chapters of this book, we will need to ensure that the technical requirements from both *Chapter 1, Exam Overview and the Current State of Cloud Workflows*, and *Chapter 2, Securing the Windows Server Operating System*, have been completed in full. We will be using both the domain controller and the file server **virtual machines** (**VMs**) to complete the exercises and tasks throughout this chapter to align with the *AZ-801* exam objectives.

To begin, let's gain some understanding surrounding what is meant by securing Windows Server networking, including the management and configuration tools available for use to achieve networking security in your organization.

Introduction to Windows Defender Firewall

Microsoft's firewall history reaches all the way back to Windows XP and Windows 2003, where a *disabled-by-default* introductory firewall approach called **Internet Connection Firewall** was introduced. Over time, versions of **Windows Firewall** and now **Windows Defender Firewall** were changed to an *on-by-default* approach whereby all unsolicited traffic is blocked, and improved the feature set by providing the following for management and increased security:

- A management console that can be configured using the user interface or **Group Policy** (within an enterprise, advanced tools to manage separate firewall policies based on device types and specific workloads)

- Introduction of **Windows Filtering Platform** to allow both inbound and outbound packet filtering and inspection, including source and destination IPv4 and IPv6 details (ports and port ranges, for example)

- A rule-based approach with the ability to select rules and protections from a list of services and applications

- Advanced connection security rules to allow or deny connections based on authorization, authentication, or certificate and ensure encryption when needed

Windows Defender Firewall generally listens for incoming requests to your Windows server and then determines which applications should be permitted to access the local server, denying connection attempts that are rejected by the firewall rulesets. In addition, **Windows Defender Firewall** has three available scenarios or profiles (depending on the network connectivity type, meaning only one profile can be applied at a time) for network connectivity rules, as shown in *Figure 5.1*, and outlined as follows:

- **Domain Profile** is typically used when computers are logged in to a domain and is applied by a network determination process to inspect whether a computer is currently attached to a domain (can connect to a domain controller), involving a check for a last **Group Policy** update and validation of whether any connection-specific **Domain Name Service** (**DNS**) suffixes have been configured.

- **Private Profile** is typically used for private or home networks.

- **Public Profile** is typically used for public network access such as hotspots, airports, hotels, stores, or your favorite coffee shop down the street:

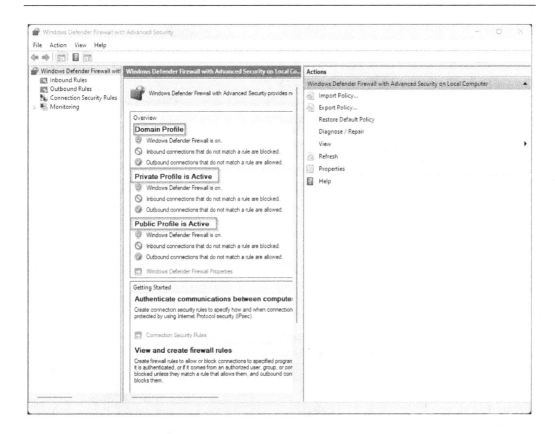

Figure 5.1 – Windows Defender Firewall default profiles

Default Windows Defender Firewall settings and best practices

As a general best practice, utilize the default firewall rules and settings, as these will provide ample protection and have been designed with a high level of security for most network designs. Do note that the default behavior is to block all inbound or incoming connections.

Now that we have a brief introduction to **Windows Defender Firewall** and the available profiles, let's discuss how to manage **Windows Defender Firewall** with the existing tools.

Managing Windows Defender Firewall

Windows Defender Firewall is managed by creating rules that are specific to the program or services being hosted and identifying all details surrounding network-facing traffic for the application, services, and drivers used. These firewall policies can be configured and customized per the organization's granular network, application, and security requirements.

This ultimately boils down to four types of **Windows Defender Firewall** rules that can be created based on specific application dependencies:

- A **Program** rule, which requires knowledge of the full path of the application

- A **Port** rule, which requires knowledge of any/all **Transmission Control Protocol** (**TCP**) or **User Datagram Protocol** (**UDP**) port connections

- A **Predefined** rule, which requires knowing only the name of an application already installed on the host computer

- A **Custom** rule, which requires advanced knowledge of the application regarding the installation path, protocols and ports used for network communication, source and destination IP addresses, connection actions, and the destined security profile

Custom rules and individualized firewall policies for **Windows Defender Firewall** can be managed via the **Windows Defender Firewall** user interface (simply by launching `wf.msc` from the **Start** menu), via **Group Policy**, or by onboarding the device into **Microsoft Defender for Business** and managing the firewall policies from within the **Microsoft Defender** portal at `https://security.microsoft.com`. As we will not be managing firewall rules via **Microsoft Defender for Business** for the *AZ-801* exam objectives, more details on managing **Windows Defender Firewall** in this manner can be further reviewed at this URL: `https://docs.microsoft.com/microsoft-365/security/defender-business/mdb-custom-rules-firewall?view=o365-worldwide`.

Let's walk through how we can manage the **Windows Defender Firewall** settings for an organization using **Group Policy**:

1. Begin by opening **Hyper-V Manager** on your device hosting the VMs.

2. Right-click on the **AZ801PacktLab-DC-01** VM, select **Start**, and ensure that the VM is in a running state.

3. Right-click on the **AZ801PacktLab-DC-01** VM and select **Connect**.

4. Use the **Action | Ctrl+Alt+Delete** menu option to begin, then log in to the **AZ801PacktLab-DC-01** VM as **Administrator**, using `Packtaz801guiderocks` as the password.

5. Open the **Start** menu, then navigate to **Windows Administrative Tools | Group Policy Management,** and select Group Policy Editor to open the management console.

6. If not already expanded, select **Domains | AD.az801.com | Group Policy Objects,** and then right-click to select **New** to begin creating a new **Group Policy Object (GPO)**.

7. In the **Name:** field, enter `AZ801Lab-CH5-FS_Firewall` and select **OK** to continue.

8. Right-click on the new `AZ801Lab-CH5-FS_Firewall` GPO and select **Edit**.

9. Expand the following path in **Group Policy Management Editor**, as shown in *Figure 5.2*: **Computer Configuration | Windows Settings | Security Settings | Windows Defender Firewall with Advanced Security | Windows Defender Firewall with Advanced Security – LDAP...**, and then select **Windows Defender Firewall Properties**:

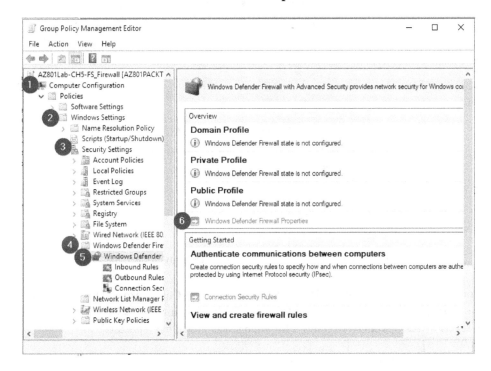

Figure 5.2 – Managing Windows Defender Firewall with GPO configuration

10. While on the **Domain Profile** tab, configure the following values and then select **Apply** to commit the changes:

*   **Firewall state**: Set this to a value of On (recommended)

*   **Inbound connections**: Set this to a value of Block (default)

*   **Outbound connections**: Set this to a value of Allow (default)

11. Remaining on the **Domain Profile** tab, select the **Customize** button for the **Settings** section, and review the settings for **Firewall settings**, **Unicast response**, and **Rule merging**, but do not make any changes, as shown in *Figure 5.3*. Note that the section for local rule merging allows for the configuration of the central management of firewall rules only or merging central policies with those individually configured on the endpoint. Select **OK** when your review is done to close the **Customize Settings for the Domain Profile** pane:

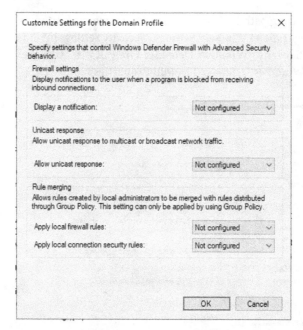

Figure 5.3 – The Customize Settings for the Domain Profile pane

12. For the **Logging** section on the **Domain Profile** tab, select **Customize** and then uncheck both **Not configured** checkboxes, change the **Size limit (KB)** setting to **32,767,** and **Log dropped packets** to **Yes**. Select **OK** to commit the changes, as shown in *Figure 5.4*:

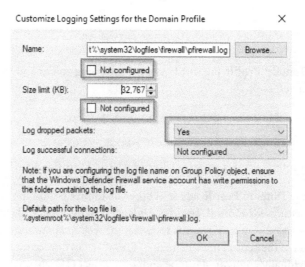

Figure 5.4 – The Customize Logging Settings for the Domain Profile pane

13. Select the **IPsec Settings** tab, as shown in *Figure 5.5*, then select the **Customize…** button for **IPsec defaults**, and then select **Advanced** and **Customize…** to review the settings. While we do not have a **Public Key Infrastructure** (**PKI**) environment created in our lab environment such as **Active Directory Certificate Services** (**AD CS**), this is a necessary review in concert with our *AZ-801* exam objectives:

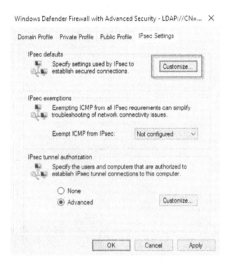

Figure 5.5 – Customizing the IPsec settings for the domain profile

14. Note that **Customize Advanced Key Exchange** Settings on the left indicate the defaults, while the **Add Security Method** pane on the right indicates the recommended best practice configuration, as shown in *Figure 5.6*. Select **Remove** twice on the existing security methods, then add a new security method as shown, and select **OK** twice to return to the **Customize IPsec Defaults** pane:

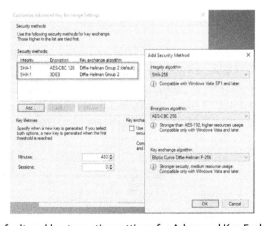

Figure 5.6 – Default and best practice settings for Advanced Key Exchange Settings

15. Under the **Data protection (Quick Mode)** section, select the radio button for **Advanced** and then select **Customize** to review the settings, noting that the values at the top of *Figure 5.7* represent the default values, and the bottom values represent the best practices configuration. Select **Remove** for all values under both **Data integrity algorithms** and **Data integrity and encryption algorithms**, then select **Add,** and use the values shown to add new algorithms for both. Select **OK** twice to accept the changes and close the pane:

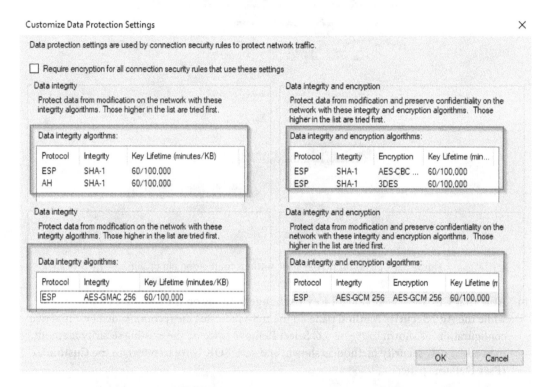

Figure 5.7 – Default and best practice configurations for Data Protection Settings

16. Under the **Authentication method** section, select the **Advanced** radio button and then select **Customize** to open the **Customize Advanced Authentication Methods** pane. To establish the best practices configuration as shown in *Figure 5.8*, select **Add** under **First authentication methods:**, select **Computer (Kerberos V5)**, and then select **OK**. Select **Add** under **Second authentication methods:** and select **User (Kerberos V5)**, then select **OK** twice to commit the changes:

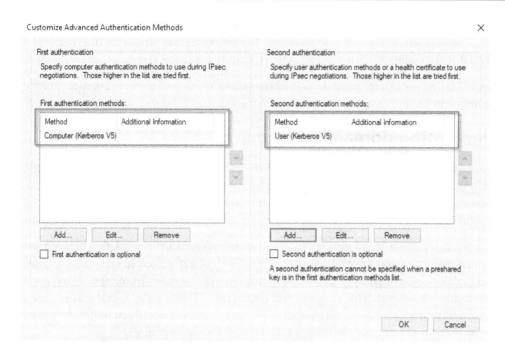

Figure 5.8 – Best practices setting for Advanced Authentication Methods

17. Select the **OK** button one last time to commit the changes to the **Windows Defender Firewall GPO,** and then close **Group Policy Management Editor** window that is currently editing the AZ801Lab-CH5-FS_Firewall GPO.

18. Locate the **File Servers Organizational Unit (OU)** from the domain tree and right-click on **File Servers,** then select **Link an existing GPO…** from the dropdown. Select the AZ801Lab-CH5-FS_Firewall GPO and then select **OK** to apply.

For this exercise, we have completed all the necessary configuration steps. Please keep the console open for this VM to complete the next exercise.

Bonus examples – implementing a basic firewall policy design

Knowing that there is no one-size-fits-all approach for many of the examples in this book, I find myself referring to this checklist approach and example firewall policy design not only to ensure that the right questions are asked to gather the proper data/information but also to refer to and utilize incredible naming conventions in my policy designs:

- https://docs.microsoft.com/windows/security/threat-protection/ windows-firewall/checklist-implementing-a-basic-firewall- policy-design

- https://docs.microsoft.com/windows/security/threat-protection/ windows-firewall/firewall-policy-design-example

In this section, we learned how to create and link a GPO to establish the base **Windows Defender Firewall** configuration settings. We also established a best practices approach for **Internet Protocol Security (IPsec)** communications to enhance network security and enable encryption where appropriate.

Next, we will learn how to implement domain isolation for a Windows Server environment to achieve our requirements surrounding network traffic design and enhanced network security.

Implementing domain isolation

IT pros and network administrators are most likely familiar with the concept and application of network isolation, where networks have unlimited physical and logical network segmentation or micro-segmentation. There are a variety of reasons to do this, and a few examples include attempting to keep traffic localized to devices for speed and efficiency and it being a great way to increase network security.

Within a Windows Server-based networking environment, you can achieve isolation between server and domain resources, thus limiting access to authorized and authenticated computers to prevent unauthorized devices from gaining access to server and domain resources. The design of this approach typically includes a network with a firewall and connection security policies that ensure expected traffic and authentication requests are allowed while unexpected or unsolicited traffic is dropped or rejected by both the firewall and the configured connection security policies.

There are two main isolation approaches:

- **Server isolation** is achieved when a server is configured to require secure authenticated and encrypted communications only using IPsec. The available controls ensure that the server will only respond to certain types of requests from a specific computer or computers.

- **Domain isolation** is achieved when domain computers are configured to require secure authenticated and encrypted communications (using IPsec) only from devices that are authorized and authenticated from the same isolated domain. Domain membership in an AD domain is required, and the design may not be limited to a single AD domain but may also include all domains in a specific forest or cross-forests where a two-way trust relationship is present between the trusted domains/forests.

Let's learn how to configure a basic firewall rule that permits access to a domain-isolated server in our *AZ-801* exam guide lab environment:

1. Begin by opening **Hyper-V Manager** on your device hosting the VMs and copy *steps 2 to 5* from the *Managing Windows Defender Firewall* section.

2. If not already expanded, select **Domains | AD.az801.com | Group Policy Objects** and then locate the `AZ801Lab-CH5-FS_Firewall` GPO, then right-click, and select **Edit**.

3. Expand the following path in **Group Policy Management Editor: Computer Configuration | Windows Settings | Security Settings | Windows Defender Firewall with Advanced Security | Windows Defender Firewall with Advanced Security – LDAP... | Inbound Rules**, and then select **Windows Defender Firewall Properties**.

4. Right-click on **Inbound Rules** and select **New Rule**.

5. Select **Custom** from the list and then select **Next >**.

6. For the **Program** section, select **All programs** and then select **Next >**.

7. For the **Protocol and Ports** section, keep the defaults and select **Next >**.

8. For the **Scope** section, keep the defaults and select **Next >**.

9. For the **Action** section, select **Allow the connection if it is secure**. Select the **Customize** button, and then select **Require the connections to be encrypted** and select **OK**, as shown in *Figure 5.9*:

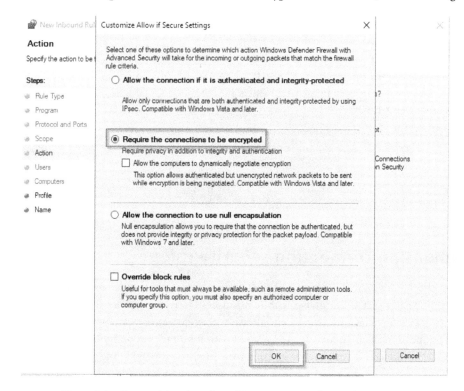

Figure 5.9 – Customizing the Allow if Secure settings for our firewall rule

10. On the **Users** page, select **Next >** to continue.

11. On the **Computers** page, select **Only allow connections from these computers:**, then select **Add**. Enter `az801` under **Enter the object names to select**, and then select **Check Names**. Select `AZ801Lab-FS-01` and `AZ801Lab-DC-01` (by holding down *Ctrl* when making multiple selections), then select **OK**. Select **Next >** to continue. Note that the recommended best practice is to utilize AD groups so that members of the groups can be easily added and removed as needed to apply different firewall rulesets.

12. On the **Profile** page, deselect **Private** and **Public**, then select **Next >** to continue.

13. For the **Name:** field, enter `AZ801-Lab-DomainIsolation` and enter `This rule is used for Domain Isolation in the AZ801 lab` for the **Description** field, then select **Finish**.

In this section, we learned about domain isolation and the features that this design approach brings to an enterprise for rigid network security and the isolation of servers and services based on standards-based authorization and authentication.

Additional reading for IPsec implementation

While we reviewed how to establish a basic IPsec approach to network security, the following article covers a greater depth of information and is a recommended additional read in preparation for the *AZ-801* exam:

```
https://docs.microsoft.com/windows/security/threat-protection/
windows-firewall/securing-end-to-end-ipsec-connections-by-using-
ikev2
```

We will now continue to our final section of this chapter, learning how to configure connection security rules within **Windows Defender Firewall**.

Implementing connection security rules

This final section focuses on layering additional connection security rules onto the inbound and outbound traffic rules that are available within **Windows Defender Firewall**. While firewall rules allow or deny traffic through the firewall configuration, they do not enforce connection security. The creation of connection security rules in conjunction with inbound and outbound rules ensures that appropriate connection security between two computers has been applied to the communication layer.

There are five main types of connection security rules:

- **Isolation**, where you can configure connection restrictions based on domain membership or device health status

- **Authentication exemption**, allowing any specified computers to bypass authentication

- **Server-to-server**, ensuring that authentication is enforced between specified computers

- **Tunnel** ensures that connections are authenticated between two computers

- **Custom**, where you can apply individually crafted connection security rules

During the previous *Implementing domain isolation* section, we adjusted our **Windows Defender Firewall** policy configuration for IPsec algorithm and authentication best practices, leading us into an additional configuration with connection security rules. Let's begin by walking through the high-level process of creating an authentication request rule for our existing Windows Defender Firewall GPO:

1. Begin by opening **Hyper-V Manager** on your device hosting the VMs and copy *steps 2 to 5* from the *Managing Windows Defender Firewall* section.

2. Expand the following path in **Group Policy Management Editor: Computer Configuration | Windows Settings | Security Settings | Windows Defender Firewall with Advanced Security | Windows Defender Firewall with Advanced Security – LDAP... | Connection Security Rules**, then right-click, and select **New Rule**.

3. On the **Rule Type** page, we will select **Server-to-server** and then select **Next >**.

4. For the **Endpoints** page, under **Which computers are in Endpoint 1?**, select **These IP addresses**, then select the **Add...** button, and add the IP address of 10.10.10.1. Additionally, for **Which computers are in Endpoint 2?**, select **These IP addresses**, select the **Add...** button, and add the IP address of 10.10.10.2, as shown in *Figure 5.10*. Select the **Next >** button to continue:

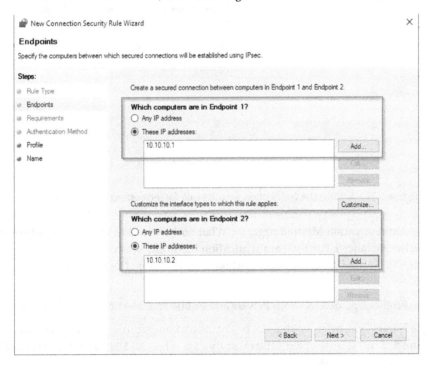

Figure 5.10 – Adding explicit endpoint IP addresses for our connection security rule

5. As a best practice, we will select **Request authentication for inbound and outbound connections**, as shown in *Figure 5.11*. In your organization configuration, you will want to follow this *crawl > walk > run* approach by starting with **Request authentication for inbound and outbound connections**, then **Require authentication for inbound connections and request authentication for outbound connections**, and finally **Require authentication for inbound and outbound connections** when all network access has been fully validated and tests cleanly in your environment:

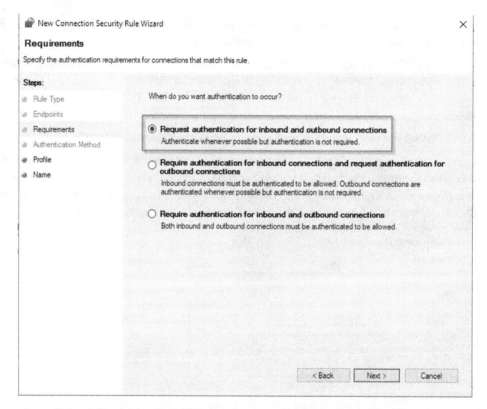

Figure 5.11 – Setting the connection security rule for requesting authentication

6. On the **Authentication Method** page, for **What authentication method would you like to use?**, select **Advanced**. For **First authentication methods:**, select Computer (Kerberos V5), and for **Second authentication methods:**, select User (Kerberos V5), then select the **OK** button.

7. On the **Profile** page, deselect both **Private** and **Public** and select **Next >** to continue.

8. For the **Name:** field, enter AZ801-Lab-ConnectionSecurity and enter This rule is used for Connection Security in the AZ801 lab for the **Description** setting, then select **Finish**.

9. (*Optional*) Feel free to power up the **AZ801PacktLab-FS-01** VM and validate that the **Windows Defender Firewall** settings are applied and working as expected.

10. Once complete with your validation steps, within **Group Policy Management**, navigate to the **File Servers** OU and locate the `AZ801Lab-CH5-FS_Firewall` GPO, then right-click, and select **Link Enabled** so that the GPO is no longer applied to the machine(s) in this OU.

In this section, we learned that creating connection security rules in conjunction with inbound and outbound rules ensures that appropriate connection security between two computers has been applied to the communication layer. We also worked through the process of enabling an authentication request rule for a GPO and learned that we can also convert the rule from a request only to require authentication as part of a phased deployment approach within our organizations.

Summary

In this chapter, we learned how to secure Windows Server networking on workloads, adding security to the communications layer for shared workloads. We also learned how to configure and manage **Windows Defender Firewall**, how to plan for and configure domain isolation, and how to implement connection security and authentication request rules for your Windows servers in a hybrid environment.

In the next chapter, we will learn how to properly secure our Windows Server storage to help mitigate data theft, data exposure, and ransomware. We will work through exercises that reinforce skills surrounding the management and configuration of Windows BitLocker drive encryption for storage volumes, enable storage encryption on **Infrastructure-as-a-Service (IaaS)** VMs using **Azure Disk Encryption**, and how to manage and monitor disk encryption keys using **Azure Key Vault (AKV)** for IaaS VM and VM scale sets.

6
Secure Windows Server Storage

We will now move into the final chapter of this section by covering security and encryption for Windows Server storage. Continuing the security theme, we will now discover how to properly secure our Windows Server storage to help protect against data theft, exposure, and ransomware. We will discuss and complete exercises surrounding planning, managing, and configuring security for Windows **Infrastructure-as-a-Service (IaaS)** VMs using **Network Security Groups (NSGs)** and **Application Security Groups (ASGs)**, the **Azure Storage firewall**, and **Windows BitLocker Drive Encryption (BitLocker)**.

This chapter will include exercises on how to manage and recover **BitLocker** encrypted volumes, enable storage encryption utilizing **Azure Disk Encryption (ADE)**, and successfully manage and secure disk encryption keys using **Azure Key Vault** for IaaS VMs and VM scale sets.

In this chapter, we will cover the following topics:

- Introduction to NSGs, ASGs, service tags, and the Azure Storage firewall
- Technical requirements and lab setup
- Managing BitLocker
- Managing and recovering encrypted volumes
- Enabling storage encryption using **ADE**
- Managing disk encryption for IaaS VMs

To begin, let's gain some understanding surrounding NSGs in **Microsoft Azure** and how they can be used to secure your cloud and hybrid workloads.

Introduction to NSGs

Let me start by saying that NSGs are arguably one of the most critical components in the Azure network design and architecture. NSGs enable the overall flow and filtering of both inbound and outbound traffic within a virtual network, between subnets and VMs, internet traffic, and additional Azure services, such as storage.

Individual security rules can allow or deny inbound or outbound traffic to or from various Azure resources, and are grouped into NSGs that can then be assigned to the following:

- A single network interface, allowing filtered network traffic on a singular interface

- An entire subnet, allowing filtered traffic on all the network interfaces in the subnet

- Both the network interface and the subnet, allowing each NSG to be independently evaluated for the application of security rules

The security rules are based on the following properties:

- A **Name** value that is unique within the NSG

- **The source/destination IP address or range**, which can be any, an individual IP address, a **Classless Inter-Domain Routing** (**CIDR**) block, a service tag (covered in the upcoming *Overview of Azure service tags* section), or an ASG

- **The priority** by which the rules are processed in order, from a value between 100 and 4,096

- **The protocol** that the traffic is using, be it TCP, UDP, ICMP, ESP, AH, or any

- **The direction** of traffic, determining inbound or outbound traffic

- **The port range**, which allows you to specify a range of ports (or a wildcard *) used to minimize the overall amount of security rules created or managed

- **The action**, which results in either **Allow** or **Deny**

> The evaluation of security rules
>
> Note that security rules are evaluated and applied based on five-tuple information, using the source, source port, destination, destination port, and protocol. If the network traffic is a match for an existing security rule, that specific rule is used for the traffic, and the remaining security rules are not evaluated for that network traffic. Additional traffic that has not yet matched will evaluate against additional security rules.

NSGs notably have three inbound and three outbound default security rules that cannot be modified (as outlined in this Microsoft documentation: `https://learn.microsoft.com/azure/virtual-network/network-security-groups-overview`), but you can create your own higher priority rules to override these defaults based on your organizational requirements. These six default rules allow outbound communication to the internet, allow communication within a virtual network and **Azure Load Balancer**, and deny all inbound traffic from the internet. The default rules are described in the following table:

Rule name	Direction	Priority	Description
`AllowVnetInbound`	Inbound	65000	Allow inbound traffic from any VM to any VM in the subnet
`AllowAzureLoadBalancerInbound`	Inbound	65001	Allow traffic from the load balancer to any VM in the subnet
`DenyAllInbound`	Inbound	65500	Deny traffic from any external source to any VM in the subnet
`AllowVnetOutbound`	Outbound	65000	Allow outbound traffic from any VM to any VM in the subnet
`AllowInternetOutbound`	Outbound	65001	Allow outbound traffic to the internet from any VM in the subnet
`DenyAllOutbound`	Outbound	65500	Deny traffic from any internal VM to any system outside of the subnet

Now that we understand what NSGs bring to your overall security design, let's learn about what ASGs bring to the design table.

Introduction to ASGs

In a nutshell, ASGs follow the application's structure or tiers and allow you to group both network security policies and VMs based on specific application groups. These ASGs can then be used as sources or destination rules within NSGs. This is an incredibly powerful feature, as this allows you to automatically apply security rules to network interfaces, no matter the IP address or overall subnet membership!

Both NSGs and ASGs bring multiple benefits to the area of network security, greatly simplifying and unifying the management experience while increasing the flexibility and agility of your architecture. For a deeper understanding of ASGs, be sure to review this additional reading: `https://docs.microsoft.com/azure/virtual-network/application-security-groups`.

Now that we have completed an overview of NSGs and ASGs, let's continue by gaining an overview of another incredible feature – **Azure service tags**.

Overview of Azure service tags

The simplest definition of an **Azure service tag** is the grouping of IP address space from a specific Azure service. This makes management flexible and scalable, as the address prefixes groups are created only by Microsoft and are automatically updated when services change or update, thus reducing the need to manually update your tenant's network security rules.

Service tags can be utilized to create inbound and/or outbound NSG rules to allow or deny internet traffic and allow or deny Azure cloud traffic, traffic to other available Azure service tags, or to achieve network isolation by prohibiting internet access to your resources while still allowing access to Azure service public endpoints.

One of the best features of service tags is that these can also be limited to specific service regions, providing even greater control over and flexibility in your network configuration. Reviewing an existing NSG in the Azure tenant, as seen in the following *Figure 6.1* example, the **Source service tag** list available shows some services, indicating the regional service tags:

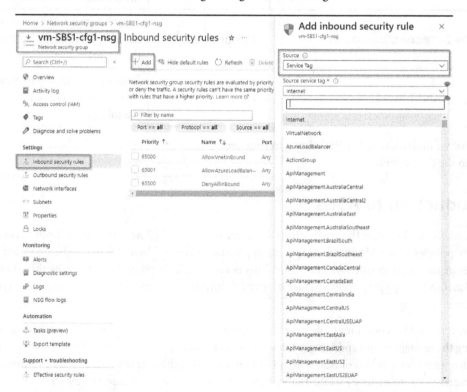

Figure 6.1 – Reviewing the application of Azure service tags with an inbound security rule

For additional review and learning surrounding Azure service tags, I highly recommend reviewing this incredible online resource, which defines all the current service tags and allows you to explore, search, and download (via JSON file) service tags and IP ranges: `https://azservicetags.azurewebsites.net/`. In addition, you can read further about the concept of virtual network service endpoint policies providing additional granularity for network and access control here: `https://docs.microsoft.com/azure/virtual-network/virtual-network-service-endpoint-policies-overview`.

Now that we have completed an overview of Azure Service tags, let's continue by gaining an overview of the **Azure Storage firewall** capabilities.

Overview of the Azure Storage firewall

The **Azure Storage firewall** provides basic access control for the public endpoint of your storage accounts and, by default, allows public access. Enabling selected IP addresses or virtual networks allows you to configure the Azure Storage firewall with known IP addresses or IP address ranges based on other cloud-based services or on-premises networks. The following configuration options, as shown in *Figure 6.2*, depict **Public network access**, **Virtual networks**, **Firewall**, and **Exceptions** settings:

Figure 6.2 – Overview of firewall and virtual network settings for an Azure Storage account

In addition, you can completely block public network access when using a feature called **Private Endpoints** (note that enabling **Private Endpoints** will bypass the Azure Storage firewall). While **Private Endpoints** may be currently out of scope for the AZ-801 exam, this is a security topic that can't be missed, so additional recommended reading can be accessed here: `https://docs.microsoft.com/azure/private-link/private-endpoint-overview`.

Now that we have gained an overview of the Azure Storage firewall, let's continue by working through the technical requirements and progressive lab setup for our hybrid lab dedicated to the exam objectives for the AZ-801 exam.

Technical requirements and lab setup

To successfully follow along and complete the tasks and exercises throughout this chapter and the following chapters in this book, we will need to ensure that the *Technical requirements* sections from *Chapter 1, Exam Overview and the Current State of Cloud Workflows*, has been fulfilled in full. We will primarily utilize our Azure tenant to complete exercises and tasks throughout this chapter to align with the AZ-801 exam objectives.

Let's begin with the setup and walkthrough steps to establish the necessary Azure resources needed for the exercises in this chapter:

1. To begin, visit `https://portal.azure.com` utilizing the Global Administrator account created in *Chapter 1, Exam Overview and the Current State of Cloud Workflows*.

2. Select **Resource groups** from the list of Azure services or simply search for `Resource groups` and select it to continue.

3. Select + **Create** to create a new resource group, selecting the **Subscription** option you created in *Chapter 1, Exam Overview and the Current State of Cloud Workflows*. Use `rg-AZ801-ExamGuide` for the **Resource Group** name and select the given **Region** you are geographically closest to, as shown in *Figure 6.3*:

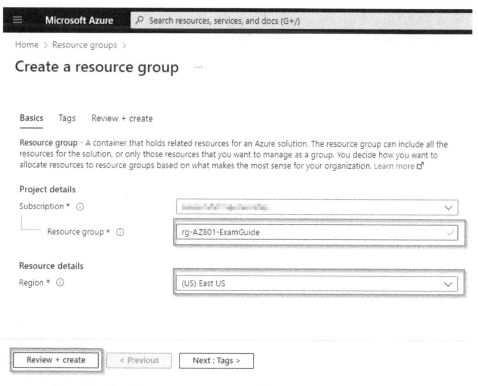

Figure 6.3 – Creating a new resource group for our Chapter 6 Azure resources

4. Click on the **Review + create** button to continue creating a new resource group.

5. Select **Virtual Networks** from the list of Azure services or simply search for `Virtual Networks` and select it to continue.

6. Select **+ Create** to create a new virtual network, and on the **Basics** tab, select the **Subscription** option you created in *Chapter 1, Exam Overview and the Current State of Cloud Workflows*. Use `rg-AZ801-ExamGuide` for the **Resource Group** name, use `vm-AZ801-vnet` in the **Name** field, and select the given **Region** you are geographically closest to, then select **Next: IP Addresses >** to continue.

7. On the **IP Addresses** tab, simply leave **IPv4 address space** as its default, then under **Subnet name**, select **default**, and change **Subnet Name** to Public, then select **Save** and **Review + create** to continue, as shown in *Figure 6.4*:

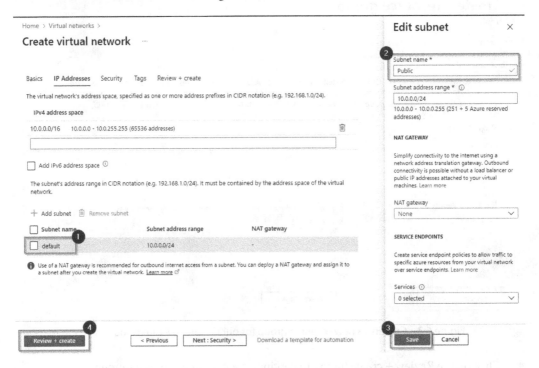

Figure 6.4 – Creating a new virtual network in our Azure tenant

8. When the validation checks have successfully passed, simply click on the **Create** button and then wait for the deployment to finish, and then click on the **Go to resource** button once the deployment has been completed successfully.

9. Select **Subnets** under the **Settings** section and then click on the **+ Subnet** button to create a new private subnet.

10. Enter Private for the **Name** field and leave the default **Subnet address range** for this exercise. Under **Service Endpoints**, for **Services**, select both Microsoft.KeyVault and Microsoft.Storage, and then click on the **Save** button.

11. Next, select **Network Security Groups** from the list of Azure services or simply search for Network security groups and select it to continue.

12. Select **+ Create** to create a new NSG, and on the **Basics** tab, select the **Subscription** option you created in *Chapter 1, Exam Overview and the Current State of Cloud Workflows*. Use rg-AZ801-ExamGuide for the **Resource Group** name, use vm-AZ801-nsg in the **Name** field, and select the given **Region** you are geographically closest to, then select **Review + create** to continue.

13. When the validation checks have successfully passed, simply click on the **Create** button and then wait for the deployment to finish, then click on the **Go to resource** button once the deployment has been completed successfully.

14. Under the **Settings** section, select **Outbound security rules**, then select **+ Add**.

15. Begin creating a new outbound security rule with the following settings as shown in *Figure 6.5* and select **Add** when completed:

    - **Source** set to **Service Tag**

    - **Source service tag** set to **VirtualNetwork**

    - **Destination** set to **Service Tag**

    - **Destination service tag** set to **Storage**

    - **Destination port ranges** set to **445**

    - **Name** updated to **AZ801-Allow-Storage-All-Regions**

Figure 6.5 – Adding a new outbound security rule for storage

16. Create a second new outbound security rule with the following settings and select **Add** when completed:

   - **Source** set to **Service Tag**
   - **Source service tag** set to **VirtualNetwork**
   - **Destination** set to **Service Tag**
   - **Destination service tag** set to **AzureKeyVault**
   - **Service** set to **HTTPS**
   - **Name** updated to **AZ801-Allow-AzureKeyVault-All-Regions**

17. Create a third new outbound security rule with the following settings (to override the default outbound rule, **AllowInternetOutbound**) and select **Add** when completed:

   A. **Source** set to **Service Tag**
   B. **Source service tag** set to **VirtualNetwork**
   C. **Destination** set to **Service Tag**
   D. **Destination service tag** set to **Internet**
   E. **Destination port ranges** set to *
   F. **Action** set to Deny
   G. **Name** updated to AZ801-Deny-Internet-All

18. Select **Inbound security rules** under the **Settings** section and begin creating a new inbound security rule for **Remote Desktop Protocol (RDP)** access with the following settings and select **Add** when completed, as shown in *Figure 6.6*:

   A. **Source** set to **Any**
   B. **Destination** set to **Service Tag**
   C. **Destination service tag** set to **VirtualNetwork**
   D. **Service** set to **RDP**
   E. **Name** updated to AZ801-Allow-RDP-All

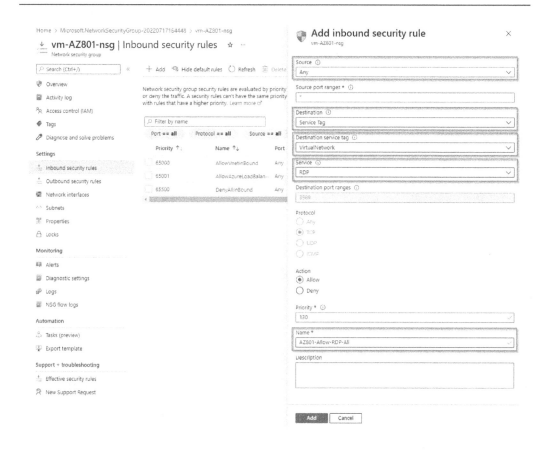

Figure 6.6 – New inbound security rule for RDP connectivity

19. Next, select **Subnets** under the **Settings** section and select + **Associate**.

20. For **Virtual networks**, select **vm-AZ801-vnet**, and for **Subnets**, select **Private**, and then click on the **OK** button to continue.

21. Let's continue by now creating a new **Azure Key Vault** setup by selecting **Key Vaults** from the list of Azure services, or simply search for Key Vaults and select it to continue.

22. Select + **Create** to begin the Key Vault creation and select the **Subscription** option you created in *Chapter 1, Exam Overview and the Current State of Cloud Workflows*. Use rg-AZ801-ExamGuide for the **Resource Group** name, use vm-AZ801-akv{uniqueID} as the **Key vault name** value (adding your own personal touch with the uniqueID value), select the given **Region** you are geographically closest to, change **Purge Protection** to **Enable**, and then select **Next : Access Policy >** to continue.

23. Check the checkbox for **Azure Disk Encryption for volume encryption** beneath the **Enable Access to:** section and then select the **Next : Networking >** button.

24. For **Connectivity method**, select **Selected networks**, and then select **Add existing virtual networks** and select your subscription. Select **vm-AZ801-vnet** for **Virtual networks** and then select **Private** for **Subnets** on the **Add networks** page, as shown in *Figure 6.7*. Select **Add** and then click on the **Review + create** button twice to initialize the deployment:

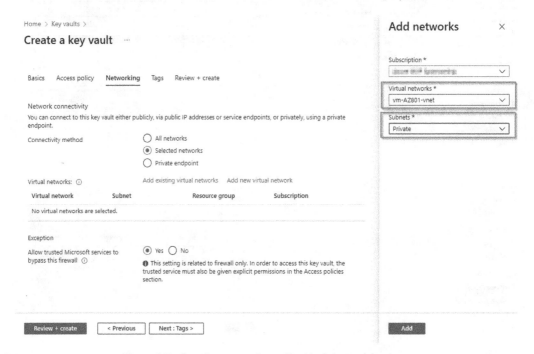

Figure 6.7 – Creating a new Azure Key Vault in our lab tenant

25. Click on the **Go to resource** button once the deployment has been completed successfully to review your newly created Key Vault resource.

26. Let's continue by creating a new storage account setup by selecting **Storage accounts** from the list of Azure services or simply searching for `Storage accounts` and selecting it to continue.

27. Click on **+ Create** to create the storage account and select the **Subscription** option you created in *Chapter 1*, *Exam Overview and The Current State of Cloud Workflows*. Use **rg-AZ801-ExamGuide** for the **Resource Group** name, provide `az801{uniqueID}` as the **Storage account name** value (where the `uniqueID` value is your first and last initial appended with `yyyymmdd`, such as `az801cg20220717`), select the given **Region** you are geographically closest to, change **Redundancy** to **Geo-redundant storage (GRS)**, and then select the **Networking** tab to continue.

28. Change **Network access** to **Enable public access from selected virtual networks and IP addresses**, then under **Virtual networks**, change **Virtual network** to **vm-AZ801-vnet**, and select **Private (10.0.1.0/24)** for the **Subnets** dropdown, as shown in *Figure 6.8*. Click on **Review + create** to continue:

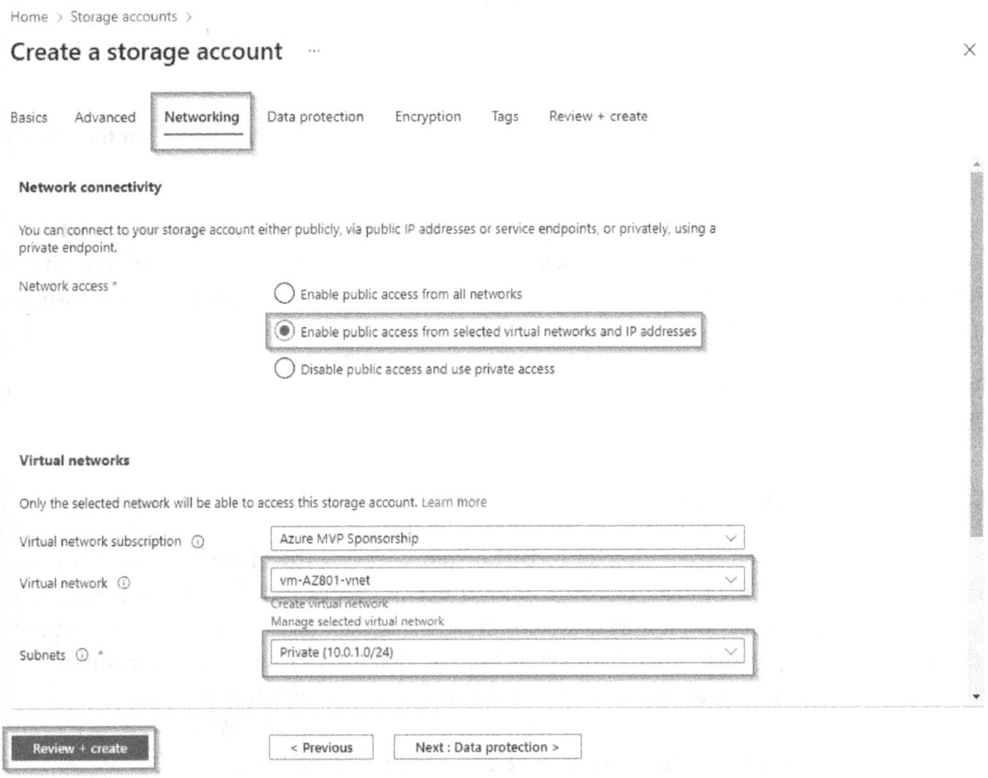

Home > Storage accounts >

Create a storage account

Basics   Advanced   **Networking**   Data protection   Encryption   Tags   Review + create

Network connectivity

You can connect to your storage account either publicly, via public IP addresses or service endpoints, or privately, using a private endpoint.

Network access *

○ Enable public access from all networks

◉ Enable public access from selected virtual networks and IP addresses

○ Disable public access and use private access

Virtual networks

Only the selected network will be able to access this storage account. Learn more

Virtual network subscription ⓘ   Azure MVP Sponsorship

Virtual network ⓘ   vm-AZ801-vnet

Create virtual network
Manage selected virtual network

Subnets ⓘ *   Private (10.0.1.0/24)

[Review + create]   [< Previous]   [Next : Data protection >]

Figure 6.8 – Creating a new storage account in our lab tenant

29. For the last few steps of this lab, we will create two VMs by first selecting **+ Create a resource** and then selecting **Create** under **Virtual machine**.

30. Select the **Subscription** option you created in *Chapter 1, Exam Overview and the Current State of Cloud Workflows*. Use **rg-AZ801-ExamGuide** for the **Resource Group** name; enter `vm-az801-svr1` as the **Virtual machine name** value; select an appropriate region; select **Windows Server 2022 Datacenter: Azure Edition – Gen2** for **Image**; for **Size**, select **Standard_B2s**; set **Username** to `az801admin`; and use `Packtaz801guiderocks` under **Password** and **Confirm password**.

31. Select the **Networking** tab and change **Subnet** for this VM to **Public**, change **NIC network security group** to **vm-AZ801-nsg**, click on the **Review + create** button to begin VM validation, and then click on the **Create** button to create the VM.

32. After the VM has been successfully created, click on the **Create another VM** button to create the second IaaS VM.

33. Repeat *steps 30-31* to create a second VM, using the following changed values:

    A. For *step 30*, use `vm-az801-svr2` under **Virtual machine name**.

    B. For *step 31*, ensure that **Subnet** is set to **Private** and **NIC network security group** is set to **None**.

34. Click on the **Review + create** button to begin VM validation and then click on the **Create** button to create the VM.

In this section, we established a new resource group in our lab tenant and created a virtual network, NSGs, and both inbound and outbound security rules. We then associated subnets with our virtual network and established an Azure Key Vault resource, an Azure Storage account, and two VMs for use in the latter part of this chapter.

In the next section, we will review the approaches and best practices for managing BitLocker for on-premises devices.

Managing BitLocker

We could easily dedicate an entire chapter to **BitLocker**, covering the history, deployment, and management aspects. For the purposes of the AZ-801 exam objectives, let's focus on a high-level overview of what BitLocker is, how it works, and how it can be used for data protection and encryption, as well as the system requirements and basic setup. For additional information on the requirements, practical applications and use, and deployment and management with BitLocker, be sure to check out the following article: `https://learn.microsoft.com/windows/security/information-protection/bitlocker/bitlocker-overview`.

BitLocker protects user data against unauthorized access and exposure on devices running Windows by providing full-volume encryption features. The present-day BitLocker features help protect the following data protection needs of organizations worldwide:

- BitLocker provides encryption for full drives, portable drives, and virtualized drives for VMs running Windows

- Most modern devices are now protected by default with BitLocker encryption

- Self-encrypting hard drives are supported and allow for the same unified set of configuration and management tools for administrators

- The option to encrypt the full hard drive, pre-provision encryption, or only encrypt the space used provides flexibility in terms of administration and provisioning time for administrators and end users

- Advances in BitLocker technology over the years mean that only a recovery key is required when encountering disk corruption or a lost PIN or password or when removing encryption – Windows Update/servicing passes automatically to handle any necessary decryption and re-encryption.

- A wide variety of **BitLocker Key Protectors** allows for unlocking or auto-unlocking

- BitLocker, while traditionally managed on-premises within **Active Directory Domain Services**, can be deployed and managed from within **Microsoft BitLocker Administration and Monitoring (MBAM)**, **Microsoft Endpoint Configuration Manager**, **Microsoft Intune**, and **Microsoft Azure**

BitLocker general system requirements

When launching BitLocker on any Windows system, there is a verification process that takes place to ensure that the following minimum system requirements are met before encrypting a volume:

- The device must meet the minimum requirements for the specific Windows **Operating System (OS)** version

- For some OS versions, particularly Windows Server versions, BitLocker is an optional feature that will need to be installed using PowerShell, Server Manager, or Admin Center

- While a hardware TPM of 1.2 or 2.0 is not required, additional pre-boot verification and multi-factor authentication rely on TPM for additional integrity checks

- Trusted BIOS or UEFI firmware is a must, and the hard drive must be set as the primary boot device before USB, CD, or network boot

- Regarding the filesystem, when using UEFI, at least one FAT32 partition is used for the system drive and one NTFS partition for the OS; when using legacy BIOS, at least two NTFS disk partitions are present, one for the OS and one for the system drive

BitLocker basic setup

The general management of BitLocker on devices (workstations and servers) can be done simply by searching for `BitLocker` from the **Start** menu and selecting **Manage BitLocker**, as shown in *Figure 6.9*:

Figure 6.9 – Managing BitLocker Drive Encryption settings on a device

For an organization managing both workstations and servers, users may be empowered to self-service and backup, restore, or even recreate their BitLocker keys for data protection purposes, as an automatic backup to AD is not enabled. However, it is recommended that the **BitLocker** settings be managed via **Group Policy**, especially the **Store BitLocker recovery information in Active Directory Domain Services** setting, and passwords and key packages stored within either Active Directory, Microsoft Intune, or Azure AD.

Additionally, it is not recommended to manage and reconfigure BitLocker from within a VM, as this is typically managed by a system administrator or an extension. Additional reading for achieving appropriate levels of information protection using BitLocker with **Group Policy** can be found here: `https://docs.microsoft.com/windows/security/information-protection/bitlocker/bitlocker-group-policy-settings`.

Now that we have covered the basics of BitLocker, let's step into how we can manage and recover BitLocker encrypted volumes.

Managing and recovering encrypted volumes

Now that we have an understanding surrounding **BitLocker** setup and know that recovery information is saved inside of **Active Directory Domain Services** (or **Azure AD** prior to the looming April 2026 mainstream end of support for **Microsoft BitLocker Administration and Monitoring**), let's walk through an overview of how we can manage BitLocker in a hybrid environment.

As with all Microsoft tools, there is a legacy configuration approach and a related **PowerShell** management approach. BitLocker is no exception here, as there is a myriad of PowerShell cmdlets to help you manage this task across an organization. A full list of PowerShell cmdlets for BitLocker administration can be found here: `https://docs.microsoft.com/windows/security/information-protection/bitlocker/bitlocker-basic-deployment#encrypting-volumes-using-the-bitlocker-windows-powershell-cmdlets`.

To reduce frustration for administrators and users, a feature called **BitLocker Key Protector** allows for various ways to unlock and allow data to be read from disk. The following parameters from the **Add-BitLockerKeyProtector**, **Backup-BitLockerKeyProtector**, and **Remove-BitLockerKeyProtector** PowerShell cmdlets give some insight as to what protectors are available and can be combined to meet strict environmental and security regulations:

Key Protector Parameter	Protector Details
`ADAccountOrGroupProtector`	BitLocker can use an AD user or group as domain authentication to unlock data drives, but this cannot be used for OS volumes
`PasswordProtector`	BitLocker simply uses a password
`RecoveryKeyProtector`	BitLocker uses a recovery key stored on a USB device
`RecoveryPasswordProtector`	BitLocker simply uses a recovery password
`StartupKeyProtector`	BitLocker uses the input of an external key from a USB device
`TpmAndPinAndStartupKeyProtector`	BitLocker uses a combination of the TPM, user PIN, and the input of an external key from a USB device
`TpmAndPinProtector`	BitLocker uses a combination of the TPM and user PIN
`TpmAndStartupKeyProtector`	BitLocker uses a combination of the TPM and the input of an external key from a USB device
`TpmProtector`	BitLocker simply uses the device TPM for encryption key protection

There are a few ways to retrieve the necessary BitLocker recovery information and they are as follows, depending on environment setup and organizational requirements:

- **BitLocker Recovery Password Viewer** allows Domain Administrators and delegated administrators to search for or view the BitLocker recovery information for AD objects, which is available as an additional Windows Server feature, as shown in *Figure 6.10*:

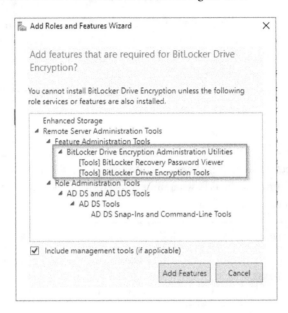

Figure 6.10 – BitLocker Recovery Password Viewer optional Windows Server feature

- A **PowerShell** command run by a delegated administrator, which allows you to search for a specific computer object using the following example:

```
$objComputer='computerNameOfWorkstationOrServer'
Get-ADObject -Filter {objectclass -eq 'msFVE-
RecoveryInformation'} -SearchBase $objComputer.
DistinguishedName -Properties 'msFVE-RecoveryPassword'z
```

- **MBAM** is a tool for organizations that are in extended support and have access to **Microsoft Desktop Optimization Pack** (**MDOP**), allowing for the management of device policies, self-service recovery and the retrieval of key packages, and administrator management and reporting insights.

- **Microsoft Endpoint Configuration Manager** can configure a BitLocker management control policy, allowing you to configure granular Client Management policies for devices and giving BitLocker compliance reporting.

- **Microsoft Endpoint Manager** involves the BitLocker base configuration settings and fixed drive, removable, and OS drive settings, as well as BitLocker device compliance reporting, as shown in *Figure 6.11*:

Figure 6.11 – BitLocker Disk Encryption profile in Microsoft Endpoint Manager

A full list of detailed recovery options can be reviewed here: `https://docs.microsoft.com/previous-versions/windows/it-pro/windows-server-2012-r2-and-2012/dn383583(v=ws.11)`.

Now that we have covered some of the BitLocker recovery approaches currently available to administrators, let's learn more about storage encryption in the cloud using ADE for IaaS VMs.

Enabling storage encryption by using ADE

When it comes to Azure VMs running as IaaS in Microsoft Azure, storage-level protection is ultimately provided in the form of encryption on the VM disk files, and can be handled through ADE using BitLocker Drive Encryption for Windows systems and **DM-Crypt** for Linux-based systems. ADE can automatically encrypt the OS disk, any data disks, and the temporary disks and will support both managed and unmanaged disks.

A few scenarios where you can utilize ADE are as follows:

- Enabling encryption on existing Azure VMs that are already in Azure
- Enabling encryption on new Azure VMs that were created from Azure Marketplace pre-created images
- Enabling encryption on new Azure VMs that were established from a customer-encrypted virtual hard drive file using existing encryption keys

In addition, there are key requirements that need to be met for ADE regarding other OSes, networking, memory, VM generation, Group Policy, and encryption key storage. The requirements are as follows:

- In terms of the **VM size**, ADE is not available on Basic A-series VMs, nor on Lsv2-series VMs (for a full list of support VM sizes and generations, including unsupported scenarios for ADE, please review the following Microsoft article: `https://learn.microsoft.com/azure/` `virtual-machines/windows/disk-encryption-overview#supported-` `vms-and-operating-systems`).

- For the **VM generation**, ADE is NOT available on Generation 2 VMs.

- For **memory**, ADE is NOT available on VMs with less than 2 gigabytes of RAM.

- For **networking**, the VM must be able to authenticate to an **Azure AD** endpoint, a managed service identity must be able to write to the **Azure Key Vault**, and the Windows VM must be able to connect to the Azure Key Vault endpoint.

- For **Group Policy**, it is recommended to NOT push any TPM protectors to the Windows VMs, and you must allow a 256-bit recovery key, as well as the AES-CBC algorithm, for any domain-joined VMs.

- For **key storage**, ADE requires an Azure Key Vault for the management and security control of disk encryption keys and secrets. The key vault and VMs must reside in the same Azure subscription and region, and the key vault access policy for **Azure Disk Encryption for volume encryption** must be enabled.

Server-side encryption is enabled by default for data encryption at rest and cannot be disabled, whereas ADE is not enabled by default, but can be enabled on both Windows and Linux Azure VMs. Note that server-side encryption does support Generation 2 Azure VMs and all Azure VM sizes, but does not, however, support temporary or unmanaged disks. For a detailed comparison of the managed disk encryption options, be sure to review this resource: `https://aka.ms/` `diskencryptioncomparison`.

Now that we have a basic understanding of what is required to enable storage encryption for ADE, let's walk through managing disk encryption keys for IaaS VMs running in **Microsoft Azure**.

Managing Disk Encryption keys for IaaS VMs

Earlier in this chapter, we completed the lab setup in our Azure tenant to establish a resource group, an Azure Key Vault, and two Azure VMs. Let's begin with this walkthrough of managing disk encryption keys by completing the following tutorial steps:

1. To begin, visit `https://portal.azure.com` utilizing your Global Administrator account created in *Chapter 1, Exam Overview and the Current State of Cloud Workflows*.

2. Open a new browser tab and search for whats my ip, then record your public IP address for later use, and close this open browser tab.

3. Select **Key vaults** from the list of Azure services or simply search for Key vaults and select it to continue.

4. Use the **Settings** section, select **Networking**, then add your public IP address under the **IP address or CIDR** section, and select **Save** to commit the change.

5. Select **Disk Encryption Sets** from the list of Azure services or simply search for Disk Encryption Sets and select it to continue.

6. On the **Disk Encryption Sets** page, click on **+ Create** to begin, as shown in *Figure 6.12*:

Figure 6.12 – Creating new disk encryption sets in your Azure tenant

7. On **Create a disk encryption set**, under **Key**, select **Create New**, and supply a **Name** value of vm-AZ801-kek, then accept the remaining defaults, and click on **Create**, as shown in *Figure 6.13*:

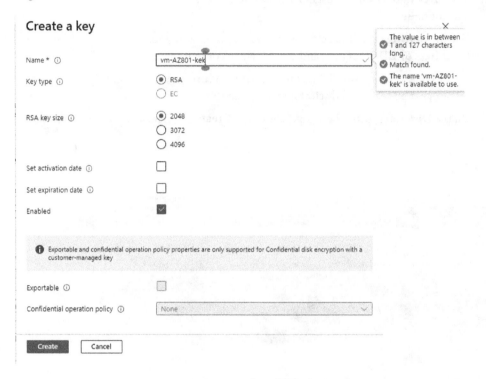

Figure 6.13 – Creating a new key for our disk encryption set in Azure Key Vault

8. Still on the **Create a disk encryption set** page, select your **Subscription** option, select **rg-AZ801-ExamGuide** under **Resource Group**, supply a **Disk encryption set name** value of vm-AZ801-des, select the **Region** in which you placed your VMs and Key Vault, find and select your Key Vault for the **Key Vault** entry, and then click on **Review + create**, as shown in *Figure 6.14*. Then, select **Create** on the verification page:

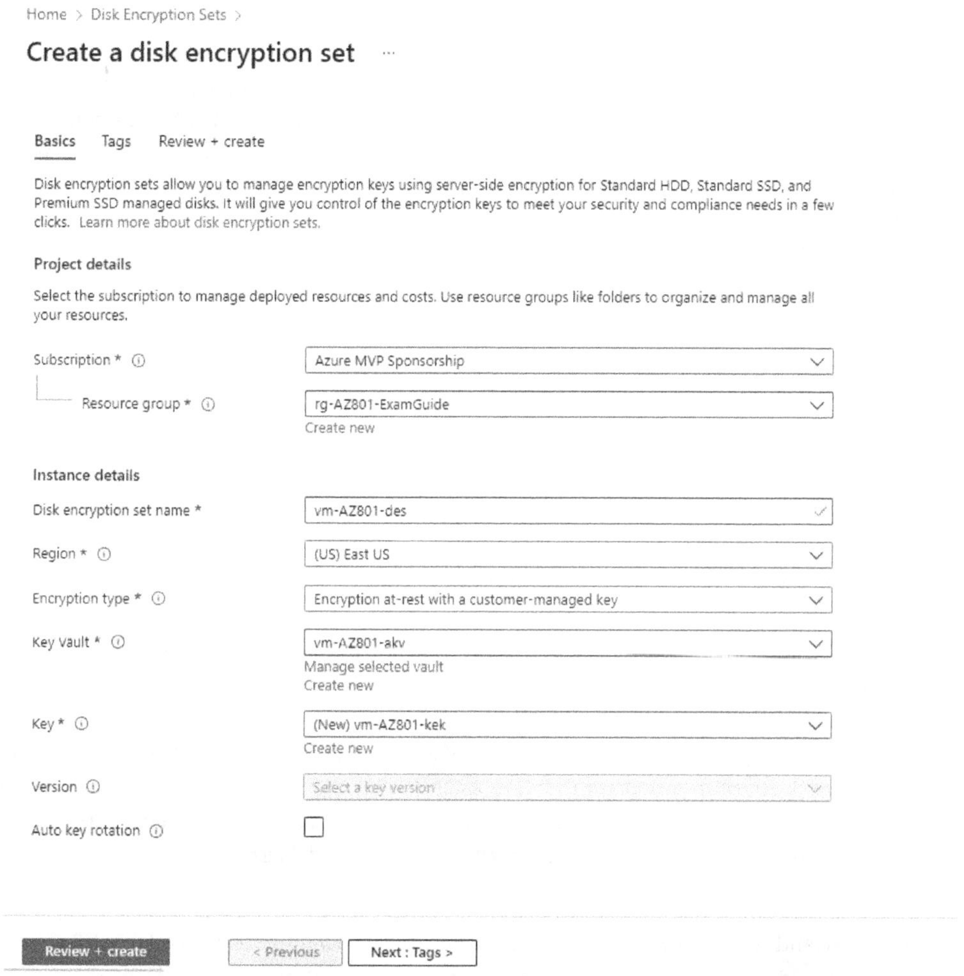

Home > Disk Encryption Sets >

Create a disk encryption set   ···

Basics   Tags   Review + create

Disk encryption sets allow you to manage encryption keys using server-side encryption for Standard HDD, Standard SSD, and Premium SSD managed disks. It will give you control of the encryption keys to meet your security and compliance needs in a few clicks. Learn more about disk encryption sets.

Project details

Select the subscription to manage deployed resources and costs. Use resource groups like folders to organize and manage all your resources.

Subscription * ⓘ

> Resource group * ⓘ

Azure MVP Sponsorship	∨

rg-AZ801-ExamGuide	∨

Create new

Instance details

Disk encryption set name *

vm-AZ801-des	✓

Region * ⓘ

(US) East US	∨

Encryption type * ⓘ

Encryption at-rest with a customer-managed key	∨

Key vault * ⓘ

vm-AZ801-akv	∨

Manage selected vault
Create new

Key * ⓘ

(New) vm-AZ801-kek	∨

Create new

Version ⓘ

Select a key version	∨

Auto key rotation ⓘ   ☐

Review + create		< Previous	Next : Tags >

Figure 6.14 – Creating a new disk encryption set in our Azure tenant

9. Select **Virtual machines** from the list of Azure services or simply search for `Virtual machines` and select it to continue.

10. Check either **vm-az801-svr1** or **vm-az801-svr2** and select **Stop** from the action bar.

11. Under the **Settings** section, select **Disks**, and then select the disk for which to change the encryption.

12. Under the **Settings** section, select **Encryption**, then change the **Encryption** selection to **Encryption at rest with a customer-managed key**, and select the **Disk encryption set** option that we recently created named `vm-AZ801-des`.

> **Key Encryption Key compatibility with Windows Server 2022**
>
> While we have deployed our VMs in this lab exercise with Windows Server 2022, it should be noted that this version of Windows Server has a newer version of BitLocker that does *not* support RSA 2048-bit Key Encryption Keys. This is a known issue, and it is recommended to use either 3,072- or 4,096-length keys for this particular OS version. More details can be found here: `https://docs.microsoft.com/azure/virtual-machines/windows/disk-encryption-faq#what-size-should-i-use-for-my-key-encryption-key--kek--`.

To complete a quick recommended tutorial on how to encrypt the OS and data disks for a VM scale set, be sure to check out this resource: `https://docs.microsoft.com/azure/virtual-machine-scale-sets/disk-encryption-powershell`.

In this section, we learned how to manage **ADE** on IaaS VMs, store keys in **Azure Key Vault**, establish a **disk encryption set** with a **Key Encryption Key**, and adjusted our IaaS VM to utilize the new customer-managed encryption set.

Next, we will complete a review of all we have learned in this chapter and set the stage for the next chapter of this book.

Summary

In this chapter, we learned some basics surrounding NSGs, ASGs, **Azure service tags**, and the **Azure Storage Firewall** as a recommended approach to establishing a layered security model. We also learned how to manage and recover BitLocker encrypted volumes, enable storage encryption using **ADE**, and successfully manage and secure disk encryption keys using **Azure Key Vault** for IaaS VMs and VM scale sets.

In the next section, and specifically in *Chapter 7, Implementing a Windows Server Failover Cluster*, we will learn how to design, configure, and manage a **Windows Server failover cluster**. We will cover how to successfully establish the building blocks for a Windows Server failover cluster, configure various storage options, and successfully configure the appropriate network settings for the failover cluster. Finally, we will learn how to configure cluster workload options, including details on when and how to use cluster sets and **scale-out file servers** to achieve continuously available application storage.

Part 3: Implement and Manage Windows Server High Availability

In this section, you will get valuable insights, tips, and tricks on how to design, implement, and manage a Windows Server failover cluster to achieve high availability for your workloads both on-premises and in the cloud.

This part of the book comprises the following chapters:

- *Chapter 7, Implementing a Windows Server Failover Cluster*
- *Chapter 8, Managing Failover Clustering*
- *Chapter 9, Implementing and Managing Storage Spaces Direct*

7

Implementing a Windows Server Failover Cluster

In this chapter, we will learn how to design and configure **Windows Server Failover Clustering (WSFC)**. We will cover how to successfully establish the building blocks for WSFC, how to configure various storage options, and how to successfully design and configure the appropriate network settings for the Failover Cluster. Finally, we will learn how to configure cluster workload options, including details on when and how to use Cluster Sets and **Scale-Out File Server** (**SOFS**) to achieve continuously available application storage.

This chapter will include exercises on how to gain familiarity with the designs, techniques, and tools to create and manage WSFC components and workloads on-premises, with hybrid components and services, and in the cloud consistent with the AZ-801 exam objectives.

In this chapter, we will cover the following topics:

- Overview of Windows Server Failover Clustering
- Technical requirements and lab setup
- Introduction to Windows Admin Center
- Creating and managing Windows Server Failover Clustering
- Configuring network settings for failover clustering
- Configuring the Windows Server Failover Clustering storage options
- Configuring cluster workload options
- Configuring Stretch Clusters
- Configuring Scale-Out File Server
- Configuring Cluster Sets

Overview of Windows Server Failover Clustering

Throughout this book, we have learned how to use Hyper-V to host our lab machines and configure Hyper-V to utilize physical machine resources that are mapped to virtual equivalent resources (think the disk, CPU, RAM, and network), thus abstracting the physical machine and hardware from the intended hosted workloads. However, with this single-host approach, Hyper-V does not provide any additional resilience or availability for protecting workloads – this is where WSFC delivers and excels.

Let's continue learning more about WSFC by introducing the services provided by this important collection of components.

What is WSFC?

WSFC is a cooperative group of independent computers that work together to enable the high availability of clustered roles to deliver applications and services called clustered roles, provide distributed configuration, and coordinated failover orchestration, and allow centralized resource management and health/insights for all hosts in the cluster. Failover Clustering is primarily considered active-passive, where all workloads and roles are running on a preferred or active node, and the passive node, which is not running any current workloads or roles, is awaiting a failover from the active node. However, active-active Failover Clusters can be created for replication purposes.

Typically, at least two hosts and up to 64 hosts total can be combined into one logical unit called a Failover Cluster. It should also be mentioned that both physical and virtual cluster nodes are supported by Microsoft and these clusters can be stretched across locations to provide additional disaster recovery and business continuity protection.

These hosts are recommended to be identical in configuration, should be certified with the Windows Server version that you intend to run, must pass failover configuration tests, and share both one or more shared storage resources and two or more collective networks (used for system management, heartbeat, and so on). A detailed list of clustering hardware requirements, network requirements, and storage options can be reviewed at `https://docs.microsoft.com/windows-server/failover-clustering/clustering-requirements`.

Failover Clusters allow these clustered roles to be continually monitored and proactively respond to environmental issues and quickly move to another available host or node in the cluster to minimize service interruptions. The cluster itself is represented in the environment by at least one entity object called a **Cluster Name Object** (**CNO**). In addition, Failover Clusters provide a feature called **Cluster Shared Volumes** (**CSV**), which provide a distributed Namespace clustered filesystem, allowing hosts to achieve simultaneous read-write operations to the same disk or **Logical Unit Number** (**LUN**).

Another major component of Failover Clustering is the deployment of shared storage that is compatible with the version of Windows Server that you are using within the Failover Cluster.

Finally, when designing and configuring a Failover Cluster, you must carefully consider the possibility that individuals can and will lose communication with the cluster at times. WSFC utilizes a quorum-based approach to determine node-level availability and overall fault tolerance during operations and failover situations. Additional supporting details on the quorum and voting methodology can be found at `https://docs.microsoft.com/windows-server/failover-clustering/manage-cluster-quorum`. Given this information, it is recommended that every Failover Cluster configuration includes a cluster quorum witness, which can be configured as a disk witness, a file share witness, a USB drive attached to a network switch, or a Cloud Witness using an Azure storage account.

What workloads can WSFC be used for?

Primarily speaking, WSFC can be used to apply the following workloads in highly available or continuously available configurations:

- Hyper-V virtual machine and application file share storage
- Windows **Distributed File System (DFS)** Namespace server
- File servers (shared disk failover)
- Microsoft SQL Server (including SQL Always On availability groups)
- Microsoft Exchange **Database Availability Groups (DAGs)**

What is the difference between a Failover Cluster and a hyper-converged cluster?

Simply put, failover is having built-in redundancy for an environment so that when and if a server fails, another will immediately take its place.

While Failover Clustering can be achieved in various ways, utilizing a software-defined hyper-converged infrastructure or cluster is the best option to provide additional hardware in a consolidated solution to provide scalable, cost-conscious, and high-performance reliability for highly available services and workloads. In addition, hyper-converged infrastructure such as **Azure Stack Hyper-Converged Infrastructure** (HCI) provides a familiar management and administration experience, while working with on-premises, hybrid, and cloud servers to provide a best-of-breed experience tailored to the needs of your organizations and customers.

To get a better sense of what **Azure Stack HCI** brings to the table, be sure to check out this additional reading, which provides a high-level overview of Azure Stack HCI: `https://docs.microsoft.com/azure-stack/hci/overview`.

Now that we have provided an overview of the features and requirements of WSFC, let's continue by walking through creating a basic Failover Cluster in our virtualized lab environment to review the requirements and gain visibility into the modern setup process. Note that these steps have been provided to demonstrate establishing a test learning environment and should not be used as a recommendation architecture for a production or enterprise deployment for your organizations or customers.

Technical requirements and lab setup

To successfully follow along and complete the tasks and exercises throughout this and the following chapters in this book, we will need to ensure that the technical requirements from *Chapter 1*, *Exam Overview and the Current State of Cloud Workflows*, have been completed in full. We will be primarily utilizing our Hyper-V virtual machine environment to complete exercises and tasks throughout this chapter to align with the AZ-801 exam objectives.

Let's begin by introducing **Windows Admin Center** and how this tool can be utilized to create and manage resources no matter where they are located.

Introduction to Windows Admin Center

In late 2019, Microsoft announced the availability of a new complementary management experience for Windows machines without requiring the use of Azure or the cloud, and was ultimately released as Windows Admin Center. Windows Admin Center is not a part of the Windows Server image by default and can be downloaded for free and installed on either Windows Server (in a gateway mode for managing multiple servers) or installed on Windows 10/11. The management solution is entirely a web-based UI tool that allows administration from virtually anywhere, given there's proper configuration and firewall access. The installation requires the use of a TLS certificate. Note that the temporary self-signed certificate that Windows Admin Center can create for use/evaluation expires in 60 days.

In addition to complementing existing Windows Server management tools, Windows Admin Center can also integrate with Azure to provide greater control for authentication and authorization. It supports many integrations directly with Azure (such as **Azure AD**, **Azure Backup**, **Azure Site Recovery**, **Azure Network Adapter**, **Azure Monitor**, **Azure Arc**, and **Azure Security Center**). The list continues to grow and all Azure hybrid service integrations with Windows Admin Center can be reviewed at `https://docs.microsoft.com/windows-server/manage/windows-admin-center/azure`. As we will soon learn, we can administer both individual Windows servers and different types of cluster management across environments!

Let's continue with the setup and walkthrough steps to establish the resources needed for the exercises in this chapter:

1. Begin by opening **Hyper-V Manager** on your device hosting the virtual machines.

2. Right-click on the **AZ801PacktLab-DC-01** virtual machine and click **Start**, and ensure that the virtual machine is running. Do the same for the **AZ801PacktLab-FS-01** virtual machine.

3. Use the **Action > Ctrl+Alt+Delete** menu option to begin, then log into **AZ801PacktLab-FS-01** as an Administrator using `Packtaz801guiderocks` as the password.

4. Use the **Start** menu to locate Microsoft Edge. Then, navigate to `https://aka.ms/wacdownload` to begin downloading the latest **Windows Admin Center** version.

5.   Locate `WindowsAdminCenterXXXX.X.msi` (where `XXXX.X` is the latest version automatically downloaded) in the `Downloads` folder and double-click the file to begin the installation. When prompted, select **I accept these terms** and then click **Next** on the screen to continue.

6.   On the **Configure Gateway Endpoint** screen, select **Required diagnostic data** and then click **Next** to continue, as shown in *Figure 7.1*:

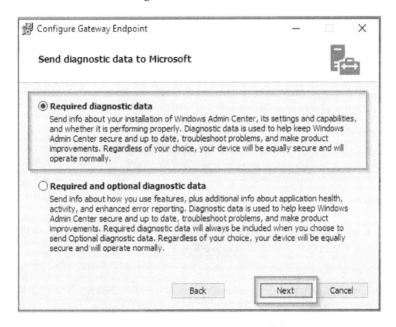

Figure 7.1 – Selecting Required diagnostic data on the Configure Gateway Endpoint screen

7.   Ensure that **I don't want to use Microsoft Update** is selected and click **Next** to continue.

8.   Accept the defaults on the second **Configure Gateway Endpoint** screen and click **Next** to continue.

9.   On the third **Configure Gateway Endpoint** screen, ensure that you have selected **Generate a self-signed SSL certificate** and ensure that **Redirect HTTP port 80 traffic to HTTPS** is selected. Then, click **Next** to continue, as shown in *Figure 7.2*:

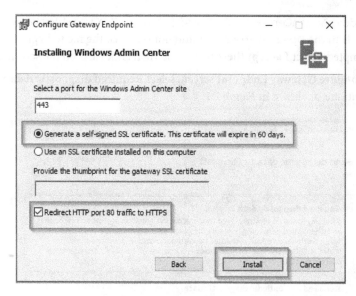

Figure 7.2 – Using a self-signed SSL certificate for this lab environment

Windows Admin Center will complete the installation and present you with a URL that can be opened (it will be `https://AZ801Lab-FS-01.AD.az801.com:443`). Select this new link to open **Windows Admin Center**. Click **Finish** on the setup screen to complete the setup.

10. Within the Windows Admin Center browser session, click **+ Add** and then click **Search Active Directory**. Enter `*` in the search field and click the **Search** button to return all lab machines available in AD. Then, check the checkbox shown in *Figure 7.3* to select all our lab devices and click **Add** to finish adding these machines to Windows Admin Center:

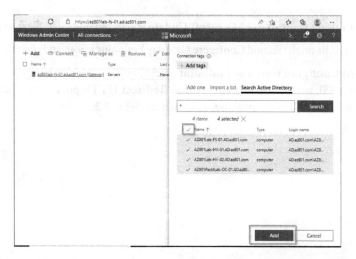

Figure 7.3 – Adding all lab machines from AD.az801.com AD to Windows Admin Center

11. In this exercise, we have completed all the necessary configuration steps. Please keep **Windows Admin Center** and the virtual machine console open to complete the next exercise.

In this section, we learned about Windows Admin Center and the features that this tool brings to any enterprise. We also learned how to download, install, and initially configure Windows Admin Center for system administration.

Next, we will learn how to complete a basic **Windows Server Failover Cluster** configuration to achieve the requirements surrounding Failover Clustering for the AZ-801 exam objectives.

Creating and managing a Windows Server Failover Cluster

Let's continue with the setup and walkthrough steps to establish the resources needed for the Failover Clustering exercises in this chapter:

1. Begin by opening **Hyper-V Manager** on your device hosting the virtual machines.

2. Right-click on the **AZ801PacktLab-HV-01** virtual machine and click **Settings**. Under **Add Hardware**, leave **SCSI Controller** as the default and click the **Add** button.

3. When prompted with **type of drive you want to attach to the controller**, select **Hard Drive** and click **Add**. Then, under **Virtual hard disk:**, click the **New** button and choose the following settings, clicking **Next** on each page:

   A. On the **Choose Disk Type** screen, select **Dynamically expanding**.

   B. On the **Specify Name and Location** screen, set the following:

      iii. **Name:** Set to AZ801PacktLab-HV-01-Data.vhdx

      iv. **Location:** Set to C:\AZ801PacktLab\VMs

   E. On the **Configure Disk** screen, set **Size:** to 10 GB and click **Finish**.

4. Select the **Add Hardware** option at the top of the **Settings** screen and this time, select **Network Adapter**.

5. Under the **Virtual switch:** dropdown, select **AZ801PacktLabInternal** and then click the **OK** button, as shown in *Figure 7.4*, to commit the changes to the virtual machine:

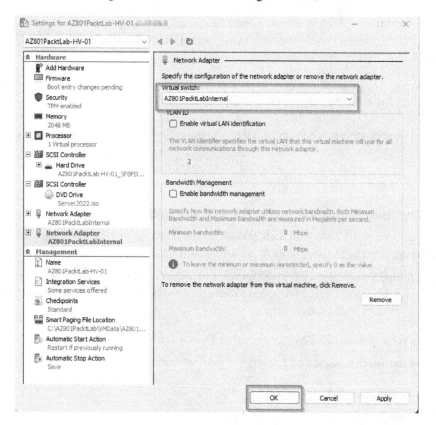

Figure 7.4 – Adding a new internal network adapter to our Hyper-V virtual host

6. Complete *Steps 2* to *5* for the **AZ801PacktLab-HV-02** virtual machine, substituting AZ801PacktLab-HV-02-Data.vhdx for *Step 3b*.

7. Finish installing Windows Server 2022 on both **AZ801PacktLab-HV-01** and **AZ801PacktLab-HV-02**, as per the instructions located in *Chapter 1*, *Exam Overview and the Current State of Cloud Workflows*, in the *Setting up a lab environment using Hyper-V* section, starting with *Step 10*.

8. Right-click on both the **AZ801PacktLab-HV-01** and **AZ801PacktLab-HV-02** virtual machines and click **Connect**.

9. Log into **AZ801PacktLab-HV-01** as an administrator and open a PowerShell session from the **Start** menu.

10. Enter the following commands for the **AZ801PacktLab-HV-01** virtual machine and enter the AZ801\Administrator credentials when prompted:

```
New-NetIPAddress -IPAddress 10.10.10.3 -DefaultGateway
10.10.10.1 -PrefixLength 24 -InterfaceIndex
(Get-NetAdapter).InterfaceIndex
Set-DNSClientServerAddress -InterfaceIndex
(Get-NetAdapter).InterfaceIndex -ServerAddresses
10.10.10.1
Rename-Computer -NewName AZ801Lab-HV-01 -Restart -Force
```

11. Enter the following commands for the **AZ801PacktLab-HV-02** virtual machine and enter the
    AZ801\Administrator credentials when prompted:

```
New-NetIPAddress -IPAddress 10.10.10.4 -DefaultGateway
10.10.10.1 -PrefixLength 24 -InterfaceIndex
(Get-NetAdapter).InterfaceIndex
Set-DNSClientServerAddress -InterfaceIndex
(Get-NetAdapter).InterfaceIndex -ServerAddresses
10.10.10.1
Rename-Computer -NewName AZ801Lab-HV-02 -Restart -Force
```

12. Once both virtual machines have successfully been rebooted, we can close out of the VM console
    for both and connect to **AZ801PacktLab-FS-01** to complete the remaining steps in this chapter.

13. Returning to Windows Admin Center, from the home page, click + **Add**. Under **Add or create
    resources**, scroll to the **Server clusters** section, then click **Create new**, as shown in *Figure 7.5*:

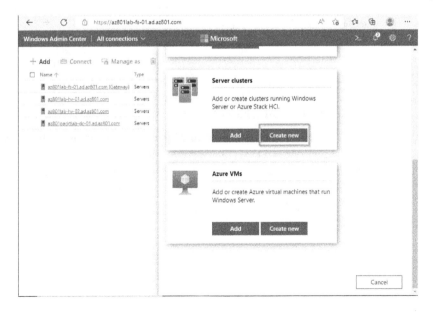

Figure 7.5 – Using Windows Admin Center to begin a Failover Cluster configuration

14. On the initial **Cluster Creation** page, we will **Select the workload type** as **Virtual Machines** and then **Select server locations** as **All servers in one site**. Click **Create** to continue.

15. On the **Deploy a Windows Server Cluster** screen (cluster creation **Step 1.1** for **Check the prerequisites**), carefully review the base requirements for creating a Windows Server Failover Cluster, as shown in *Figure 7.6*. When completed, click **Next** to continue the setup:

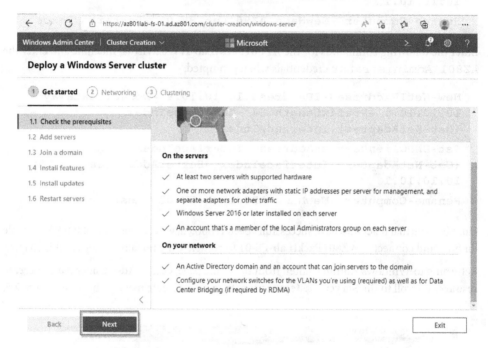

Figure 7.6 – Deploy a Windows Server Cluster prerequisites review screen

16. In setup **Step 1.2** for **Add servers** for cluster creation, supply a **Username** value of administrator and a **Password** value of Packtaz801guiderocks.

17. For **Enter the computer name, IPv4 address, or fully qualified domain name of each server**, add the following machines one at a time, clicking the **Add** button to add them, as shown in *Figure 7.7*. Click **Next** when both virtual machines have been successfully added:

A. az801lab-hv-01.ad.az801.com

B. az801lab-hv-02.ad.az801.com

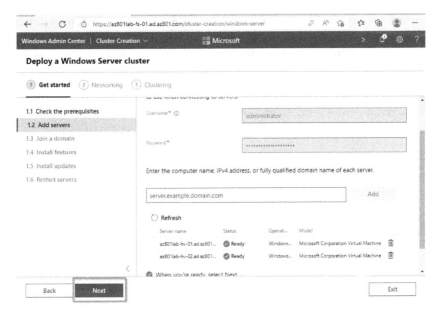

Figure 7.7 – Adding servers to the initial Cluster Creation manager

18. In setup **Step 1.3** for **Join a domain** for cluster creation, for **Domain username**, use a value of az801\administrator and for **Domain password**, use a value of Packtaz801guiderocks. Click **Next** when the validation has been completed, as shown in *Figure 7.8*:

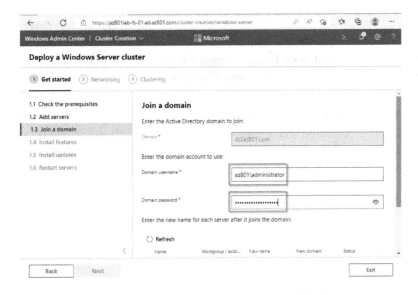

Figure 7.8 – Supplying a domain join account to be used for cluster creation

19. In setup **Step 1.4** for **Install features** for cluster creation, click **Install features** to automatically configure all the necessary features to the two Hyper-V hosts. After a few minutes, you will see **Status** showing **Installed** for both machines. Now, we can click **Next**, as shown in *Figure 7.9*:

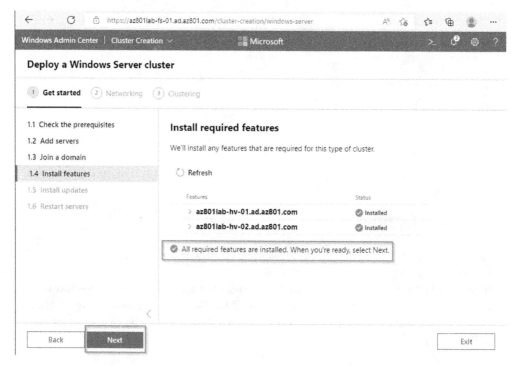

Figure 7.9 – Installing the required features for Windows Failover Cluster configuration

20. In setup **Step 1.5** for **Install updates** for cluster creation, both Hyper-V servers will not have internet access and will report errors regarding Server status. Simply click **Next** to continue to the next step.

21. In setup **Step 1.6** for **Restart servers** for cluster creation, both Hyper-V servers will complete any final feature installation and update installations and will reboot to complete any pending activities. Click the **Restart servers** button, as shown in *Figure 7.10*, and monitor the **Status** column for both servers showing a Ready status. Then, click the **Next: Networking** button:

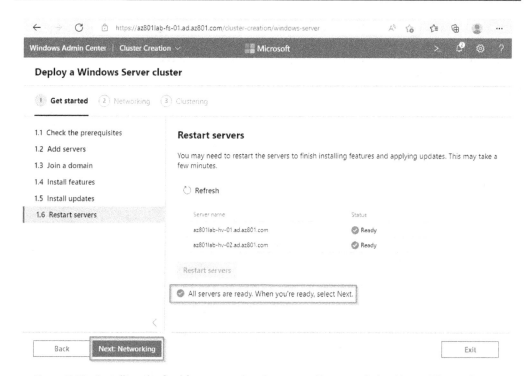

Figure 7.10 – Installing the final features and updates, as well as completing Hyper-V host reboots

22. In this exercise, we have completed all the initial steps for the **Get Started** section of the Windows Server cluster configuration. Please keep **Windows Admin Center** and the virtual machine console open to complete the next exercise.

In this section, we learned how to use Windows Admin Center to begin creating a new Windows Server cluster and reviewed the prerequisites and initial configurations needed on Hyper-V hosts to set the building blocks for our Failover Custer.

Next, we will continue our basic **Windows Server Failover Cluster** configuration with the appropriate network settings surrounding Failover Clustering for the AZ-801 exam objectives.

Configuring network settings for Failover Clustering

Unsurprisingly, WSFC heavily relies on the proper network design and configuration of the Failover Cluster. It is highly recommended to ensure that your network configuration avoids single points of failure by using redundant switches, routers, and teamed network adapters where possible and that you reduce both packet loss and latency wherever possible to ensure a high-quality service. For additional reading on just how crucial networking is with Failover Clustering, be sure to review this optional article on additional requirements to take into consideration for various environmental requirements: `https://techcommunity.microsoft.com/t5/failover-clustering/failover-clustering-networking-basics-and-fundamentals/ba-p/1706005`.

Let's continue with the setup and walkthrough steps to establish the networking resources needed for the Failover Clustering exercises in this chapter:

1. While on setup **Step 2.1** for **Check network adapters** for cluster creation, ensure that you see four verified network adapters listed, two for each of the Hyper-V virtual hosts, as shown in *Figure 7.11*. Click **Next** to continue:

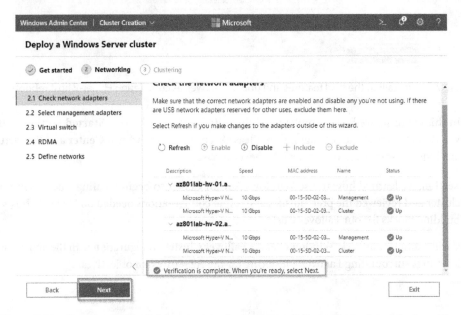

Figure 7.11 – Checking the available network adapters in our Failover Cluster lab virtual machines

2. In setup **Step 2.2** for **Select management adapters** for cluster creation, ensure that you select the network adapter **Name** that indicates Ethernet for both servers listed, as shown in *Figure 7.12*. Then, click **Apply and test** to begin the test. Once the successful changes have been applied, click **Next** to continue:

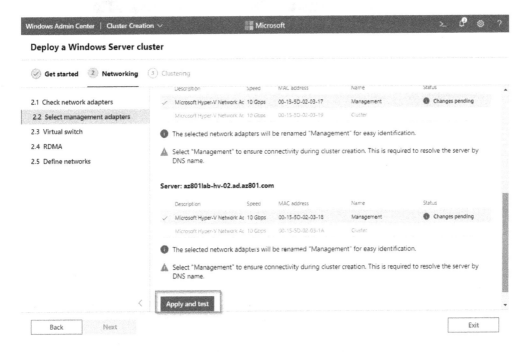

Figure 7.12 – Selecting management adapters for our Failover Cluster

3. In setup **Step 2.3** for **Virtual switch** for cluster creation, confirm that your configuration screen looks like what's shown in *Figure 7.13* (the **Name** field may be something other than **Cluster** and we will be able to change that in an upcoming step):

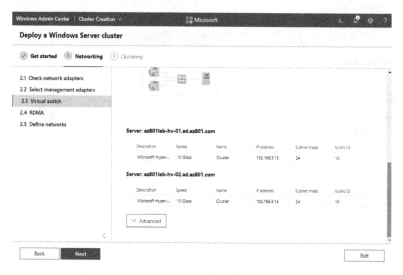

Figure 7.13 – Creating a virtual switch for cluster management

4. In setup **Step 2.4** for **Optionally configure RDMA** for cluster creation, we will not be choosing any settings on this screen, so click **Next**, as shown in *Figure 7.14*:

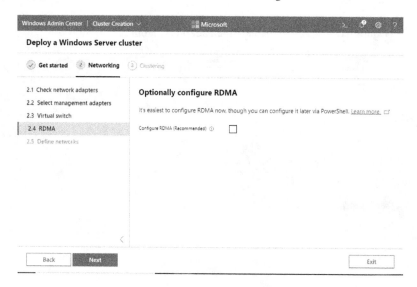

Figure 7.14 – Skipping the optional RDMA configuration

5. In setup **Step 2.5** for **Define networks** for cluster creation, update the following values for each of the Failover Cluster servers, as shown in *Figure 7.15*:

A. For **Server: az801lab-hv-01.ad.az801.com**, set **Name** to Cluster, **IP address** to 192.168.3.13, **Subnet mask** to 24, and **VLAN** to 0.

B. For **Server: az801lab-hv-02.ad.az801.com**, set **Name** to Cluster, **IP address** to 192.168.3.14, **Subnet mask** to 24, and **VLAN** to 0:

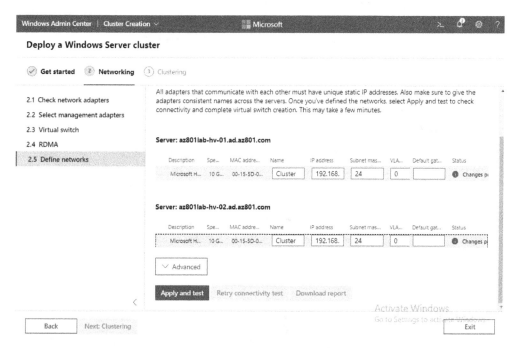

Figure 7.15 – Defining the cluster network configuration

6. Click the **Apply and test** button to complete the networking section of this Failover Cluster configuration and start cluster validation.

7. In setup **Step 3.1** for **Validate cluster** for cluster creation, click the lone **Validate** button shown in *Figure 7.16* to begin the validation process. Note that this will take a few minutes to complete. If any errors are reported, you will want to review the previous setup steps for consistency:

Figure 7.16 – Validating the cluster before it is created

8. In setup **Step 3.2** for **Create cluster** for cluster creation, set **Cluster name** to AZ801WSFC, while for **IP address**, set a **Cluster IP address** of 10.10.10.100. Then, click the **Create cluster** button to continue, as shown in *Figure 7.17*. Note that this cluster IP is also known as the **floating IP address**:

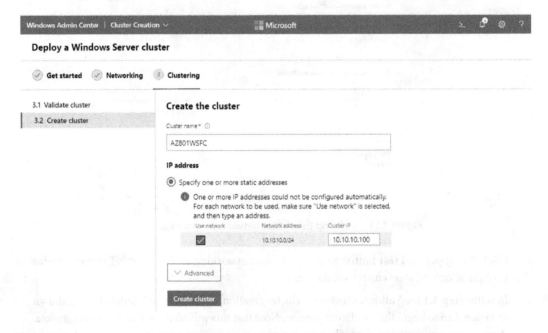

Figure 7.17 – Creating the Windows Server Failover Cluster

9. Once the cluster has been created, you will see the screen shown in *Figure 7.18*. Feel free to click the **Finish** button once you've reviewed it:

Figure 7.18 – Finalizing the cluster's creation

10. In this exercise, we have completed all the necessary steps for our Windows Server cluster configuration. Please keep **Windows Admin Center** and the virtual machine console open to complete the next exercise.

In this section, we learned how to use Windows Admin Center to create a Windows Server Failover Cluster while covering the basic network requirements for a successful lab deployment.

Next, we will continue learning about the Windows Server Failover Cluster storage options consistent with the AZ-801 exam objectives.

Configuring Windows Server Failover Cluster storage options

Any clustering technology must incorporate support for a distributed filesystem that allows for efficient operations and access to shared storage across all cluster nodes while reducing data corruption and improving failover speed and workload distribution.

The options that can be used for Failover Cluster shared storage are as follows:

- **Storage Spaces Direct (S2D)** or internal/direct attached storage enables a highly available and incredibly scalable replication of storage across nodes by providing an easy approach to pooling and distributing locally attached storage across multiple cluster nodes.

- iSCSI, shared **Serial-Attached SCSI (SAS)**, **Storage Area Network (SAN)**, or **Fibre Channel (FCoE)** storage.

- CSV with **NT File System (NTFS)** or **Resilient File System (ReFS)** provides a multi-host read/write access filesystem to a shared disk. Hosted applications can then read/write to the same shared data location from any node within the Failover Cluster. This shared block volume can then be provided to the Failover Cluster by other varying storage technologies, such as traditional SANs, iSCSI, or S2D.

- **SMB 3.0** file shares as shared storage or SOFS with the ability to map an SMB remote share to a mapped drive letter, allowing for remote mount point traversal for container workloads.

As you can see, there is a multitude of opportunities and configurations available to achieve the requirements of any Failover Cluster design, regardless of where the workloads are running. Let's walk through how to create a small S2D storage pool for our Failover Cluster to gain some familiarity with a modern storage configuration:

1. Using **Hyper-V Manager**, connect to either `AZ801PacktLab-HV-01` or `AZ801PacktLab-HV-01` as the local administrator.

2. Using the **Start** menu, launch a new **Windows PowerShell as administrator** session.

3. Enter the `Enable-ClusterS2D -CacheState Disabled` command and press *Enter*.

   The PowerShell results shown in *Figure 7.19* will indicate that process of creating the Storage Spaces Direct cluster has begun or that the S2D pool has already been successfully created:

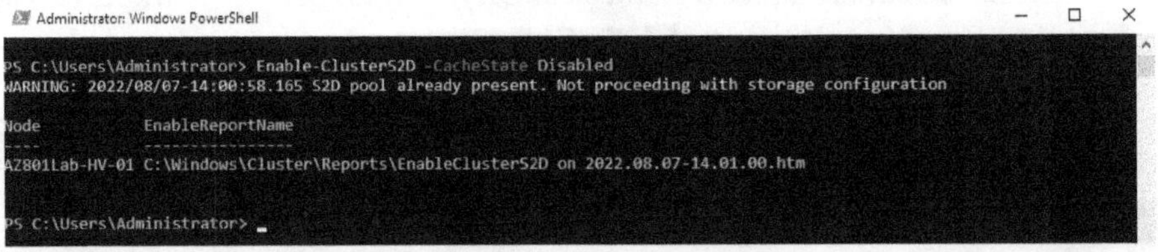

Figure 7.19 – Creating an S2D storage pool

4. Wait while the disks are scanned across all hosts and the cluster is created, as shown in *Figure 7.20*:

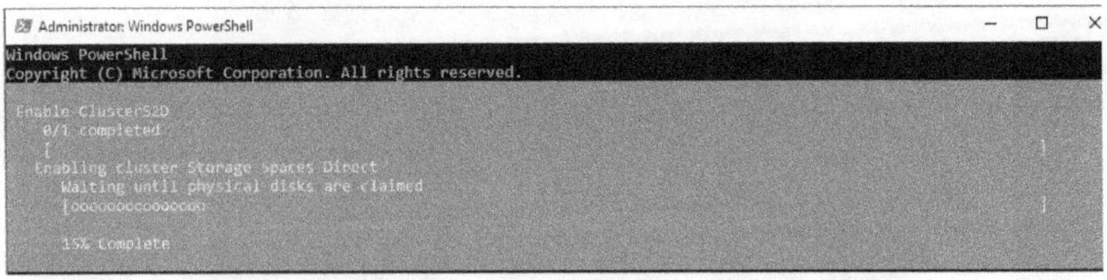

Figure 7.20 – Validation that the S2D cluster is being created

5. Once the S2D cluster has been created, return to **Windows Admin Center** on the **AZ801PacktLab-FS-01** virtual machine and refresh the **Cluster Manager Dashboard** page. This time, you will notice that a new **Storage Spaces Direct** cluster has been identified and additional setup time will be needed for the cluster, as shown in *Figure 7.21*:

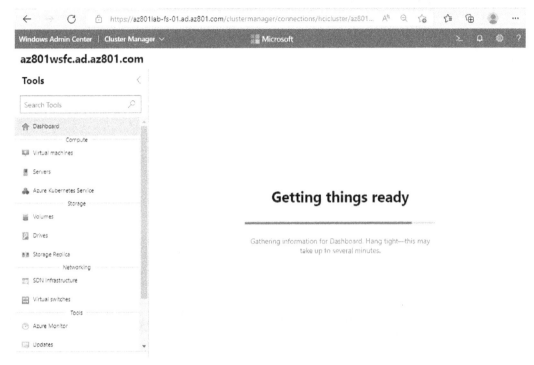

Figure 7.21 – Windows Admin Center Cluster Manager dashboard detecting a new S2D configuration

6. Feel free to navigate the updated **Cluster Manager** dashboard, noting that you will now see additional cluster health statistics surrounding disk and storage metrics and overall performance. This completes all the necessary steps for our Windows Server S2D cluster configuration.

Earlier in this chapter, when you finished creating the Failover Cluster, there was a recommendation to ensure you have a cluster witness, which helps to ensure that you have more than half of the nodes available. This is called a quorum. Let's review some high-level steps to use **Azure Storage** as our **Cloud Witness**:

1. Sign in to `https://portal.azure.com` and use either an existing Storage account or create a new one to be used as a **Cloud Witness**.

2. Locate your **Storage account** and under **Security + networking**, select **Access keys**. Copy the **Storage account name** value and the value of either **key1** or **key2**, as shown in *Figure 7.22*:

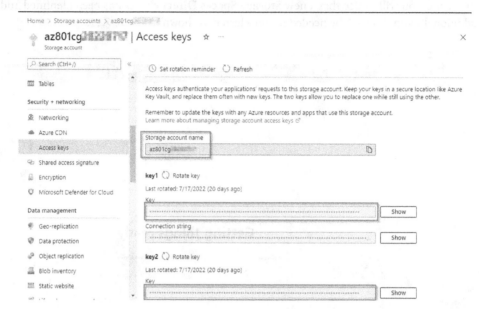

Figure 7.22 – Gathering the storage account name and access keys for Cloud Witness configuration

3. Under the **Settings** section, select **Endpoints** and then scroll to locate **Primary endpoint**. Copy this **Blob service** value for later use, as shown in *Figure 7.23*:

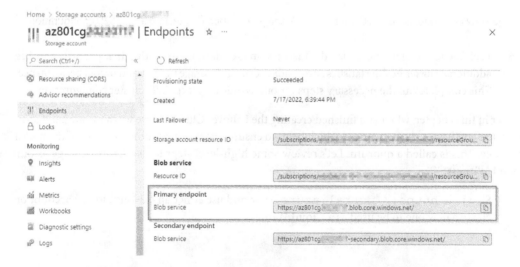

Figure 7.23 – Gathering the primary blob service endpoint for our storage account

4. Go back to **Windows Admin Center**, return to our **Cluster Manager**, and select **Settings**. Then, under **Cluster**, select **Witness**. Change **Witness type** to **Cloud witness**, then paste in the **Azure storage account name** value, the **Azure storage account key** value, and the **Azure service endpoint** value, as shown in *Figure 7.24*. Click **Save** to finish configuring the Cloud Witness:

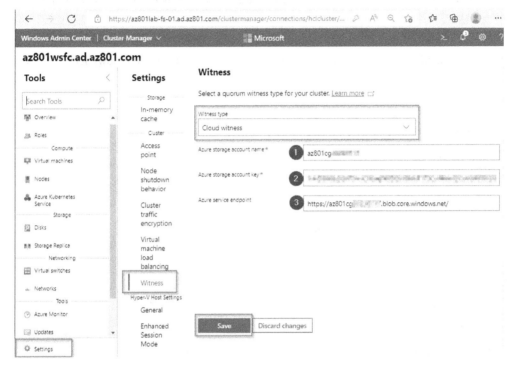

Figure 7.24 – Completing the Cloud Witness configuration in Windows Admin Center

5. In this exercise, we have completed all the necessary steps for our **Cloud Witness** configuration. Please keep **Windows Admin Center** and the virtual machine console open to complete the next exercise.

Note on Cluster Witness and internet connectivity

It is recommended to set up a Cloud Witness, but an SMB file share can also be used. However, when using a Cloud Witness, all server nodes in the cluster must have a reliable and always-on internet connection to provide a proper quorum.

In this section, we learned how to use Windows Admin Center to create a Cloud Witness for our WSFC.

Next, we will continue learning about Windows Server Failover Cluster workload options, consistent with the AZ-801 exam objectives.

Configuring cluster workload options

Once the Failover Cluster has been configured according to the architectural requirements for a business or customer, the intended workload(s) as part of the cluster design must then be configured and loaded onto the cluster for proper workload handling. Each Failover Cluster is different based on the workload's characteristics, which means that the possibilities are endless for the design and architecture of a Failover Cluster.

A list of the clustered roles that you can configure with high availability on a Windows Server Failover Cluster are as follows:

- Generic Application
- Generic Script
- Generic Service
- Namespace Server
- DFS Namespace Server
- **Distributed Transaction Coordinator** (DTC)
- File Server
- Hyper-V Replica Broker
- iSCSI Target Server
- iSNS Server
- Message Queuing
- Other Server
- Virtual Machine
- WINS Server

To make these workloads more relatable to real-world use, let's take what we have learned in this chapter and apply it to one of the more popular ways to utilize Hyper-V Failover Clustering: hosting Hyper-V machines. For most organizations, it's imperative that you not only design and architect an environment for resiliency but also address highly available resources in the event of a failure, general management or configuration changes, and monthly patching of environments.

Consider for a moment that one running VM encounters a failure such as a blue screen during startup – the remaining virtualization and host layers of the design will receive a signal that health has been degraded and the current running state of the VM can then be seamlessly started on another node in the cluster, thus reducing the amount of overall downtime for an application or service. The environment that was built in this lab, to meet the hosted Hyper-V virtual machine requirements, could extend this host clustering model to put each service or hosted application into a running VM that is backed by shared storage and replicated data and is highly available across the host cluster nodes.

As we will learn in *Chapter 8, Managing Failover Clustering*, these services can then be easily and quickly migrated from one running host to another (even via automation) to provide the necessary availability requirements while allowing for administrative and patch management efforts. In *Chapter 9, Implementing and Managing Storage Spaces Direct*, we will learn how to provide resilient storage for hosting virtual machines. As the last missing piece, we will learn how to effectively manage the virtual machines for replication and failover in *Chapter 12, Protecting Virtual Machines by Using Hyper-V Replicas*.

Now that we have reviewed the available cluster workloads and cluster design considerations, let's review the topic of **Stretch Clusters** for disaster recovery purposes.

Configuring Stretch Clusters

While Failover Clusters generally provide localized high availability for workloads in the same physical location, Stretch Clusters provide similar functionality across multiple physical locations. Stretch clusters are generally more complex to design, implement, and manage and require unidirectional and synchronous replication (called Storage Replicas) between disaster recovery sites. However, once the cluster solution has been properly configured, this solution provides a way to keep the replication of volumes and data with all cluster servers in sync to ensure minimal (active-passive configuration) or zero (active-active configuration) data loss during a system failure.

For optional reading on Stretch Cluster replication with shared storage, check out this incredible article: `https://docs.microsoft.com/windows-server/storage/storage-replica/stretch-cluster-replication-using-shared-storage`.

Now that we have reviewed what stretch clusters provide for disaster recovery planning and architectural design for critical business workloads, let's review the topic of **SOFS**.

Configuring Scale-Out File Server

The power behind SOFS is inherently to provide active-active clustered storage as continuously available file shares for server application storage. This, in turn, allows you to share the same folder at the same time on all the nodes in the cluster, and the active-active feature ensures that the file share is always available, even when a node is down because of maintenance or in the event of a failure. In addition, the bandwidth for the share can always be increased simply by adding another node to the cluster, thus reducing bandwidth constraints.

General use of SOFS includes, but is not limited to, the following:

- **Virtual Machine Manager** (**VMM**) library shares of files and/or virtual machine templates
- SQL Server live database files (including database, log, and backup files)

- Hyper-V live virtual disk and configuration information
- Clustered shared volume caching
- **Internet Information Services (IIS)** configuration and general data

A great overview of when to use SOFS, along with additional overview information, can be found at `https://docs.microsoft.com/windows-server/failover-clustering/sofs-overview`.

Now that we have reviewed what SOFS provides for disaster recovery planning and Failover Clustering, let's review the topic of **Cluster Sets**.

Configuring Cluster Sets

For some business-critical needs where a higher level of availability and scalability is needed for clustered workloads, you can configure Cluster Sets. The idea surrounding Cluster Sets is that you implement a cluster of clustered resources or workloads, ultimately combining hyper-converged clusters into a consistent resource fabric across the environment. Many of the benefits of Cluster Sets include, but are not limited to, the following:

- Overall increased resiliency in the event of multiple node failures
- Live migration of virtual machines between clusters
- Ability to seamlessly change the hardware and compute life cycle when adding or retiring clusters over time
- No requirement for identical compute and storage between available cluster nodes
- A unified storage Namespace is provided across all individual clusters, thus providing enhanced flexibility to the VM architecture and its design

A cluster set consists of the following high-level resources:

- A **management cluster**, which is considered the Failover Cluster that hosts and is considered the management plane of the entire cluster set.
- A **cluster set Namespace referral SOFS**, which provides an instance of SOFS for the entire management cluster.
- A **cluster set master**, which is a resource responsible for orchestrating communications between member clusters.
- **Member clusters**, which are the clusters hosting VMs, S2D workloads, SOFS shares, and additional workloads.

- A **cluster set worker**, which is an individual resource hosted on each of the member clusters. It is responsible for interacting with and monitoring local cluster resources or resource placements within the clusters.

- A **fault domain**, which is essentially a group of hardware that could fail together. Each node of the cluster can participate in a fault domain within an availability set.

- **Availability sets**, which allow you to configure the redundancy level of the clustered workloads across fault domains, providing a logical workload grouping.

With that, we have reviewed what Cluster Sets provide for disaster recovery planning and the architectural design of critical business workloads. Next, we will review what we have learned in this chapter and set the stage for the next chapter of this book.

Summary

In this chapter, we learned how to design and configure a **Windows Server Failover Cluster**. We also learned how to successfully establish the building blocks for a Windows Server Failover Cluster by learning how to configure various storage options and successfully design and configure appropriate network settings for the failover cluster. We also gained familiarity with the designs, techniques, and tools to create and manage Windows Server Failover Cluster components and workloads on-premises, with hybrid components and services, and in the cloud consistent with the AZ-801 exam objectives.

In the next chapter, we will learn how to manage a Windows Server Failover Cluster. We will cover topics such as cluster-aware updating, how to recover failed cluster nodes, and how to upgrade existing cluster nodes to a newer Windows Server version (Windows Server 2022, in this example set). Finally, we will learn about additional features for managing Failover Clusters while utilizing Windows Admin Center.

8

Managing Failover Clustering

In this chapter, we will continue with the failover clustering theme and build out our experience to include how to configure components, such as cluster-aware updating, how to recover failed cluster nodes, and how to upgrade existing cluster nodes to a newer Windows Server version (Windows Server 2022, in this example set). Finally, we will learn how to manage failover clusters by utilizing Windows Admin Center.

This chapter will include exercises on how to manage **Windows Server Failover Cluster** (WSFC) components and workloads on-premises, with hybrid components and services, and in the cloud consistent with the AZ-801 exam objectives.

In this chapter, we will cover the following topics:

- Technical requirements and lab setup
- Managing failover clusters using Windows Admin Center
- Implementing cluster-aware updating
- Installing Windows updates on cluster nodes
- Recovering a failed cluster node
- Upgrading a node to Windows Server 2022
- Failing over workloads between nodes

Technical requirements and lab setup

To successfully follow along and complete the tasks and exercises throughout this chapter and the following chapters in this book, we will need to ensure that the technical requirements from *Chapter 1, Exam Overview and the Current State of Cloud Workflows*, and *Chapter 7, Implementing a Windows Server Failover Cluster*, have been completed in full. We will be primarily utilizing our Hyper-V virtual machine environment to complete the exercises and tasks throughout this chapter so that we align with the AZ-801 exam objectives.

Managing failover clusters using Windows Admin Center

Sure, we now know how to create a WSFC using **Windows Admin Center**. However, we can also use Windows Admin Center to manage existing failover clusters by simply adding them as known cluster resources (this also includes hyper-converged clusters such as **Azure Stack HCI**).

Now that the cluster has been added, the cluster management tools from Windows Admin Center allow us to manage the following (as part of an ever-growing list):

- **Dashboard**, which displays the current cluster status and any alerts at a high level
- **Compute**, which covers any hosts, virtual machines, or containers in the cluster while providing the ability to pause nodes, simulate a failure, validate the cluster, remove the cluster, and create and review validation reports for the following cluster workloads:

  - **Virtual machines**
  - **Servers**
  - **Azure Kubernetes Service**

- **Storage**, which provides the management features of any attached or local storage, the presented volumes, and a **storage replica** which provides replication of data and volumes between clusters and servers for recovery purposes:

  - **Volumes**
  - **Drives**
  - **Storage Replica**

- **Networking**, which provides some management features surrounding **software-defined networking (SDN) infrastructure**, and cluster **virtual switches**
- **Tools** for remote monitoring and cluster notifications for anomalies and/or failures, cluster-aware updating operations, cluster insights, and performance monitoring, and security configuration for the cluster based on core cluster security and encryption of data between cluster nodes and storage:

  - **Azure Monitor**
  - **Updates**
  - **Diagnostics**
  - **Performance Monitor**
  - **Security**

Now that we have introduced the tools available to us for failover cluster management when using Windows Admin Center, let's dig into the utilization of these tools to implement **cluster-aware updating**, as well as general management cluster roles.

Implementing cluster-aware updating

In *Chapter 7, Implementing a Windows Server Failover Cluster*, we learned about failover clustering regarding the network, storage, and overall requirements. Now, we wish to focus on how to keep failover clustering up to date and operating at peak performance.

Let's continue learning more about **Windows Server Failover Clustering** features by introducing cluster-aware updating.

Introduction to cluster-aware updating for Windows Server failover clusters

Cluster-aware updating (**CAU**) is a feature introduced with Windows Server 2012 that allows administrators to manage Windows Updates for an entire cluster of hosts and services. CAU attempts to reduce or eliminate the amount of service and guest downtime while providing an automated approach for applying orchestrated updates and maintenance across an entire failover cluster. During what's called an updating run, CAU puts each node into a temporary maintenance mode and fails over any clustered roles onto other available nodes. Updates are then installed on the maintenance node(s), any necessary server restarts are completed, and the node is automatically placed back into active rotation when healthy and CAU moves onto the next node in the cluster.

There are two modes that CAU can work in, and those are as follows:

1. **Self-updating mode**, where CAU can run directly on a cluster node that is meant to be updated, based on a predetermined update schedule, giving full automation features to administrators and businesses.

2. **Remote updating mode**, where CAU runs on a server that is not part of the cluster (is not a cluster node), whereby CAU remotely connects to any of the failover cluster nodes and completes the necessary cluster updates on demand.

While cluster-aware updating can be configured and managed via **Server Manager** or **PowerShell** on each Hyper-V node or via the failover cluster node, the preferred method is to utilize **Windows Admin Center** to complete the configuration and orchestration from a centralized and browser-accessible administration tool. However, note that not everything can be completed from Windows Admin Center – as each organization has its own set of security requirements, there may be certain configurations, such as proxy setup or a **virtual computer object** (**VCO**) configuration such as CAUaz801mgx, as shown in *Figure 8.1*, that must be taken into consideration before completing the CAU setup:

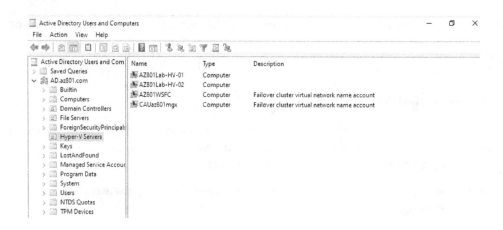

Figure 8.1 – Identifying a VCO created by cluster-aware updating in AD

For additional information on pre-staging a cluster name account in AD, go to `https://docs.microsoft.com/previous-versions/windows/it-pro/windows-server-2008-R2-and-2008/cc731002(v=ws.10)#steps-for-prestaging-the-cluster-name-account`.

Now, let's review the requirements for CAU:

- Use a repeatable process or script to install the failover clustering feature and clustering tools for all the necessary cluster operations

- Ensure that you have a proper administrator account that allows local administrative permission to the cluster nodes for updates and management, including remote-updating mode for CAU on the update coordinator and any cluster validation or best practice analyzer tools

- All cluster nodes must be in the same AD domain and the cluster name must have proper DNS resolution

- Enough cluster nodes must be online so that the cluster has a proper quorum

- All cluster nodes must have the cluster service running and set to automatically start

- Any pre-update or post-update scripts for CAU must be installed and available to all cluster nodes

- All cluster nodes must have remote management enabled and configured, including PowerShell Remoting, Windows Management Instrumentation, .NET Framework 4.5, and proper firewall rules to allow automatic restarts

An exhaustive list of all requirements and best practices for CAU can be found at `https://docs.microsoft.com/windows-server/failover-clustering/cluster-aware-updating-requirements`.

Analyzing cluster readiness before CAU configuration

Three tools are available to complete a readiness check to ensure that you are correctly set up before configuring and installing the CAU role. These tools are a PowerShell tool called Test-CAUSetup, the **Analyze Cluster updating readiness** link within the CAU tool on the host, and the readiness tool within Windows Admin Center. More details on the full set of tests to run, as well as common resolutions, can be found at `https://docs.microsoft.com/previous-versions/windows/it-pro/windows-server-2012-R2-and-2012/jj134234(v=ws.11)?redirectedfrom=MSDN#test-cluster-updating-readiness`.

An example of the cluster updating readiness tool from within the CAU tool directly on a host node is shown in *Figure 8.2*:

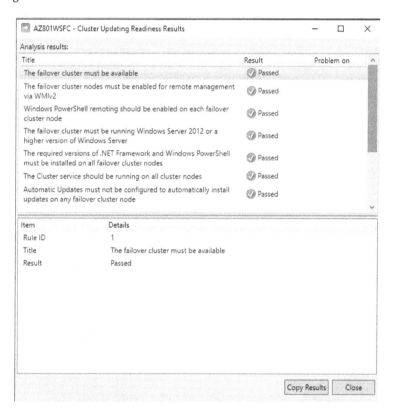

Figure 8.2 – Cluster Updating Readiness Results from the CAU tool

Now that we have provided an introduction and covered the requirements needed to successfully implement CAU, let's walk through the CAU implementation using **Windows Admin Center**:

1. Ensure that all four virtual machines in your Hyper-V lab for AZ801 are currently in the **Running** state.

2. Right-click on the **AZ801PacktLab-HV-01** virtual machine. Then, select **Settings**, and under **Add Hardware**, select **Network Adapter**. Finally, click the **Add** button.

3. Under the **Virtual Switch:** dropdown, select **AZ801PacktLabExternal** and then click the **OK** button to commit the changes to the virtual machine.

4. Repeat *steps 2* and *3* while using **AZ801PacktLab-HV-02** for the virtual machine.

5. Begin a virtual machine connection to AZ801PacktLab-FS-01 and log in as administrator.

6. Open **Microsoft Edge** and navigate to our Windows Admin Center instance at https:// az801lab-fs-01.ad.az801.com and if prompted, log in as administrator.

7. Navigate to and select our **server cluster** named az801wsfc.ad.az801.com and if prompted, log in as administrator.

8. Once **Cluster Manager** has successfully loaded, scroll down to the **Tools** section, select **Updates** to begin, as shown in *Figure 8.3*, and select **Add Cluster-Aware-Updating role**:

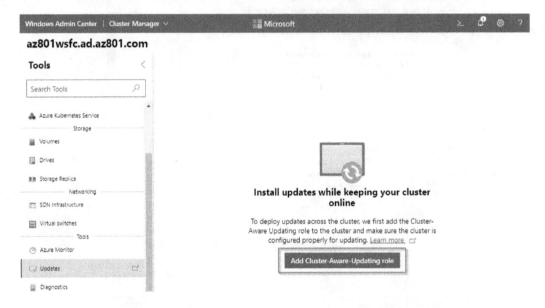

Figure 8.3 – Adding the CAU role to our failover cluster

9. These steps may take a few minutes to complete; you can view notifications in Windows Admin Center for **Configuring Cluster-Aware Updating**. Once the validation steps for CAU begin, you will see a **Checking a few things** notification, as shown in *Figure 8.4*:

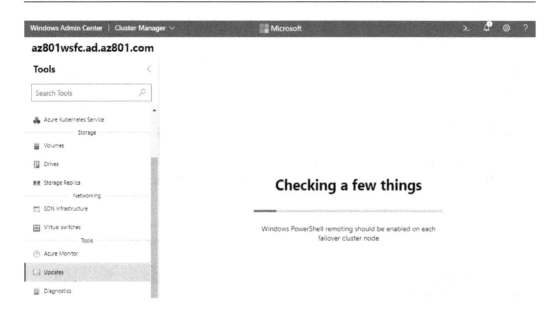

Figure 8.4 – Beginning the CAU validation and implementation steps

10. After a few more minutes, you will see the screen change to **Checking for updates**, as shown in *Figure 8.5*:

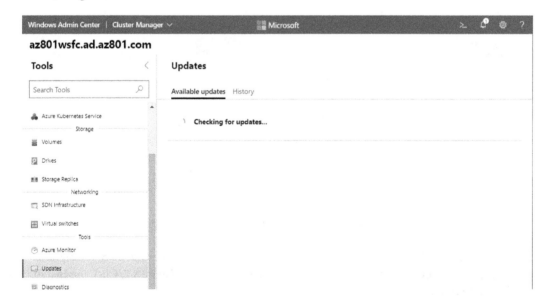

Figure 8.5 – Checking for updates for CAU

11. This part of the process may take some time as it must check in and evaluate which updates are necessary for the cluster nodes involved, so this will be a natural break in our walkthrough.

Now that we have completed the initial CAU implementation, let's move on to installing Windows updates on cluster nodes, where we will install the Windows updates gathered by the CAU process.

Installing Windows updates on cluster nodes

Now that we have given the CAU process enough time to check in, evaluate each node of our cluster, and determine available updates for our cluster, let's continue learning how to effectively install Windows updates on our cluster nodes:

1. Given good connectivity to the internet from your Hyper-V hosts, you will soon see an indicator that **Updates are available**, as shown in *Figure 8.6*. Click the **Install** button to continue:

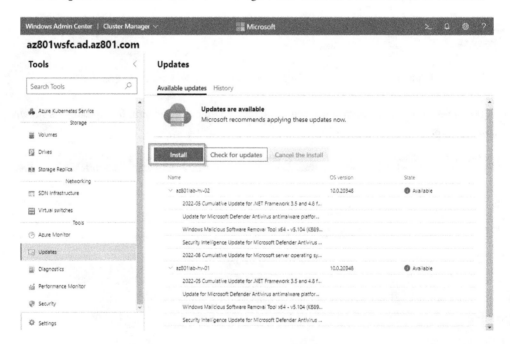

Figure 8.6 – Reviewing available updates via the CAU tool

2. You will be asked to review the updates once again, noting that cluster roles will be migrated between hosts and hosts will be selectively updated, as shown in *Figure 8.7*. Click the **Install** button to continue:

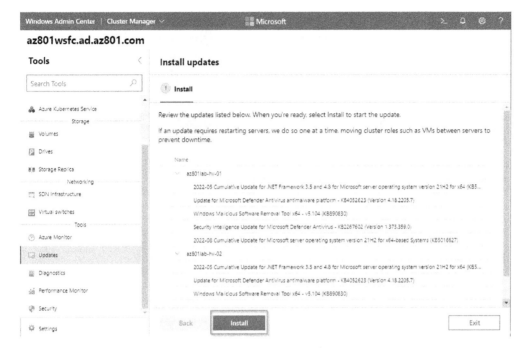

Figure 8.7 – Installing the available updates within CAU

3. At this point, the update run will commence, and you will soon see a screen indicating that the update evaluation for the hosts has begun, as shown in *Figure 8.8*:

Figure 8.8 – Starting the CAU update run

4.   After a few minutes, the screen will refresh again, indicating that updates are being installed, moving from **Starting** to **Waiting**, and then to **Staging**, as shown in *Figure 8.9*. Note that only one host can be selected at a time, meaning that the cluster node will be paused in the background, current connections will be drained, and the clustered roles will be evicted from this host before being placed into maintenance mode:

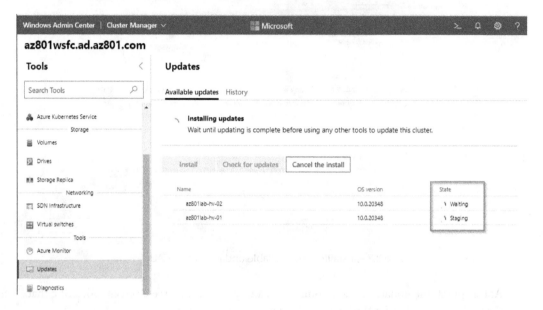

Figure 8.9 – Installing updates via CAU and monitoring the staging and installation progress

5.   After a few minutes, you will notice the node has completed an automated restart, is indicating that updates have been applied, as shown in *Figure 8.10*, and has notified the cluster that it is once again available and healthy for receipt of clustered roles. The second cluster node, az801lab-hv-02, now gets paused as a cluster node, its current connections are drained, and the clustered roles are evicted from this host and moved to az801lab-hv-01 so that CAU can complete update orchestration on the second cluster node:

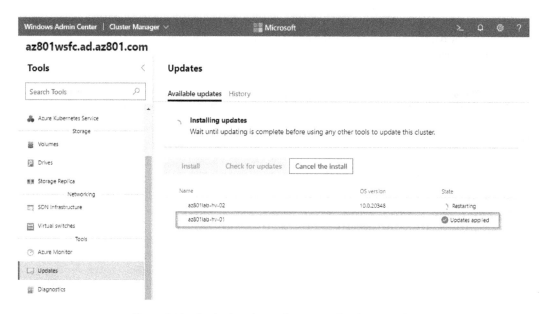

Figure 8.10 – Reviewing the update status for cluster nodes

6. Once both cluster nodes have completed all the necessary steps for update orchestration, you will be presented with a screen refresh, indicating that the cluster is up to date with quality updates, as shown in *Figure 8.11*. Note that many Azure Stack **Hyper-Converged Infrastructure (HCI)** offerings can also check for available hardware updates, including but not limited to drivers and BIOS updates:

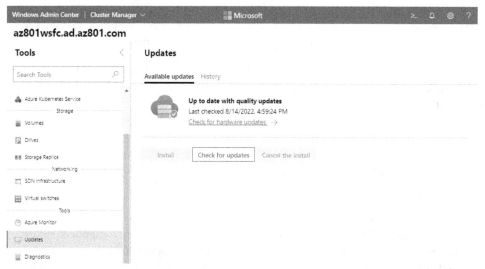

Figure 8.11 – Confirmation of successful CAU update run completion

7.   Finally, selecting the **History** tab will allow you to review your most recent CAU attempts grouped by date stamp and review the updates applied, as well as any successes/failures during the update run, as shown in *Figure 8.12*:

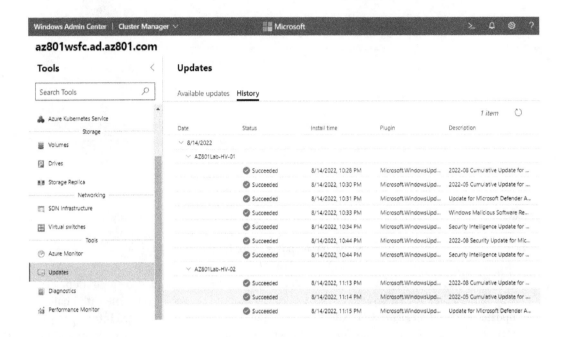

Figure 8.12 – Reviewing the successfully applied updates via the CAU method

8.   With that, we have completed all the necessary configuration steps. Please keep Windows Admin Center and your virtual machine console open to complete the next exercise.

In this section, we learned how to implement CAU using Windows Admin Center. We also learned how CAU works by completing the download, install, and cluster roles and reboot orchestration within a WSFC.

Next, we will explore how to recover a failed cluster node for a WSFC.

Recovering a failed cluster node

Try as we might to architect and design a system that is resilient, always-on, and always available, we must also plan for recovery in the context of inevitable system failure. Typically, there are two prominent types of cluster failures: an irreparable failure such as a hardware component failure, or a reparable failure that could be a temporary system failure such as a system fault, an operating system error, or another hardware failure. However, every environment is different, and some are vastly more complex than others, so your mileage and requirements may vary.

For a general approach to recovering from failures that apply to most workloads, the following high-level steps can be followed to complete recovery in most cases:

1. Identify the failed node and validate that the cluster roles have been moved to another available node.
2. Locate the failed node, then pause and evict the node from the failover cluster configuration from a currently available/functioning cluster node.
3. Confirm that the failed cluster node has been evicted from the failover cluster configuration.
4. Replace any of the failed hardware components.
5. If needed, perform a disk or volume restore on the failed node to rebuild the node.
6. Using a failover cluster management tool (either Windows Admin Center or the Failover Cluster Manager management console), add the previously failed node back into the existing cluster.
7. Confirm that the same cluster administrator accounts are present on all cluster nodes.
8. Complete any necessary setup steps to add the node back to the proper failover cluster instance and rejoin the cluster.

There is also the potential of a much wider systems or communications failure, or a misconfiguration present within the failover cluster. The following scenarios come to mind, but are not limited to them:

- Cluster quorum corruption
- Cluster quorum disk failure
- Cluster disk data loss
- Cluster disk corruption or failure
- Cluster quorum rollback
- Majority node set cluster failure
- Complete cluster failure

This type of disaster recovery approach is achieved by completing a forced quorum process that is entirely manual and is well documented at `https://docs.microsoft.com/sql/sql-server/failover-clusters/windows/wsfc-disaster-recovery-through-forced-quorum-sql-server?view=sql-server-ver16` for optional reading.

Now that we have reviewed some steps for recovering a failed cluster node, let's learn how to upgrade a cluster node to Windows Server 2022.

Upgrading a node to Windows Server 2022

While the cluster that we have configured in our lab is already utilizing Windows Server 2022 as the operating system, cluster node OS upgrades can be completed using a **Cluster OS Rolling Upgrade** approach to upgrade the cluster node operating system without stopping the cluster workloads.

The Cluster OS Rolling Upgrade approach allows for great flexibility, allowing you to upgrade the cluster node by node from Windows Server 2019 to Windows Server 2022 for all nodes in the cluster without any downtime. A new cluster does not need to be created during this process, and the upgrade process itself is fully reversible at any point except the final stage step (when the `Update-ClusterFunctionalLevel` cmdlet has been issued). In addition, if you must stay in a mixed OS mode for a brief time, patching and general cluster operations can be supported for both OS versions.

Before you begin planning your Cluster OS Rolling Upgrade process, you will want to run the following PowerShell commands to determine the current state of the cluster:

```
Get-Cluster | Select ClusterFunctionalLevel
Get-ClusterNode
Get-ClusterNodeSupportedVersion
```

Then, you must ensure you meet the requirements before proceeding with the Cluster OS Rolling Upgrade process:

- You must start with a cluster that is running Windows Server 2012 R2 or newer. Note that this upgrade approach is an N-1 for support levels, meaning that you can only upgrade to one OS version newer than where you begin and cannot skip versions (for instance, you cannot directly upgrade from Windows Server 2012 R2 to Windows Server 2022).
- You must verify that the Hyper-V nodes utilize CPUs that support **second-level addressing tables (SLATs)**.

The Cluster OS Rolling Upgrade process can be completed by following these high-level upgrade steps:

1. The cluster database must be backed up.
2. The cluster workload data and configuration must be backed up.
3. You must confirm that CAU is not currently applying updates to the cluster (it is recommended that you suspend any scheduled updates).
4. For each cluster node intending to be upgraded:

    A. The cluster node must be paused, and all roles drained from the node.

    B. The cluster node must then be evicted from the active cluster.

    C. The cluster node will then undergo a clean OS install and Failover Clustering features reinstalled to the cluster node.

D. The cluster node will have its storage and networking reconfigured.

E. The cluster node can then be rejoined to the active cluster.

F. The cluster role(s) will then be reinstalled on the cluster node and workload failover can resume for this cluster node.

5. Once all the cluster nodes have been successfully updated, the overall Cluster Functional Level can be upgraded.

The following is a list of gotchas and limitations to consider when planning for a Cluster OS Rolling Upgrade for your failover cluster:

- You must ensure that you have sufficient workload capacity on the cluster during this process as one node will be removed from the cluster at a time

- When using Storage Spaces Direct with local storage, there will be bandwidth degradation during this upgrade process until the nodes have been added back to the cluster

- You must ensure that no backups are running against the cluster while completing these OS upgrades for the cluster nodes

- You must use a node that is running the newer version of Windows Server to add new nodes to the cluster or utilize Windows Admin Center to complete this task

- You must avoid storage changes to the cluster while running in mixed OS mode

- In-place upgrades for the cluster node OS are strongly discouraged

For a full detailed article explaining the workflow, additional benefits, and insightful limitations, be sure to check out the following URL: `https://docs.microsoft.com/windows-server/failover-clustering/Cluster-Operating-System-Rolling-Upgrade`.

Now that we have reviewed how to upgrade a cluster node to Windows Server 2022, let's learn how to fail over workloads between cluster nodes.

Failing over workloads between nodes

Throughout this chapter, we learned about the types of workloads that a failover cluster can host, as well as how best to manage the clusters in general. Let's continue our learning objectives by discussing how to fail over workloads between cluster nodes using Windows Admin Center:

1. From the **Cluster Manager** area within **Windows Admin Center**, select Servers under the **Compute** section and review the options for the clustered nodes, as shown in *Figure 8.13*, including the additional options under the ellipses for cluster node management:

Figure 8.13 – Server node management within Cluster Manager

2. Selecting just one of the cluster nodes not only allows us to review the specific details and configuration of the server node but also allows us to visualize what happens when we select **Pause** to place the cluster node into maintenance, noting that workloads are automatically moved to other nodes in the cluster, as shown in *Figure 8.14*:

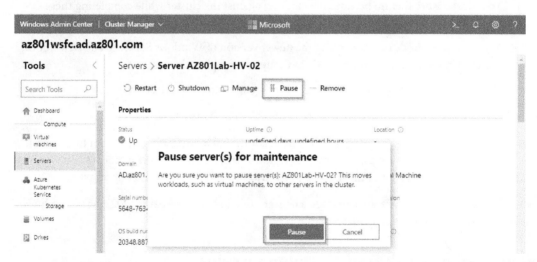

Figure 8.14 – Selecting Pause to place the cluster node

3. You will notice in *Figure 8.15* that this selected cluster node has been placed into maintenance, and all connections are draining while workloads are queued to be moved onto another available cluster node within the failover cluster:

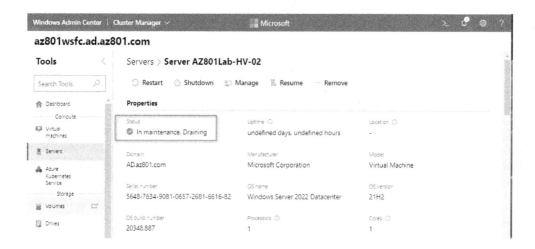

Figure 8.15 – Cluster node placed into maintenance mode and migrating workloads

4. In *Figure 8.16*, you will see that the cluster node is completely in maintenance mode and all connections and workloads have been failed over to another cluster node. To take the cluster node back out of maintenance and rebalance the cluster workloads, select the **Resume** button:

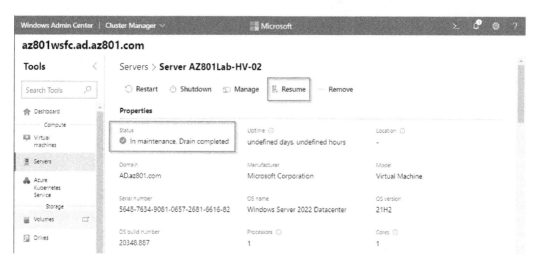

Figure 8.16 – Taking the cluster node back out of maintenance using the Resume button

5. With that, we have completed all the necessary configuration steps. Please close **Windows Admin Center** and feel free to shut down all running virtual machines within **Hyper-V Manager**.

With that, we have failed over workloads to another available node within the failover cluster.

Summary

In this chapter, we learned how to configure components such as CAU, how to recover failed cluster nodes, and how to upgrade existing cluster nodes to a newer Windows Server version. We also learned how to manage failover clusters by utilizing Windows Admin Center, consistent with the AZ-801 exam objectives.

In the next chapter, we will learn how to configure Storage Spaces Direct and then use Storage Spaces Direct within a failover cluster. We will also discuss how to upgrade a Storage Spaces Direct node, as well as implement proper security configuration for Storage Spaces Direct, discovering topics such as converged and hyper-converged deployments.

9
Implementing and Managing Storage Spaces Direct

In this chapter, we will learn how to configure **Storage Spaces Direct** (**S2D**) and then manage S2D within a failover cluster. We will also discuss how to upgrade an S2D node, as well as implement proper security and networking configurations for S2D while discovering topics such as converged and hyper-converged deployments.

This chapter will include exercises on how to implement and manage S2D components and workloads on-premises, with hybrid components and services, and in the cloud consistent with the AZ-801 exam objectives.

In this chapter, we will cover the following topics:

- Technical requirements and lab setup
- Introduction to **Storage Spaces Direct** (**S2D**)
- Implementing networking for S2D
- Deploying S2D on Windows Server
- Upgrading an S2D cluster node

Technical requirements and lab setup

To successfully follow along and complete the tasks and exercises throughout this chapter and the following chapters in this book, we will need to ensure that the technical requirements from *Chapter 1, Exam Overview and the Current State of Cloud Workflows*, and *Chapter 7, Implementing a Windows Server Failover Cluster*, have been completed in full. We will be primarily reviewing our Hyper-V virtual machine environment to complete the exercises and tasks throughout this chapter to align with the AZ-801 exam objectives.

Let's begin by introducing S2D, how it can be utilized, and what features it brings to your resilient storage design and architecture.

Introduction to Storage Spaces Direct (S2D)

In *Chapter 7, Implementing a Windows Server Failover Cluster*, we learned how to quickly create an S2D cluster for our AZ801 failover cluster lab and discussed the benefits of this S2D design for any failover cluster. We're now going to go a bit deeper into the subject of S2D to gain a better understanding of the power of **Microsoft Storage Spaces** and S2D technologies.

Microsoft S2D was first introduced in Windows Server 2016 as an evolution of the Storage Spaces technology introduced in Windows Server 2012. S2D is considered a software-defined storage technology designed to protect data from drive failures by grouping three or more drives into a common disk resource pool, allowing for multiple copies of stored data. Additionally, the technology is designed to utilize locally attached drives from different industry-standard servers, allowing for a highly scalable and resilient storage array without the traditional **network-attached storage** (**NAS**) or **storage area network** (**SAN**) management overhead and expenses.

Now that we have introduced S2D, let's learn about the various uses of S2D within environments.

Uses for S2D

S2D can be used to provide flexibility and efficiency when seeking to expand your existing network storage capacity (let's be honest here, we have all experienced the ill-fated *disk full* error message at least once in our careers). The data can be made available at separate locations, given there are fast and low-latency network connections, and S2D can most certainly be used to help scale your **Hyper-V** deployments within the business, allowing virtual machines to access the same data concurrently.

The real beauty of this S2D software is that decisions on where to store data and on which hardware are automatically made for you when you are using different storage technologies within an S2D cluster. For instance, files that are frequently used are automatically stored on faster storage drives, whereas infrequently accessed files (for example, backup and archive data) are stored on slower conventional disks. To systems and users, you are simply accessing another network share and the underlying hardware and technology are seamless and transparent to the users and systems accessing the data!

Let's continue learning more about S2D by reviewing the prerequisites for establishing and maintaining a resilient cluster environment.

Prerequisites for S2D

The first major prerequisite for S2D is that licensing for Windows Server Datacenter edition is required for this feature set. For S2D to function, you must have multiple hard drives on a single server or multiple servers that each have one or more locally attached drives where the servers can simply be connected over Ethernet with no special cables required for setup. However, the individual nodes of the cluster should have at least a 10 Gbps network connection to communicate with each other and should support **Remote Direct Memory Access (RDMA)** with either **RDMA over Converged Ethernet (RoCE)** or **iWARP** protocols to ensure low latency and reduce Ethernet overhead between cluster nodes.

There are also a vast array of component combinations that allow the use of traditional **hard disk drives (HDDs)**, **solid-state drives (SSDs)**, or **non-volatile memory express (NVMe)** drives. These drives can be attached via **Serial Attached SCSI (SAS)** or **Serial Advanced Technology Attachment (SATA)**. Microsoft requires all devices to be at least Windows Server 2016 certified, as per the **Software-Defined Data Center (SDDC)** standard to ensure a high level of compatibility and certification.

Notably, this technology utilizes features within Windows Server that we have already reviewed in this book – **Windows Server Failover Clustering (WSFC)**, **Server Message Block (SMB)**, **Cluster Shared Volume (CSV)**, and **Storage Spaces**. One additional feature named **Software Storage Bus** is also used and is at the core of S2D as it is a virtual bus responsible for allowing each server node to see all disks spanning the cluster.

A full list of components within S2D is as follows:

- **Windows Server** and **Failover Clustering** as part of the core components to allow connection between the cluster nodes

- **Networking hardware** uses SMB 3.0 and above over Ethernet to allow communication between cluster nodes with the network recommendations, as previously discussed, for the greatest performance and resilience

- **Storage hardware** that requires at least two Microsoft-approved cluster nodes and up to 16 nodes with appropriate physically attached storage drives to each server, with recommendations of at least two SSDs and at least four more drives for resiliency

- A **Software Storage Bus** that acts as a virtual storage bus across cluster nodes, allowing all server nodes to see all local drives within the cluster

- A **Storage Bus Cache** that improves both I/O and throughput by partnering the fastest drives with slower drives to provide additional server-side read-and-write caching

- A **storage pool** is automatically created as a collection of drives; the recommendation is to use one storage pool per cluster

- **Storage Spaces** provides a fault-tolerant approach to the virtual disks, providing resiliency to simultaneous drive failures with data mirroring, erasure coding, or both features, and can also participate in both rack and chassis fault tolerance for the cluster

- A **Resilient File System (ReFS)** allows for accelerated operations for virtual file operations (creation, expansion, and checkpoint merging to name a few) and introduces intelligent tiering of data, moving data between hot and cold storage, depending on its frequency of use

- **Cluster Shared Volumes** (**CSV**) is responsible for unifying all ReFS volumes into a single identifiable namespace that's accessible from any server within the cluster, presenting the namespace as a locally accessible volume

- **Scale-Out File Server** (**SOFS**) is only used in converged deployments and provides access to data using the SMB3 protocol to server nodes, allowing S2D to present itself as network-attached storage to the cluster

> **S2D hardware requirements**
>
> A full detailed list of S2D hardware requirements can be found at `https://docs.microsoft.com/windows-server/storage/storage-spaces/storage-spaces-direct-hardware-requirements`.

Now that we have covered the overall prerequisites for S2D clusters, let's dig into the deployment modes available to administrators for establishing S2D clusters.

Deployment modes for S2D

S2D was designed for two main deployment scenarios: converged mode and hyper-converged mode (also known as **hyper-converged infrastructure**, or **HCI**). *Figure 9.1* identifies the two deployment modes and indicates where S2D is supplying the software-defined storage for the cluster:

Microsoft Storage Spaces Direct (S2D) deployment modes

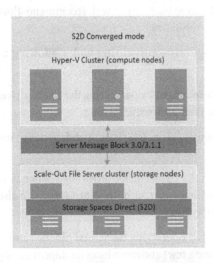
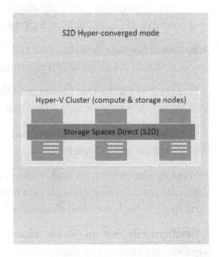

Figure 9.1 – Microsoft Storage Spaces Direct (S2D) deployment modes compared

For converged mode, the compute and storage resources reside in separate clusters to support large-scale deployments and workloads. Microsoft S2D is then configured to run directly on the storage cluster nodes within a **Scale-Out File Server** cluster to provide the storage tier. Hyper-V then runs on separate compute nodes in a separate cluster to provide the compute tier. Finally, the SMB 3.0/3.1.1 protocol provides the communications layer between the two disparate clusters.

Conversely, for hyper-converged mode, both the compute and storage resources are part of the same cluster, and storage is directly attached to each server node in the cluster to ensure connected full compute and storage nodes. Hyper-converged mode also results in a much smaller hardware footprint with lower overall administrative costs.

Many vendors offer full HCI environments preconfigured with S2D configuration that allows for a quick and painless deployment within environments. For a full list of pre-configured servers that are suitable for **Azure Stack HCI** hyper-converged hardware, please review `https://azurestackhcisolutions.azure.microsoft.com/` and `https://aka.ms/AzureStack` as recommended reading.

> **Private Cloud Simulator for Windows Server 2022**
>
> In addition, it is worth noting that as part of the Windows Hardware Lab Kit, Microsoft provides a **Private Cloud Simulator** (**PCS**) that allows vendors to validate their hardware configurations before being certified by Microsoft as an Azure Stack or Azure Stack HCI solution. Additional optional reading can be found at `https://docs.microsoft.com/windows-hardware/test/hlk/testref/private-cloud-simulator-server-2022`.

Now that we have covered the deployment modes for S2D clusters, let's dig into the various tools available to administrators for managing and monitoring S2D clusters.

Managing S2D

To successfully manage and monitor an S2D cluster, the following list of tools can be used:

- **Server Manager** and **Failover Cluster Manager** to monitor storage disks and pools of disks, as shown in *Figure 9.2*:

Figure 9.2 – Using Failover Cluster Manager to manage disks and storage pools for S2D

- **Windows Admin Center** cluster volumes and drive management, as shown in *Figure 9.3*:

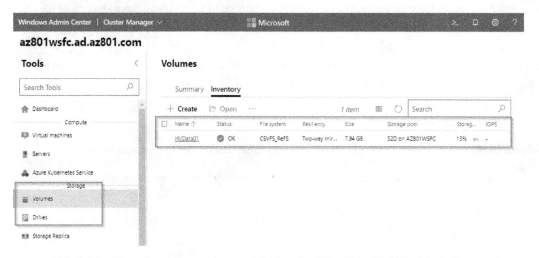

Figure 9.3 – Managing storage volumes and drives for S2D within Windows Admin Center

- **Windows PowerShell** for command-line creation, administration, and repair, as shown in *Figure 9.4*:

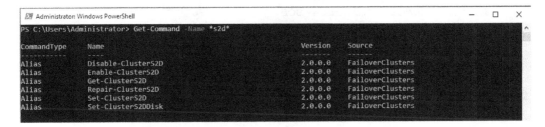

Figure 9.4 – Currently available Windows PowerShell cmdlets for S2D cluster management

As for general monitoring of storage, especially storage resync status, the following tools can be used for insights and additional troubleshooting:

- The Windows PowerShell Get-HealthFault command (for Windows Server 2019 and above), as shown in *Figure 9.5*:

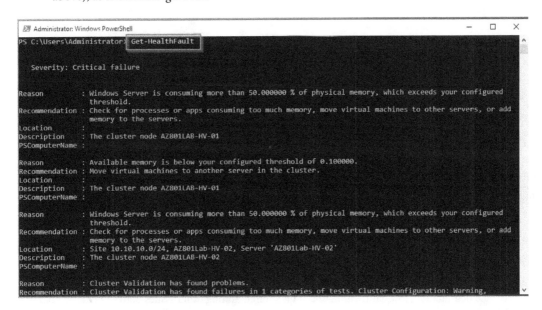

Figure 9.5 – Running the Get-HealthFault Windows PowerShell cmdlet to validate cluster health

- For an **Azure Stack HCI** cluster, the following Windows PowerShell commands can be used to complete a cluster validation with **Data Center Bridging** (**DCB**):

```
Install-Module Validate-DCB
Validate-DCB
```

- **Windows Admin Center** and the **cluster validation report**, as shown in *Figure 9.6*:

Figure 9.6 – Locating and running the Validate cluster (Preview) report within Windows Admin Center

Now that we have covered the tools and ways that S2D clusters can be monitored, let's discuss the implementation details of networking for S2D clusters.

Implementing networking for S2D

Arguably the most important part of an S2D cluster is the network. If the networking components of the cluster are not well designed or not implemented properly, you will most likely experience performance degradation, low bandwidth, and high latency, among other issues between cluster nodes.

One of the big advantages of using S2D is the simplicity of networking configuration and the reduction of CPU usage (by as much as 30% or more, given proper **remote-direct memory access** (**RDMA**) network interfaces and design). Let's review the minimum and recommended network requirements for S2D.

The following networking requirements are recommendations from Microsoft:

- Minimum network connectivity for a two to three-node cluster:

  - 10 Gbps (or faster) network interface(s)

  - Two or more network connections from each node recommended to improve both redundancy and overall performance

- Recommended network connectivity for a high-performance cluster (four or more nodes):

  - **Network interface cards** (**NICs**) that are RDMA capable, with **iWARP** (Microsoft-recommended for ease of configuration) or RoCE

  - Two or more network connections from each node recommended to improve both redundancy and overall performance

  - 25 Gbps (or faster) network interface(s)

- Switched or switchless network interconnects for the cluster nodes:

  - The switched approach requires that the network switches used in the cluster are properly configured and tuned for the expected bandwidth and networking type.

  - The switchless approach means that nodes can be directly interconnected to each other, thus avoiding the use of a network switch. It is required that every single node has a direct connection with every node of the cluster.

Switch Embedded Teaming

Switch Embedded Teaming (**SET**) can also be used to mix networking loads while using **Quality of Service** (**QoS**) or **Data Center Bridging** (**DCB**) to help control bandwidth allocation. For a general example, think of a network team that has been configured with two physical NICs and two virtual NICs. An affinity relationship can then be created between one virtual NIC and one physical NIC to ensure that both physical NICs are used in the SET configuration.

For more details on SET technology, be sure to review the following URL: `https://docs.microsoft.com/azure-stack/hci/concepts/host-network-requirements#overview-of-key-network-adapter-capabilities`.

After you have made your choices regarding designing your environment based on the networking requirements, properly selected switches, network adapters, and cabling are a must to deliver the performance and resilience of your S2D design. Equally as important is monitoring the networking for your S2D cluster. This can generally be completed within the **Failover Cluster Manager**, as shown in *Figure 9.7*, or by utilizing the Windows PowerShell `Get-StorageSubSystem {Name} | Get-StorageHealthReport` command to determine the cluster-wide health data:

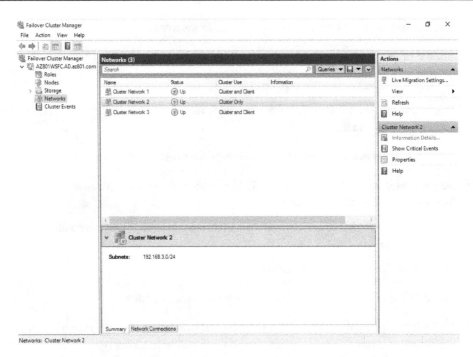

Figure 9.7 – Managing the S2D network through Failover Cluster Manager

Now that we have covered the networking considerations for deploying S2D clusters, let's discuss the deployment details for them.

Deploying S2D on Windows Server

While S2D is truly an abstracted software-defined data center technology from Microsoft, all architectures should entail a best-practices implementation that considers the physical configuration, networking configuration, a newly managed filesystem, and the latest tools provided by Microsoft for managing this S2D technology.

With that said, I recommend the following with any S2D designs to make a positive impact on the overall performance, resiliency, and stability of your clusters:

- Use certified Windows Server clustered solutions (**Azure Stack HCI** solutions)
- Utilize **Windows Admin Center** to create and manage **S2D**
- Use RDMA-enabled network configurations where appropriate
- Use SET, as discussed in the *Implementing networking for S2D* section of this chapter
- Use an ReFS where applicable

Information and design details that are needed before proceeding with the approach to deploying S2D are as follows:

- Determine the deployment option, either converged mode or hyper-converged mode

- Determine all the server/node names in advance and ensure that all naming policies, placement of files and folders, and resource names are followed

- Identify and coordinate the details for domain naming and domain joining, identifying the appropriate domain name before implementation

- Identify networking components in advance, including the type of **RDMA** being used for the network adapters, as well as any **RoCE** requirements pertinent to the environment in which you will be completing the implementation

- Identify any VLAN IDs necessary for the management network adapters on the server, as well as any additional networks required to complete the cluster

The high-level steps for deploying S2D on Windows Servers are as follows (the best experience involves setting up the S2D cluster using Windows Admin Center, as we learned in the *Creating and managing a Windows Server Failover Cluster* section of *Chapter 7, Implementing a Windows Server Failover Cluster*):

1. Deploy Windows Server to all cluster nodes:

    A. Install the intended Windows operating system on every server in the cluster.

    B. Connect to the servers to validate the version of the OS and network connectivity, that the appropriate Windows Updates are applied, the devices are ready to be joined to the appropriate domain, and that all necessary administration tools are installed.

    C. Join the devices to the domain and add all necessary administrative accounts.

    D. Install all necessary roles and features to all servers in the cluster.

2. Configure the network per best practices and recommendations as previously covered in this chapter, as well as the specifications of the device vendors.

3. Configure S2D:

    A. Clean the drives and ensure that there are no partitions and no data.

    B. Validate the cluster.

    C. Create the cluster.

    D. Configure a cluster witness.

    E. Enable S2D using the following PowerShell cmdlet:

    ```
    Enable-ClusterStorageSpacesDirect -CimSession
    {ClusterName}
    ```

F.   Create volumes.

G.   Enable the Cluster Shared Volumes cache if needed and/or intended for use within the cluster you are creating.

H.   Deploy virtual machines (or other workloads) for hyper-converged deployments.

4.   If you're using converged mode for your cluster solution, you must deploy the Scale-Out File Server:

A.   Create the Scale-Out File Server role using Server Manager, Windows PowerShell, or Windows Admin Center.

B.   Create all necessary file shares.

C.   Enable Kerberos-constrained delegation.

Now that we have covered the general considerations and high-level steps for deploying S2D clusters, let's discuss the upgrade details and approaches for them.

Upgrading an S2D cluster node

Another advantage of using S2D clusters is that you have great flexibility in handling operating system upgrades over time, using the Cluster OS Rolling Upgrade process to complete any necessary version-to-version updates. A total of four options are available to administrators, with two approaches allowing virtual machines to stay running and two approaches involving all virtual machines in a stopped state. Let's review the upgrade options, and what each option offers in terms of pros or cons, depending on the needs of your environment and workloads.

S2D upgrade options

The following options are available when updating your S2D cluster:

- **In-place upgrade while all virtual machines are running**, where each server in the cluster can be upgraded without any virtual machine downtime but the storage jobs require a waiting period to complete after each server node upgrade has been completed.

- **In-place upgrade while all virtual machines are in the stopped state**, where each server in the cluster can be quickly updated with faster storage job operations. However, this introduces virtual machine downtime.

- (Microsoft-recommended) **Complete a clean Operating System installation while virtual machines are running**, where each server in the cluster can be reinstalled without any virtual machine downtime and you must install and configure all server roles and applications again.

- (Microsoft-recommended) **Complete a clean operating system where virtual machines are in the stopped state**, where each server in the cluster can be quickly reinstalled. However, this introduces virtual machine downtime, and you must install and configure all server roles and applications again.

Now that we have reviewed the available S2D upgrade options for clusters, let's dig into the prerequisites and general limitations.

Prerequisites and general limitations

As with any upgrade options for S2D clusters, some prerequisites must be met before proceeding. These often introduce limitations, depending on the chosen option and specific environmental configurations. Let's review the prerequisites and general limitations:

- You must ensure you have proper backups that are accessible and usable in the case of any issues raised during the upgrade process.

- It is recommended that you validate your access to and support with the hardware vendor BIOS, driver, and firmware updates that are compatible and support the latest version of Windows Server.

- It is recommended that the latest Windows Updates are applied on the Windows OS that you are migrating to, to avoid any upgrade process failures.

- While upgrading, ReFS is fully supported. There are missed benefits when migrating from Windows Server 2016 to Windows Server 2019 and above, and you must create new ReFS volumes and migrate the data to benefit from the newest performance features and functionality.

- It is recommended that you put each S2D server node in storage maintenance mode during the upgrade to reduce upgrade issues.

Full details surrounding the four upgrade paths, along with all prerequisites and detailed limitations, can be found at the following URL for you to review. This is also recommended reading: `https://docs.microsoft.com/windows-server/storage/storage-spaces/upgrade-storage-spaces-direct-to-windows-server-2019`.

With that, we have upgraded an S2D cluster node. Next, we will review what we have learned in this chapter and set the stage for the next chapter of this book.

Summary

In this chapter, we learned how to configure S2D and then use it within a failover cluster to provide resiliency and redundancy for a failover cluster. We also discussed how to upgrade an S2D node, as well as implement proper security and networking configurations for S2D, while discovering topics such as converged and hyper-converged deployments consistent with the AZ-801 exam objectives.

In the next chapter, we will learn how to implement disaster recovery patterns and practices. We will discuss managing backup and recovery options for Windows Server. We will also cover how to install and use Azure Backup Server for general backup and recovery of files and folders.

After, we will discuss how to configure and use **Azure Recovery Services** vault to manage file and folder backups using backup policies. Finally, we will learn how to recover virtual machines from snapshots, how to recover a VM to a new Azure Virtual Machine, and how to successfully restore a VM to its previous running state.

Part 4: Implement Disaster Recovery

In this section, we will learn how to design, implement, and monitor disaster recovery options and approaches to keep your workloads protected in the event of an emergency or crisis so that you can meet defined recovery objectives for your business.

This part of the book comprises the following chapters:

- *Chapter 10, Managing Backup and Recovery for Windows Server*
- *Chapter 11, Implementing Disaster Recovery Using Azure Site Recovery*
- *Chapter 12, Protecting Virtual Machines by Using Hyper-V Replicas*

10

Managing Backup and Recovery for Windows Server

In this chapter, we will discuss managing backup and recovery options for Windows Server. We will cover how to install and use **Azure Backup Server** for the general backup and recovery of files and folders. We will then discuss how to configure and use **Azure Recovery Services vaults** to manage how we back up files and folders using backup policies. Finally, we will learn how to recover virtual machines from snapshots, how to recover a VM to a new Azure Virtual Machine, and how to successfully restore a VM to its previous running state.

This chapter will include exercises on how to implement and manage backup and recovery options for files, folders, and VM workloads running on-premises, with hybrid components and services, and in the cloud consistent with the AZ-801 exam objectives.

In this chapter, we will cover the following topics:

- Technical requirements and lab setup
- Using Azure Backup Server to manage, back up, and recover files
- Using Azure Recovery Services vaults to manage, back up, and restore files and folders
- Configuring backups for Azure Virtual Machines using the built-in backup agent
- Recovering a VM using temporary snapshots
- Restoring and recovering VMs to existing or new Azure Virtual Machines

Technical requirements and lab setup

To successfully follow along and complete the tasks and exercises throughout this chapter and the following chapters in this book, you need to ensure that the technical requirements from *Chapter 1, Exam Overview and the Current State of Cloud Workflows*, and *Chapter 7, Implementing a Windows Server Failover Cluster,* have been completed in full. We will be primarily reviewing our Hyper-V virtual machine environment and Microsoft Azure to complete the exercises and tasks throughout this chapter to align with the AZ-801 exam objectives. Please ensure that both the `AZ801PacktLab-DC-01` and `AZ801PacktLab-FS-01` virtual machines have been powered on and are showing a `Running` state in **Hyper-V Manager**. We will not be utilizing the failover cluster VMs during this chapter, so they can remain in a powered-off state.

Let's begin by reviewing the existing and additional cost management options for our Azure environment to give insights into managing services that will be utilized in this chapter and the chapters to follow.

Cost management review

As we have learned so far in this book, many services can be utilized and scaled between on-premises, hybrid, and cloud-native environments and there is always some cost associated. The Azure free account that this book instructs you to establish has a spending limit of $200 per month that is turned on by default. When this **Azure spending limit** is reached (per month), any of the services and resources deployed will be disabled or put into a read-only state for the remainder of the current billing period.

While this default $200 spending limit can't be increased, it can be removed (either temporarily or indefinitely) and your tenant can be converted into a **pay-as-you-go** (**PAYG**) or subscription-based model, incurring costs and billing directly to a payment method selected by you (typically a credit card).

It should go without saying that your mileage may vary here – while working through the exercises in this book from this point forward, some services may be interrupted by consumption of the free account resource allocation, where we may reach the $200 spending limit. It is wholly up to you whether you choose to wait for the next billing cycle to rehydrate any paused or disabled services or step up and convert your free account into a PAYG pricing model.

I highly recommend that you remain on the PAYG account as the exercises in this book are designed to keep costs low or utilize free services wherever possible. However, if you choose to upgrade your free account to PAYG, please carefully read through `https://docs.microsoft.com/azure/cost-management-billing/manage/spending-limit` and `https://docs.microsoft.com/azure/cost-management-billing/manage/upgrade-azure-subscription` regarding Azure's spending limit and upgrade process for moving to PAYG.

Now that we have reviewed how our Azure account manages cost by default in addition to the optional approaches for cost and resource management (as outlined in this Microsoft article: `https://learn.microsoft.com/azure/backup/azure-backup-pricing`), let's learn about the **Microsoft Azure Backup Server** (**MABS**) agent and how it can be used to back up and restore files, folders, and system state or volumes from on-premises devices to **Microsoft Azure**.

Using Azure Backup Server to manage, back up, and recover files

Every business has critical workloads that need to be protected from data loss, corruption, and ransomware, and it is the job of the IT pro and architects to ensure that data, application workloads, virtual machines, and all artifacts are adequately protected and can be recovered in a disaster scenario. Azure Backup provides these features and more as a zero-infrastructure solution that provides a predictable cost model while meeting the following business models:

- **Recovery Point Objectives** (**RPO**), where an agreed-upon and acceptable amount of data can be lost (over time) during an incident

- **Recovery Time Objectives** (**RTO**), where an asset or service must come back online after an agreed-upon time frame

Azure Backup is also easily managed from the Azure portal and provides a ton of self-service capabilities for backup policy management and backup and restore capabilities while offering flexible retention for any compliance needs of the business. A great optional read from Microsoft regarding how to plan for and protect against ransomware can be found at the following URL: `https://docs.microsoft.com/azure/security/fundamentals/backup-plan-to-protect-against-ransomware`.

Azure Backup generally supports the following backup scenarios:

- **Azure Files** shares with the availability of snapshot management

- **On-premises** backup of files, folders, and system state data using the **Microsoft Azure Recovery Service Agent** (**MARS**), or utilizing the **Microsoft Azure Backup Server** (**MABS**) or **Data Protection Manager** (**DPM**) server for on-premises virtual machines running in either Hyper-V or VMware infrastructure

- Native **Azure VMs** backup features to allow for quick and easy integrated management of backup and recovery points

- **SQL Server and/or SAP HANA databases running in Azure VMs** for covering full, differential, and t-log backups, a quick 15-minute RPO, and available point-in-time recovery options for these intelligent workload backups

We could argue that one of the more confusing things with backup features is that there are at least three different ways to utilize Azure Backup to protect your data and multiple workloads throughout any business and data center(s). To help clarify this not only in general but for our AZ-801 exam objectives, we will focus on the general features of both MABS and MARS as it pertains to appropriate backup and recovery methods for your files, virtual machines, system state, and other workloads.

It should be noted that both MARS and MABS are free downloads from Microsoft, and both allow you to configure and manage backups from on-premises and hybrid servers to the Microsoft Azure cloud. The key differentiator here is that the MARS agent is installed directly on the intended server(s) for protection and will only allow backup and restore of files for that server, whereas MABS is traditionally installed on a separate server and serves as a centralized management tool to identify and orchestrate backups for many endpoints at once. It's also important to know that the MABS protection agent package will need to be installed on each machine that will be backed up using MABS.

> **MABS protection matrix**
>
> Microsoft publishes a wonderful resource that identifies various server versions and workloads that can be protected with Azure Backup Server, so this comes as recommended reading. It is available at the following URL: `https://docs.microsoft.com/azure/backup/backup-mabs-protection-matrix`.

In addition, Azure Backup utilizes a **Recovery Services vault** to assist with managing, storing, and encrypting the backup data. The Recovery Services vault is a resource that allows for easy configuration by simply selecting the vault you want to utilize for backup or utilizing the vault credentials while abstracting the underlying Azure Backup storage account in the background. This also leads to increased security by introducing the **RBAC** security boundaries used on all Azure resources, thus helping to further protect the recovery data.

Now, let's discuss the prerequisites and recommendations for setting up and configuring **Microsoft Azure Backup Server** to meet the AZ-801 exam objectives.

Installing and managing Azure Backup Server

One of the main requirements is to determine whether you need to protect on-premises workloads or workloads running on Azure VMs. Let's discuss some of the requirements for each environment, as well as some of the nuances:

- If the requirement is to protect workloads running on-premises, here are the requirements:

  - This method is considered disk-to-disk-to-cloud

  - Your **MABS** must be located on-premises and connected to a domain

  - The server can be a Hyper-V VM, a VMware VM, or a physical host

- The minimum requirements for the server are two CPU cores and 8 GB of RAM

- The Windows Server OS can be either Windows Server 2016, 2019, or 2022

- If the requirement is to protect workloads on Azure VMs, here are the requirements:

  - This method is considered disk-to-cloud

  - You must locate the **MABS** server on an Azure VM and have that VM connected to a domain

  - It is recommended that you utilize an Azure gallery image of either a Windows Server 2016 Datacenter, Windows Server 2019 Datacenter, or Windows Server 2022 Datacenter

  - The minimum requirements for the server are **Standard_A4_V2** with four cores and 8 GB of RAM

Dedicated Microsoft Azure Backup Server

Note that MABS is intended to run on dedicated server resources, so it is not supported or recommended to install MABS on domain controllers, application servers, exchange servers, any cluster nodes, or any devices running **System Center Operations Manager**. In addition, the installation is not supported on Windows Server core or **Microsoft Hyper-V Server**.

Now that we have covered the installation requirements, let's walk through the steps of installing and managing **Microsoft Azure Backup Server** in a hybrid environment by creating a **Recovery Services vault** in **Azure Backup center**, then installing the MABS software package on our **AZ801PacktLab-FS-01** virtual machine:

1. To begin, right-click on the **AZ801PactkLab-FS-01** machine in Hyper-V Manager and select **Checkpoint** to create a new VM snapshot. Then, under **Checkpoints**, select the new checkpoint by right-clicking, then rename the checkpoint `Pre MABS Installation` or another desired checkpoint name.

2. Additionally, right-click on the **AZ801PactkLab-FS-01** machine again, and this time select **SCSI controller**, then **Hard Drive**, and proceed with adding a new dynamically expanding 20 GB hard disk to our VM in the `C:\AZ801PacktLab\VMs` folder.

3. Remotely connect to our **AZ801PactkLab-FS-01** machine as `AZ801\Administrator`. We will create a Recovery Service by visiting `https://portal.azure.com` in a browser and utilizing the Global Administrator account you created in *Chapter 1*.

4. Select **Backup center** from the list of Azure services or simply search for `Backup center` and select it to continue.

5. From the **Backup center** overview, select **+ Vault** from the tabs, as shown in *Figure 10.1*:

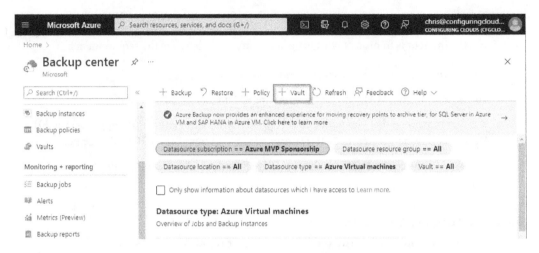

Figure 10.1 – Selecting the +Vault tab to create a new Recovery Services vault in Azure

6. Next, on the **Start: Create Vault** screen, select **Recovery Services vault** and then select **Continue**, as shown in *Figure 10.2*:

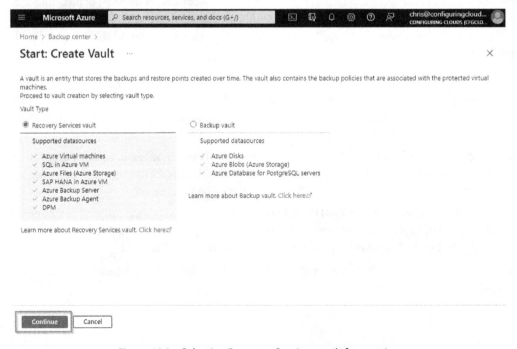

Figure 10.2 – Selecting Recovery Services vault for creation

7. On the **Create Recovery Services vault** screen, select the **Subscription** property you created in *Chapter 1*. Use rg-AZ801-ExamGuide for the **Resource Group** name, enter az801ARSVault under **Vault Name**, and select a **Region** you are geographically closest to. Then, select **Review + Create**, as shown in *Figure 10.3*:

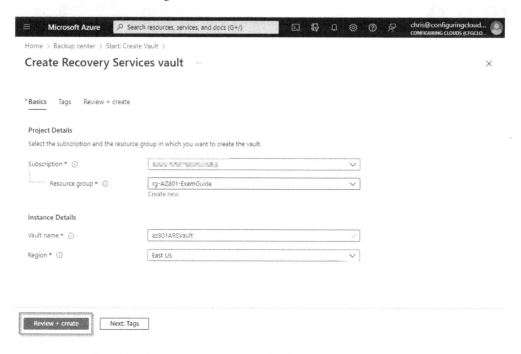

Figure 10.3 – Entering configuration details for the Recovery Services vault resource

8. Select **Create** to begin provisioning this Recovery Services vault resource. Select **Go To Resource** once the activity has been completed.

9. Under the **Getting started** blade on the left, select **Backup**.

10. For **Where is your workload running?**, select `On-Premises`, while for **What do you want to backup?**, select `Files and Folders, Hyper-V Virtual Machine,` and `System State`. Select **Prepare Infrastructure** to continue, as shown in *Figure 10.4*:

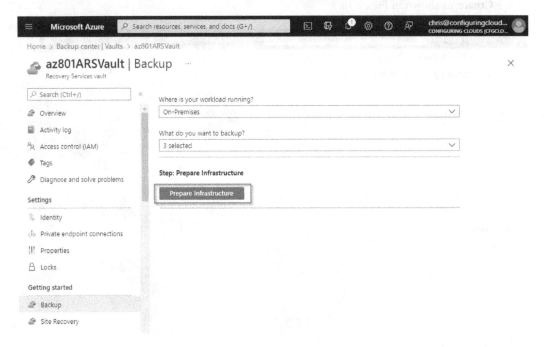

Figure 10.4 – Selecting a workload location and what to back up for Azure Backup

11. Under **Azure Backup Server step 1**, select the **Download** link to begin downloading the **Microsoft Azure Backup Server** components. Note that this package may take a while to download – please ensure that you select all available filenames to complete the download successfully and allow Microsoft.com to download multiple files at the same time.

12. Once the download has been completed, for **Step 2**, select the checkbox for **Already downloaded or using the latest Azure Backup Server installation**, which will make the **Download** button visible, as shown in *Figure 10.5*. Select the **Download** button to begin downloading a `VaultCredentials` file:

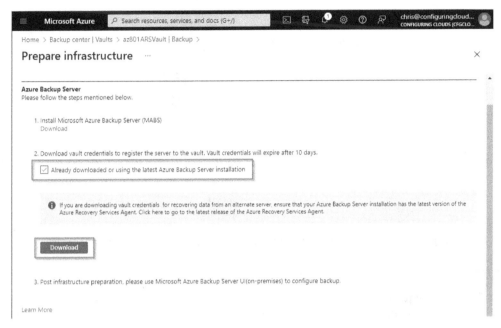

Figure 10.5 – Downloading the necessary Recovery Services Vault credentials

13. On the **AZ801PacktLab-FS-01** virtual machine, navigate to the `Downloads` folder and double-click the `System_Center_Microsoft_Azure_Backup_Server_v3.exe` file to begin the **MABS** installation.

14. Select **Next**, then **I accept** and **Next** twice, and then **Extract** to begin the extraction process, placing a new folder and files automatically in `C:\System_Center_Microsoft_Azure_Backup_Server`.

15. To avoid an agent registration error during setup for the Recovery Services vault, you will need to ensure that you have the MARS agent version 2.0.9249.0 or above downloaded and placed in the `C:\System_Center_Microsoft_Azure_Backup_Server\System_Center_Microsoft_Azure_Backup_Server\MARSAgent` folder to replace the existing agent. The updated agent can be downloaded from `https://aka.ms/azurebackup_agent`.

16. Open `C:\System_Center_Microsoft_Azure_Backup_Server\System_Center_Microsoft_Azure_Backup_Server\Setup.exe` to begin the installation, then select **Microsoft Azure Backup Server** to continue, as shown in *Figure 10.6*:

Figure 10.6 – Selecting the Microsoft Azure Backup Server installation option

17. The Microsoft Azure Backup Server setup will begin. Click **Next** to continue.

18. On the **Prerequisite Checks** page, click the **Check** button. Once all the checks have passed, click **Next** to continue, as shown in *Figure 10.7*:

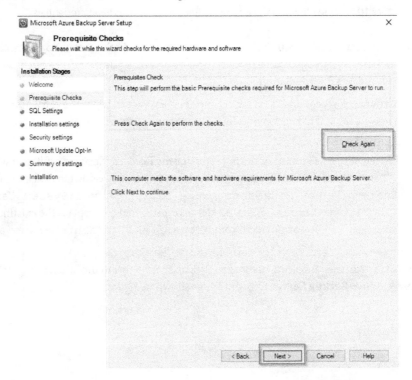

Figure 10.7 – Completing initial prerequisite checks for MABS

19. On the **SQL Settings** page, we will select **Check and Install** to validate all prerequisites for SQL and complete the necessary SQL and tools installation, as shown in *Figure 10.8*:

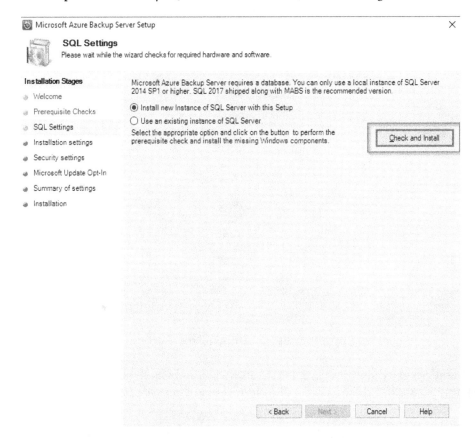

Figure 10.8 – Completing the prerequisites check and installation

Missing .NET 3.5 features warning

Note that this step may fail if you are missing the .NET 3.5 features on this server or have not yet rebooted to apply any of the prerequisite changes. If .NET 3.5 is missing from the server, utilize the `Install-WindowsFeature -Name NET-Framework-Core` PowerShell command to install the prerequisite and then restart the installation from the beginning.

20. Installing the prerequisites may take a few moments and will then switch to the next step of configuring the file and installation locations. Review the defaults on this screen and click **Next** when you're ready, as shown in *Figure 10.9*:

Figure 10.9 – Reviewing the SQL settings for the MABS installation

21. On the **Security Settings** page, for the **Password** and **Confirm password** fields, enter `Packtaz801guiderocks` and click **Next**, as shown in *Figure 10.10*:

Figure 10.10 – Entering the credentials for the restricted local user accounts

22. On the **Microsoft Update Opt-In** page, select **Use Microsoft Update when I check for updates (recommended)** and click **Next** to continue.

23. On the **Summary of settings** page, click **Install**, as shown in *Figure 10.11*:

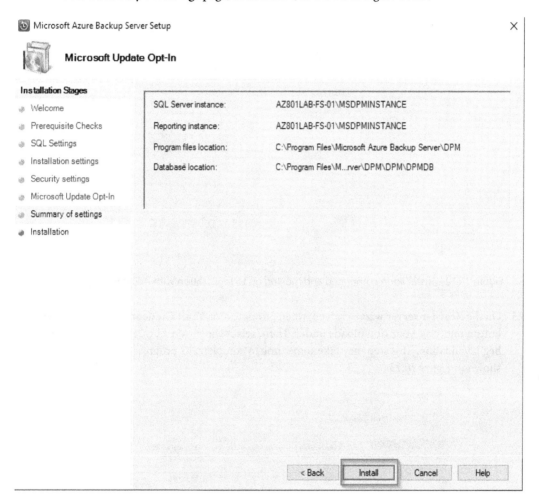

Figure 10.11 – Reviewing the summary of settings

24. Once the installation has been completed, you will see the following screen, asking you to proceed by registering the server with **Azure Backup vault**, as shown in *Figure 10.12*:

Figure 10.12 – Installation completed and moving on to registration with Azure Backup vault

25. On the **Register server wizard** screen, when prompted for **Vault Credentials**, click the **Browse** button and visit your downloads folder. Then, select the `*.VaultCredentials` file to begin validation. This step may take some time to complete; its progress can be monitored, as shown in *Figure 10.13*:

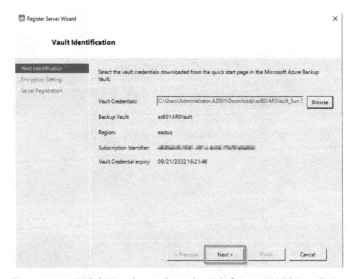

Figure 10.13 – Validating the vault credentials for our MABS installation

26. On the **Encryption Setting** page, ensure that you set **Passphrase** and **Confirm Passphrase** to `Packtaz801guiderocks` or use the **Generate Passphrase** button to generate a 16-character or more passphrase for use, enter `C:\Users\Administrator.AZ801\Downloads` as the location for saving the password to, as shown in *Figure 10.14*, and then click **Finish** to complete the agent registration in Azure:

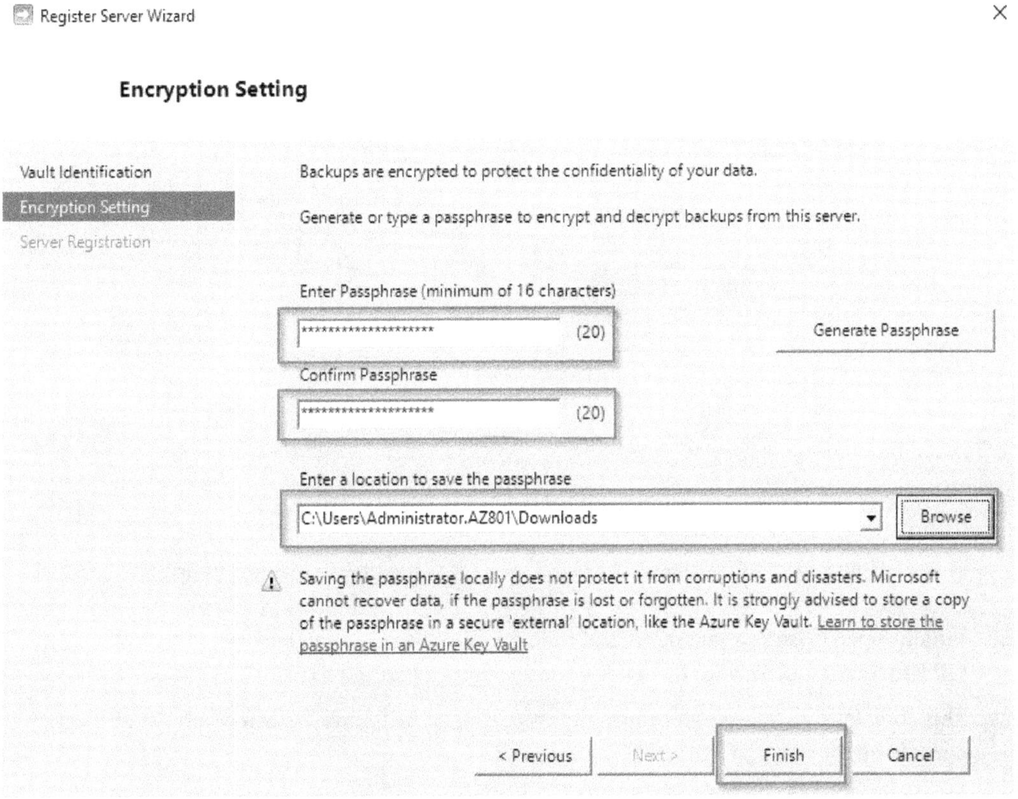

Figure 10.14 – Establishing the encryption settings for the backup agent

27. Once the registration process has been completed, you will see a confirmation screen, as shown in *Figure 10.15*. Selecting **Close** on this screen will automatically launch the **Microsoft Azure Backup** console for configuration:

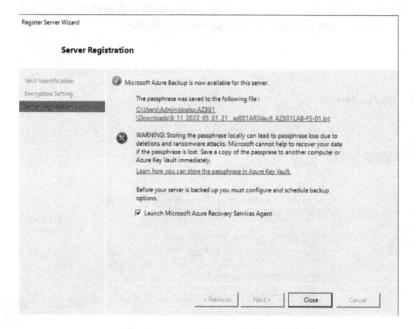

Figure 10.15 – Completing the MABS server registration with Azure Backup

28. Finally, open **Disk Management** and then **enable** and **initialize** our new 20 GB disk as the **GPT formatted** drive with the letter **E** for use in the next exercise.

29. This completes all the steps for this walkthrough, so we will leave the **Microsoft Azure Backup** console up for our next exercise.

Now that we have completed the steps for installing, configuring, and registering our on-premises Microsoft Azure Backup Server with Azure Backup, let's learn how we can back up and recover files and folders using our newly configured Microsoft Azure Backup Server.

Backing up and recovering files and folders using Azure Backup Server

While this may not truly be a one-click backup solution for installing and configuring Microsoft Azure Backup Server, doing so using Microsoft Azure Backup Server is relatively painless, allowing you to establish protection groups for your devices, quickly access recovery options for your data and devices, and access monitoring and reporting for backup schedules, jobs, data usage, and overall usage trends in your backup architecture.

Now that we have our Azure Backup Server implemented in our lab environment, let's choose a small workload to back up to demonstrate the functionality between **Azure Backup Server** and **Azure Recovery Service vault**:

1. Begin by opening Microsoft Azure Backup Server on our virtual machine, then select **Management** and then **Disk Storage**. Click **Add** and then select our new volume by selecting **Add** > and then **OK** to add a new volume, using DPM as the name when prompted.

2. Next, select **Protection** from the left pane to establish a **Protection Group**.

3. Select **Action** from the menu bar, then **Create protection group…** to continue.

4. On the **Create New Protection Group** page, click **Next** > to continue.

5. On the **Select protection group type** page, select **Servers** and then click **Next** > to continue.

6. On the **Select Group Members** page, expand AZ801LAB-FS-01 > All Volumes > C:\ > Users then select it by placing a checkbox for both the Administrator and Administrator.AZ801 folders. Click **Next** > to continue; notice that we have additional objects that can be natively protected using this approach, as shown in *Figure 10.16*:

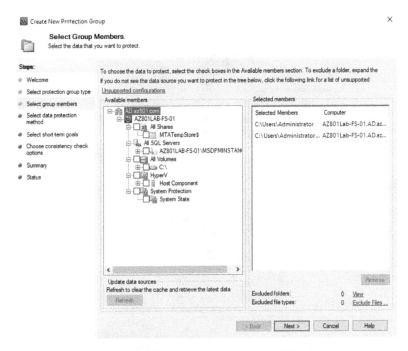

Figure 10.16 – Selecting group members from available data to protect

7. On the **Select Data Protection Method** page, accept the defaults and click **Next** > to ensure that short-term protection is stored on disk and online protection is enabled.

8. On the **Select Short-Term Goals** page, select **Just before a recovery point** to ensure an express full backup is run before any scheduled recovery points, then click **Next >** to continue.

9. On the **Review disk storage allocation** page, update **Target Storage** to our new DPM volume and click **Next >** to continue.

10. On the **Choose replica creation method** page, accept the defaults and click **Next >** to continue.

11. On the **Choose consistency check options** page, accept the defaults and click **Next >** to continue.

12. On the **Specify online protection data** page, accept the defaults and click **Next >** to continue.

13. On the **Specify online backup schedule** page, change the time to the next available time and click **Next >** to continue.

14. On the **Specify Online retention policy** page, accept the defaults shown in *Figure 10.17* and spend some time reviewing the retention policy and best practices for on-premises and cloud backup scenarios, as documented at `https://docs.microsoft.com/azure/backup/guidance-best-practices`:

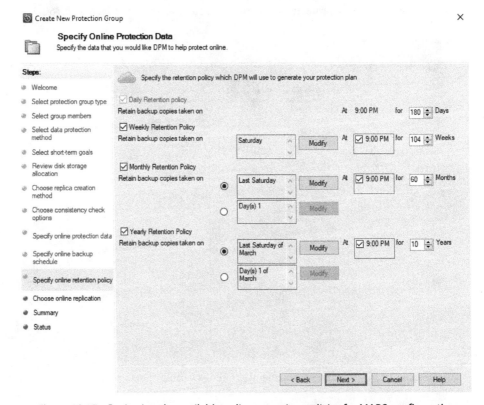

Figure 10.17 – Reviewing the available online retention policies for MABS configuration

15. On the **Choose online replication** page, accept the defaults and click **Next >** to continue. Note that offline backup settings are also available from this screen, and this option is called offline seeding for backups, which uses the **Azure Import** feature.

16. On the **Summary** page, review the settings and select **Create Group** to complete this part of the exercise. The backup process will take some time to create the local replica and online recovery points, as shown in *Figure 10.18*:

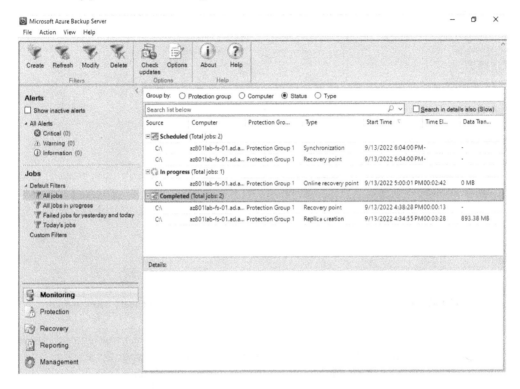

Figure 10.18 – Monitoring MABS jobs using the console

17. The online backup of the data in the **Recovery Services vault** area will also appear in Azure for additional management and monitoring, as shown in *Figure 10.19*:

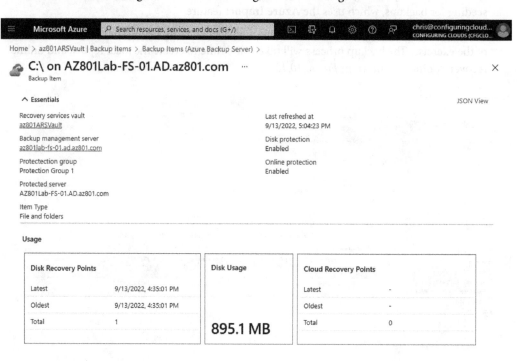

Figure 10.19 – Viewing the replicated backup data from MABS to Azure Recovery Services vault

18. To continue with this exercise and work with the recovery of files, be sure to follow the steps found in the following online documentation: `https://docs.microsoft.com/azure/backup/back-up-file-data#recover-backed-up-file-data`.

19. Once you have finished learning from this part of the lab, we need to follow additional destructive steps to revert our virtual machine to a prior state before **MABS** was installed while also removing the backup resources stored within the Azure Recovery Services vault. To proceed, first, shut down the **AZ801PactkLab-FS-01** machine in Hyper-V Manager.

20. Once the VM has been shut down, identify the **Pre MABS Installation** checkpoint (or another desired checkpoint name) from earlier in our lab, and click **Apply** to remove the **Microsoft Azure Backup Server** configuration.

21. Finally, visit `https://portal.azure.com` and locate our `az801ARSVault` resource, then select **Properties** and select **Update** for **Security Settings**. Set both **Soft Delete** and **Security Features** to **Disabled** and click **Save**.

22. Now, visit **Backup Infrastructure** > **Backup Management Servers** and select the protected server, and then select **Delete**. Type the name of the server as `AZ801LAB-FS-01.AD.AZ801.com`, select the reason as **Decommissioned**, type **Decommissioned** as the comment, then select the respective checkbox and click **Delete** to complete this operation.

23. Visit the **Overview** area for our resource, then click **Delete** to remove it. Then, select the checkbox for **I confirm that the above steps have been performed…** and select **Yes** to complete the deletion.

24. Recreate `az801ARSVault` using *Steps 4 through 10* from the *Installing and managing Azure Backup Server* section earlier in this chapter. However, do not choose `Hyper-V Virtual Machine`, as indicated in *Step 10*.

25. This completes all the steps for this walkthrough. Please keep the virtual machine running for our next exercise.

Now that we have completed the steps to install, configure, and register our on-premises **Microsoft Azure Backup Server** with **Azure Backup**, let's learn how we can use **Azure Recovery Services vault** to manage, back up, and restore files and folders.

Using Azure Recovery Services vault to manage, back up, and restore files and folders

Now, let's switch focus to the **Microsoft Azure Recovery Services (MARS)** agent or the **Azure Backup Agent**. Previously, we learned that we could back up files, folders, and system state from both on-premises machines and Azure VMs with this agent in use as part of Microsoft Azure Backup Server. This agent is meant to be lightweight without the overhead of MABS while still storing backup data in an Azure Recovery Services vault. Later in this chapter, we will cover using the **Azure Backup extension** for Azure VMs. The subtle difference between the extension versus the MARS agent is that the agent can handle specific files and folders on the VM, whereas the extension backs up the entire VM. This means that file and folder-level recovery will require both components to be configured on Azure VMs.

In addition, as this agent is wholly dependent on connection to **Azure Active Directory**, **Azure Storage**, and **Azure Backup** endpoints running in **Microsoft Azure**, any networking and firewall configuration must include a review of the necessary URL and IP addresses to ensure that no communications are inadvertently blocked or prevented. Be sure to review the following URL to ensure that internet access is allowed for the device where the **MARS** agent is installed: `https://docs.microsoft.com/azure/backup/install-mars-agent# before-you-start https://docs.microsoft.com/azure/backup/install-mars-agent#before-you-start`.

Let's walk through an overview of how the MARS agent is configured for backup policies on a VM for use with an **Azure Recovery Services vault**.

Creating a backup policy for an Azure Recovery Services vault

Using the **Microsoft Azure Backup console** for **MARS** backup, we can easily schedule a backup policy to indicate when, what data, and how long we should retain backups. Let's begin learning how to establish this backup policy for a virtual machine with the MARS agent already installed and registered with an Azure Recovery Services vault:

1. While remaining on the **AZ801PacktLab-FS-01** server, if not already open, select **Microsoft Azure Backup** to begin.

2. Select **Schedule Backup**. Then, on the **Select Items to Backup** page, select **Add Items** and then the **C:\Users** directory. Finally, click **Next >** once the validation has been completed, as shown in *Figure 10.20*:

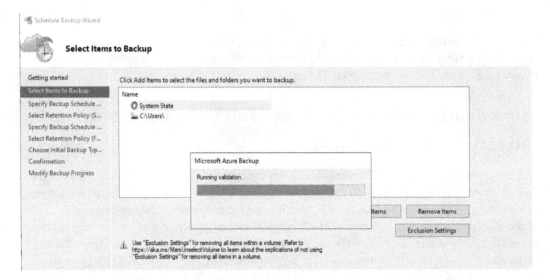

Figure 10.20 – Selecting items to back up

3. On the **Specify Backup Schedule** page, accept the defaults and click **Next >** to continue.

4. On the **Select Retention Policy (Files and Folders)** page, as we had completed for our **MABS** configuration earlier in this chapter, notice that this retention policy is identical. Accept the defaults here and click **Next >** to continue, as shown in *Figure 10.21*:

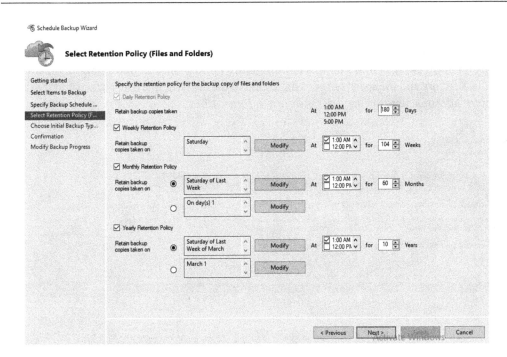

Figure 10.21 – Selecting a retention policy for files and folders

5. On the **Choose Initial Backup Type (Files and Folders)** page, accept the default of **Transfer over the network** and click **Next >** to continue, as shown in *Figure 10.22*:

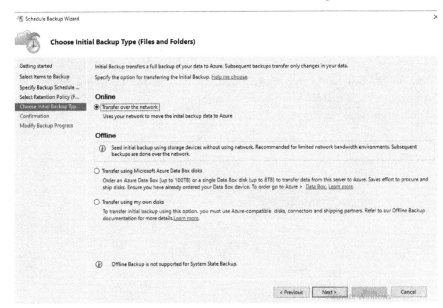

Figure 10.22 – Choosing an initial backup type for our files and folders backup

6. On the **Confirmation** page, click **Finish**, then on the **Modify backup Progress** page, click **Close**.

7. We could wait for the schedule to start the backup, but we want to begin this process now, so let's select **Back Up Now** from the **Microsoft Azure Backup** console. Depending on the size of the directory and the speed of your connection to the internet, this data backup and transfer may take some time to complete, as shown in *Figure 10.23*:

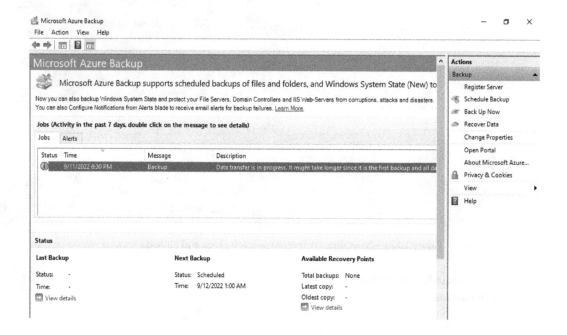

Figure 10.23 – Validating that the backup has begun

8. Once the backup task has been completed, we can validate from both the **Microsoft Azure Backup** console and the **Recovery Services vault** that the backup job has completed and has stored data in the vault, as shown in *Figure 10.24*:

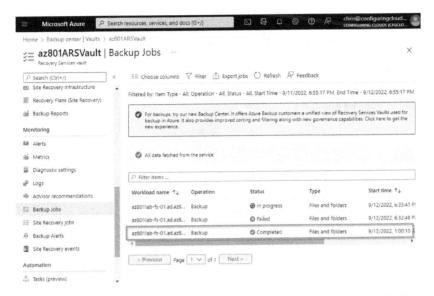

Figure 10.24 – The backup job has completed and has stored data in the Recovery Services vault

9. Now, let's attempt to recover a file from the recently completed backup using the **Recover Data** option in **Microsoft Azure Backup**. We will begin by selecting **This server**, then selecting **Next**.

10. On the **Select Recovery Mode** page, we will select **Individual files and folders** and then click **Next**, as shown in *Figure 10.25*. Note that we also have the option of recovering a full volume, as well as the system state:

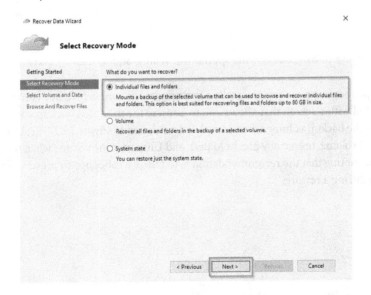

Figure 10.25 – Selecting the Individual files and folders recovery mode

11. Then, we need to **Select the volume** as **C:** and then identify the most recent backup date and time. Select **Mount** to continue, as shown in *Figure 10.26*:

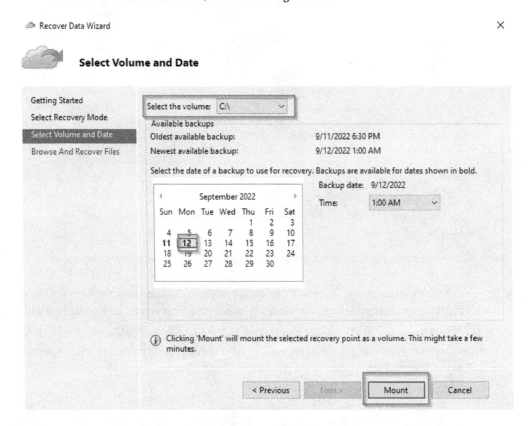

Figure 10.26 – Selecting the appropriate volume and date for recovery

12. The volume mounting process may take a few minutes to complete. Once completed, you will be presented with a screen that allows you to **Browse** the mounted backup volume, or simply **Unmount** the backup volume when completed, as shown in *Figure 10.27*. Feel free to navigate the restored volume, restore any files or folders, and **Unmount** the volume when we've completed everything, noting that the recommendation is to utilize a robocopy to preserve file and folder attributes during a restore:

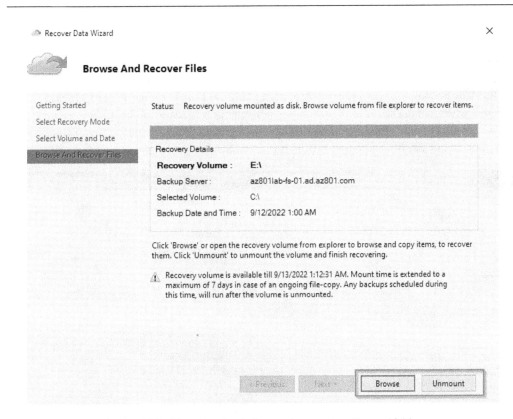

Figure 10.27 – Mounting, browsing, and recovering files and folders

13. With that, we have completed this walkthrough, so we may now close the **Microsoft Azure Backup** console for our next exercise.

Now that we have completed the steps to install, configure, and register our on-premises **Microsoft Azure Recovery Services** agent with **Azure Backup**, let's learn how we can use an **Azure Recovery Services vault** to manage, back up, and restore files and folders.

Managing backups in an Azure Recovery Services vault

Now that we have some backup resources stored in Azure as part of the centralized Backup center, let's review the available management and governance features within the Azure portal and Backup center.

Backup policies

Backup policies allow you to specify the backup schedule for VMs, a feature called Instant Restore that also retains instant recovery snapshots along with disk backup for faster recovery, and a retention range for daily, weekly, monthly, and yearly backup points, as shown in *Figure 10.28*. One thing to note is that when making changes to backup policies, retention changes will be made to all existing and future recovery points. However, the addition of new retention categories (for example, adding a weekly, monthly, or yearly backup point) will only be added to future recovery points:

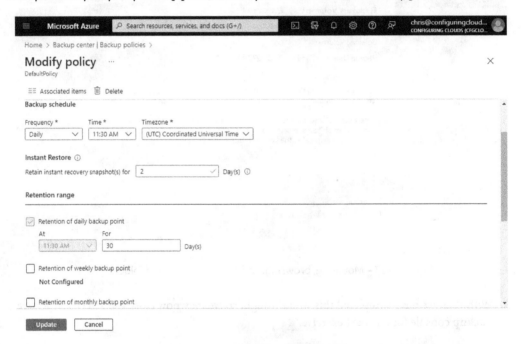

Figure 10.28 – Review of the DefaultPolicy backup policy in Backup center

Backup Jobs

Backup Jobs simply shows jobs for varying data sources across all instances of **Recovery Services vaults** for quick insights and monitoring of workloads being protected, in addition to quickly reviewing and managing errors or failed backup jobs within the environment, as shown in *Figure 10.29*:

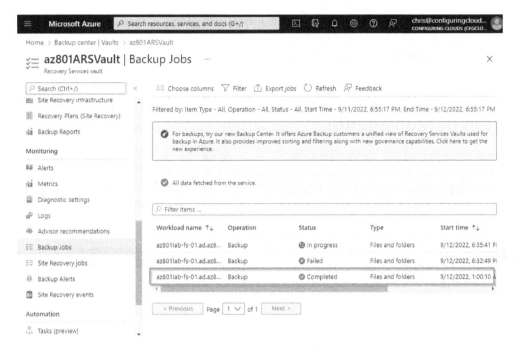

Figure 10.29 – Review of Backup Jobs in Backup center

Azure Backup PowerShell and Azure CLI examples

For those seeking either a quick repeatable approach to backups or simply looking to automate these backup processes, Microsoft provides some incredibly deep resources on both subjects at `https://learn.microsoft.com/azure/backup/powershell-backup-samples` and `https://learn.microsoft.com/azure/backup/create-manage-azure-services-using-azure-command-line-interface`, respectively.

Backup Reports

Backup Reports requires Log Analytics and workspace design to be configured so that you can receive logs from configured diagnostics settings. These reports give quick insights across all backup instances, jobs, policies, and data usage, and even provide optimization opportunities based on past and present workloads. More information about this powerful Backup Reports tool and the necessary diagnostics configuration can be found at `https://docs.microsoft.com/azure/backup/azure-policy-configure-diagnostics`.

Finally, most Microsoft tools that have a user interface typically have PowerShell resources to help with the backup and management tasks. Additional details on this collection of PowerShell tools can be found at `https://docs.microsoft.com/azure/backup/backup-client-automation`.

Now that we have covered some of the tools that can help us manage backups and configuration within **Azure Recovery Services vaults** and the **Backup center**, let's move on to configuring backups for Azure Virtual Machines using the built-in tools provided by Microsoft.

Configuring backups for Azure Virtual Machines using the built-in backup agent

Here's where the fun begins! We have read about it, and now we truly get to experience the one-click backup approach directly in Microsoft Azure. Once you have Azure VMs created and have a configured Azure Recovery Services vault, within **Backup center**, you can simply navigate to the **Policy and Compliance | Protectable datasources** section, as shown in *Figure 10.30*, to review any virtual machines that are not currently configured for backup:

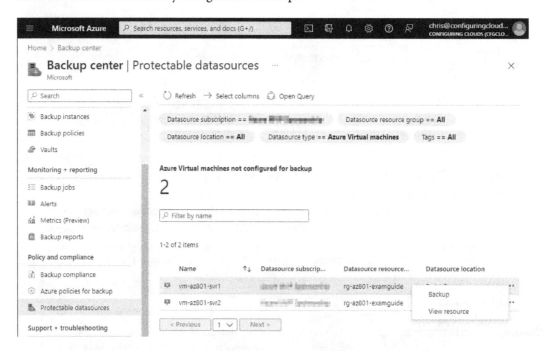

Figure 10.30 – Reviewing VM data sources that are not currently backed up

Right-clicking and selecting **Backup** from the menu provides us with a **Welcome to Azure Backup for Azure VMs** page that gives us a simple and straightforward approach to enabling backup for VMs with configurable backup policies and policy sub-types, with a simple **Enable backup** button to begin the **Volume Shadow Copy Service** (**VSS**)-based process, as shown in *Figure 10.31*:

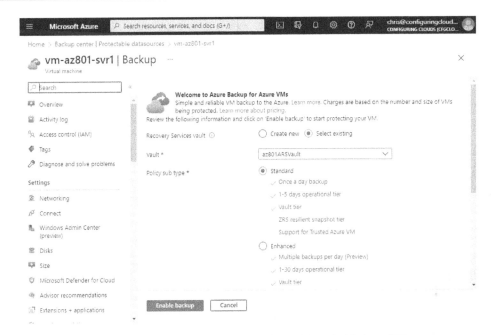

Figure 10.31 – True one-click Azure Backup achieved with Azure VMs

Azure Backup encrypts VMs by default with **Storage Service Encryption** (**SSE**) and can also handle Azure VMs that are encrypted with **Azure Disk Encryption** (**ADE**). One key aspect to note here is that when backing up VMs that are using ADE, both the **BitLocker Encryption Keys** (**BEKs**) as well as the **Azure Key Vault key encryption keys** (**KEKs**) get backed up and encrypted. This ultimately allows users with proper permissions to retain the ability to restore keys and secrets in the key vault if/when needed and can also successfully recover the encrypted VM. For additional reading on ADE, be sure to review this URL: `https://learn.microsoft.com/azure/virtual-machines/windows/disk-encryption-overview`. As for additional reading on SSE, be sure to review this URL: `https://learn.microsoft.com/azure/virtual-machines/disk-encryption`.

Another important part of snapshot creation is snapshot consistency. There are three types of snapshot consistency:

- **Application-consistent** snapshots capture both pending I/O operations and memory content and ensure the consistency of the application data before backup. This snapshot approach is the primary backup approach for Windows VMs.

- **Crash-consistent** snapshots usually occur when a VM is in the process of shutting down while a backup operation is in progress, and in this scenario, only the existing data on the disk is captured at backup time.

- **File-system consistent** snapshots ultimately take a snapshot of all available files at the same time. This snapshot approach is the primary backup approach for Linux VMs.

Azure Backup covers both Windows VMs as well as Linux VMs. To take an app-consistent snapshot on a Linux VM, both pre and post-scripts must be customized and configured for the VMs. In addition, it's beneficial to know that two extensions are managed by Microsoft to help handle snapshots, with a separate extension for each server OS:

- **VMSnapshot** is loaded into the Windows Azure VM the first time a backup is issued and uses VSS to take a snapshot of both the data on the disk as well as memory

- **VMSnapshotLinux** is loaded into the Linux Azure VM the first time a backup is issued, and only includes a snapshot of the disk

In addition to this introduction, a great resource that provides an overview of the components of Azure VM backups, as well as how it all works and recommended best practices, can be found at the following URL: `https://docs.microsoft.com/azure/backup/backup-azure-vms-introduction`.

Now that we have provides a brief introduction to backups for Azure Virtual Machines, let's review the overall benefit that VM snapshots bring to Azure and how they are utilized for backup and recovery purposes.

Recovering a VM using temporary snapshots

One of the things I believe is valuable and infrequently talked about is the ability to create incremental snapshots for managed disks. This is not only more cost-effective but also allows for the image to be used immediately after being taken, and results in the ability to create either a fully managed disk or a full snapshot to help with disk restoration purposes.

In addition, these incremental snapshots can be copied to virtually any Azure region, and only differential changes since the last snapshot in the source region are copied to the target region.

There are some great insights and recommendations, as well as restrictions, documented at the following URL for you to review: `https://docs.microsoft.com/azure/virtual-machines/disks-incremental-snapshots?tabs=azure-cli`.

Let's learn about some approaches to restoring Azure VMs using **Azure Backup**.

Restoring or recovering VMs to existing or new Azure Virtual Machines

There are several restore options available to administrators when attempting to recover Azure VMs from the recovery points stored within **Azure Recovery Services vaults**. Let's review some of the restore options:

- **Create a new VM** allows you to create a new virtual machine quickly and efficiently from a selected restore point, giving you flexibility in naming and placing the resource group, the storage location, and the virtual network for the restored VM.

- **Restore disk** recovers a disk that can be used to create a new VM or add to an existing VM. Template functionality is available for customizing the destination VM for creation.

- **Replace existing** restores the disk directly on top of the existing disk. This requires that the current VM persists/exists.

- **Cross Region (secondary region)** allows you to restore the Azure VM and selected recovery point if – and only if – the backup is completed in the secondary region.

Let's review an example of restoring an encrypted VM from within **Backup center**, as shown in *Figure 10.32*, noting that we can only restore disks for the specified **Restore point**:

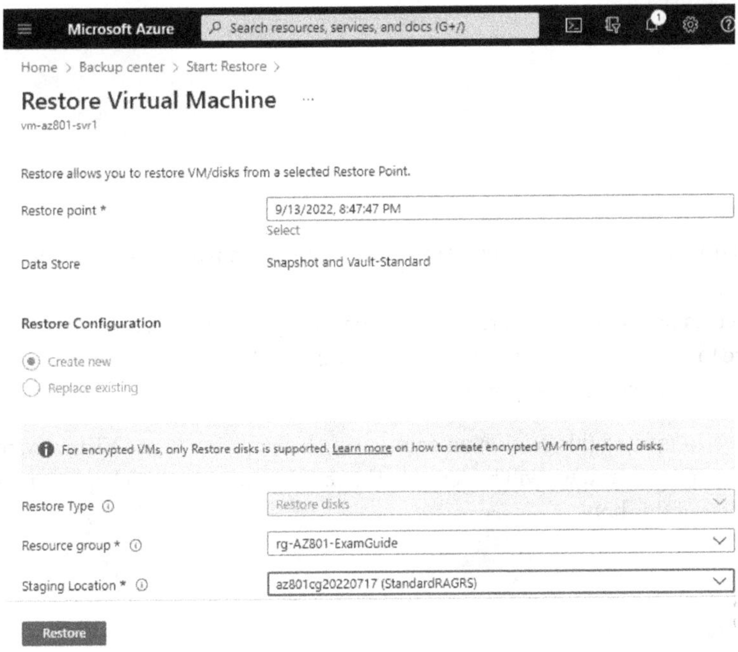

Figure 10.32 – Restoring an encrypted VM from within Backup center

For this example, this virtual machine has a crash-consistent and vault-standard snapshot available for recovery, allowing us to create a new virtual machine with great flexibility in terms of the placement and use of Azure resources, as shown in *Figure 10.33*:

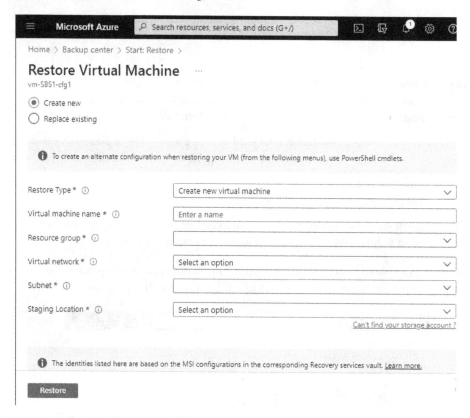

Figure 10.33 – Restoring a virtual machine from a crash-consistent snapshot to a new virtual machine

For additional details on all of the restore options, nuances, recommendations, and benefits of each approach, be sure to review this incredible article: `https://docs.microsoft.com/azure/backup/backup-azure-arm-restore-vms`.

With that, we have completed this section on configuring backups for Azure Virtual Machines using the built-in backup agent. Next, we will review what we have learned in this chapter and set the stage for the next chapter of this book.

Summary

In this chapter, we learned how to manage backup and recovery options for Windows Server. We covered how to install and use **Azure Backup Server** for general backup and recovery of files and folders. Then, we discussed how to configure and use **Azure Recovery Services vaults** to back up files and folders using backup policies. Finally, we learned how to recover virtual machines from snapshots, how to recover a VM to a new Azure Virtual Machine, and how to successfully restore a VM to its previous running state consistent with the AZ-801 exam objectives.

In the next chapter, we will learn about another valuable disaster recovery tool called **Azure Site Recovery**. We will discuss how to configure Azure Site Recovery, how to create and implement a recovery plan, and how to configure recovery policies to ensure your workload remains online in the event of a planned or unplanned outage/failure.

11

Implementing Disaster Recovery Using Azure Site Recovery

In this chapter, we will learn about another valuable disaster recovery tool called **Azure Site Recovery** (**ASR**). ASR is a tool that affords a business continuity and disaster recovery option to replicate on-premises and Azure Virtual Machines to a secondary data center location, whether it's on-premises or in the cloud. We will discuss how to configure ASR, how to create and implement a recovery plan, and how to configure recovery policies to ensure your workload remains online in the event of a planned or unplanned outage/failure.

This chapter will include walkthroughs on how to implement and manage ASR components and workloads on-premises, with hybrid components and services, and in the cloud consistent with the AZ-801 exam objectives.

In this chapter, we will cover the following topics:

- Technical requirements and lab setup
- Configuring Azure Site Recovery networking
- Configuring Azure Site Recovery for on-premises VMs
- Configuring a recovery plan
- Configuring Azure Site Recovery policies
- Configuring Azure Site Recovery for Azure Virtual Machines
- Implementing VM replication to a secondary data center or Azure Region

Technical requirements and lab setup

To successfully follow along and complete the tasks and exercises throughout this chapter and the following chapters in this book, you need to ensure that the technical requirements from *Chapter 1, Exam Overview and the Current State of Cloud Workflows*, have been completed in full. Many aspects of ASR services require a rigorous amount of detail and require infrastructure configuration time for both initial protection as well as ongoing replication and protection. We will focus on the requirements, best practices, and high-level concepts in this chapter and will be primarily reviewing areas of the Azure portal environment to follow the walkthroughs throughout this chapter that align with the AZ-801 exam objectives.

Let's begin by reviewing what ASR provides to assist with business continuity and disaster recovery services.

Introduction to Azure Site Recovery

ASR is a cloud-native Azure cloud data protection and disaster recovery service that protects both on-premises and Azure-hosted virtual machines and servers. ASR provides a single replication solution for protecting and restoring data from virtual machines running on either VMware or Hyper-V clusters and physical servers. In the event of a failure and based on your organization's disaster recovery strategy (let's be honest, this will happen at some point), ASR provides a fully automated disaster recovery service that encrypts data both in transit and at rest, while also providing the ability to fail over services to a secondary site or even another Azure region while meeting your organizational requirements and policies.

Business continuity and disaster recovery (BCDR) plans typically require **recovery time objectives (RTOs)** and **recovery point objectives (RPOs)** for your workloads and applications, identifying a maximum amount of time (in hours) that your business can survive without specific workloads and applications. With the replication of workloads that ASR provides, your organization's BCDR plans can be met by failing over the workloads and applications with ease from the primary site to a secondary site, including automatic failover of traffic.

A few of the more powerful features of ASR are as follows:

- ASR remains a cloud-native fully integrated offering and provides the flexibility for operationalized simulation and validation of failover testing with minimal configuration and clicks without interruption to your production services

- On-premises virtual machine replication can be replicated to Azure or to a secondary on-premises data center, where appropriate, and is end-to-end encrypted

- Centralized management can be orchestrated all from within the Azure portal, from initial setup and configuration to ongoing management, failover, and failback operations, including restoring systems in the order that you choose (this can also be orchestrated via the use of **Azure Automation Runbooks**, as detailed at `https://learn.microsoft.com/azure/site-recovery/site-recovery-runbook-automation`).

- Application consistency is retained during replication via the use of application-consistent snapshots

- ASR can manage full end-to-end network management during the replication and disaster recovery scenario, including orchestration of network configuration, IP address reservations, **Network Security Groups** (**NSGs**), and load balancers to ensure the virtual machine workloads can effectively communicate in the failover location site or region

- ASR is designed to be cost-effective by design, and only requires you to pay for what you use as a consumption-based pricing and scalability model

- ASR is continually improved and updated, and updates can be reviewed at `https://learn.microsoft.com/azure/site-recovery/site-recovery-whats-new`

Now that we have discussed some of the features of ASR, let's dig into the required components to ensure that ASR can be easily enabled.

The following components are required for setting up ASR for your environments and specific infrastructure:

- A valid **Azure virtual network** is required for any replicated workloads to use and communicate within Azure, including any public and private IP address reservations.

- An **Azure Recovery Services vault** within your Azure subscription that will contain the replicated virtual machines, any policies and recovery plans, and the failover information for both source and target failover/failback locations.

- The **Microsoft Azure Site Recovery Provider** and **Microsoft Azure Recovery Services** (**MARS**) agent are deployed onto the on-premises site's Hyper-V server host.

- ASR receives the replication communication over HTTPS port 443 (outgoing) and **Azure Storage** is used to store the duplicated virtual machines. The following URLs must be accessible for the replication process:

  - `*.hypervrecoverymanager.windowsazure.com`

  - `*.accesscontrol.windows.net`

  - `*.backup.windowsazure.com`

  - `*.blob.core.windows.net`

  - `*.store.core.windows.net`

  - `login.microsoftonline.com`

- ASR requires access to the Microsoft 365 IP ranges for authentication and will require the use of the **Azure Active Directory** (**Azure AD**) service tag-based **Network Security Group** (**NSG**) rule to allow authentication (more details can be reviewed at `https://learn.microsoft.com/azure/site-recovery/azure-to-azure-troubleshoot-network-connectivity#outbound-connectivity-for-site-recovery-urls-or-ip-ranges-error-code-151037-or-151072`).

- Appropriate **credentials** for Azure, requiring both the **Virtual Machine Contributor** and the **Recovery Contributor** RBAC roles allowing for proper permission to manage both the VM and the storage for ASR, and the vault registration credentials key file if/when using VMware machines, System Center **Virtual Machine Manager** (**VMM**), or Hyper-V sites/hosts.

- An appropriate **configuration server** that includes the following components:

  - A **process server** that provides the gateway for replication traffic to Azure or another site and is used for caching data, compressing replicated data, and encrypting data (be sure to review the following URL, which is specifically related to the requirements for VMware environmental setup: `https://learn.microsoft.com/azure/site-recovery/vmware-azure-set-up-process-server-scale#sizing-requirements`)

  - A **Master Target server**, which is used for failback from Azure only

  - A configuration server can be one of the following:

    - **VMware machines** -ASR replication appliances, which can be established as pre-configured virtual machine images (OVA) or by utilizing PowerShell on an existing virtual machine to establish the appliance configuration

    - **System Center Virtual Machine Manager** (**VMM**) with the **Azure Site Recovery Provider** installed and registered

    - **Hyper-V Sites/Hosts** with the **Azure Site Recovery Provider** installed and registered

- **Azure Storage** used for the replicated virtual machines must *not* have soft delete disabled so that consistency and compliance for the Recovery Services vault are achieved.

An example of the overall Site Recovery Infrastructure can be seen in *Figure 11.1*, indicating the four types of infrastructures that can be protected:

Figure 11.1 – Overall Azure Site Recovery infrastructure covering four different scenarios

Now that we have reviewed the requirements and necessary components for a successful **Azure Site Recovery** configuration, let's dive deeper into the networking requirements as these are a must-have for any successful deployment.

Configuring Azure Site Recovery networking

The hard and fast rule of any virtual machine in Azure is that the VM must have at least one network interface attached to it (called the primary network interface) and can contain multiple network interfaces as the OS and VM size supports (called secondary network interfaces). Depending on the network design, multiple networks might be configured for different workloads and security requirements. It is imperative to know that when failing over or migrating virtual machines to Azure from an on-premises environment, all network interfaces in the same VM must be connected to the same virtual network, which will require design and planning.

The great news is that the network interface settings can be flexibly modified under the replication settings for the virtual machine, to accommodate network mapping and network interface settings for both failover and test failover settings, as shown in *Figure 11.2*:

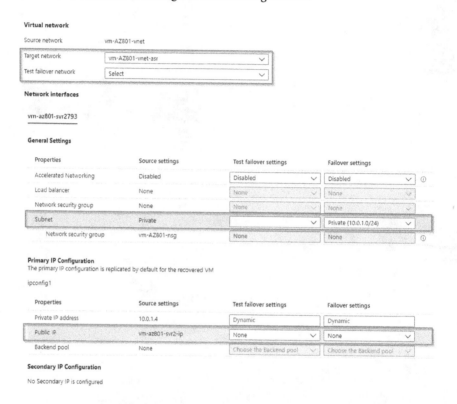

Figure 11.2 – Available network mapping changes for a replicated item in ASR

The target network is important for any workloads being hosted on Hyper-V, VMware, and physical machines as you will need to identify what the private IP address will be in the recovery network. When the target and source networks are similar, the configuration of the network mappings will be straightforward. However, when the target and source networks are different, you will need to consider translating the IP address spaces for the workloads and services.

Ultimately, networking in Azure should be an extension of the on-premises network so that applications can move from one site to the other seamlessly. Azure allows you to add site-to-site connectivity to virtual networks created in Azure. You can then add site-to-site connectivity either at the time you are creating the network or at a later point in time and include replication of both **Active Directory** and **DNS** into Azure in your design.

Finally, ASR also provides support for accelerated networking to improve networking performance with workloads compliant with single root I/O virtualization to a virtual machine. This allows for reduced latency, reduced jitter, and lower overall CPU utilization to enable high performance for demanding network workloads, and this support extends across different Azure regions.

Additional reading on how to establish connectivity to Azure VMs after failover can be reviewed at the following URL: `https://learn.microsoft.com/azure/site-recovery/concepts-on-premises-to-azure-networking`.

With that, we have completed this section on configuring ASR networking. Next, we will learn about configuring ASR for on-premises virtual machines.

Configuring Azure Site Recovery for on-premises VMs

As we have learned throughout this book, **Windows Admin Center** greatly streamlines a lot of the configuration tasks for modern infrastructure management, and the configuration of Azure Site Recovery is no exception. Windows Admin Center merely requires internet connectivity and a gateway registration to your Azure tenant and subscription, at which point a myriad of **Azure Hybrid Services** immediately become available to you.

A visual representation of the Azure Hybrid Center can be seen in *Figure 11.3*; additional details on the setup and configuration for Windows Admin Center and Azure integration can be found at `https://learn.microsoft.com/windows-server/manage/windows-admin-center/azure/`:

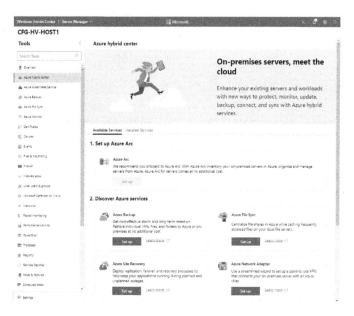

Figure 11.3 – Azure Hybrid Center as seen in Windows Admin Center

Let's review the steps for protecting our on-premises Hyper-V virtual machines with ASR:

1. To begin, it is necessary to set up and register your host with **Azure Site Recovery**. As shown in *Figure 11.4*, utilizing **Windows Admin Center**, we can sign in with an account that has the appropriate permissions to the Azure subscription, then select the resource group, appropriate **Recovery Services Vault**, the location that determines the Azure region, and the storage account to be used to save the replicated VM workloads:

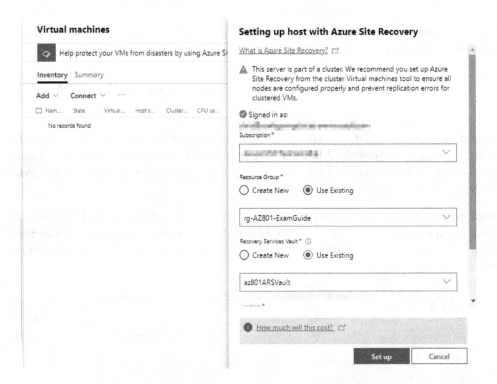

Figure 11.4 – Setting up a Hyper-V host with Azure Site Recovery

2. Replication can be set up by selecting any VM listed within the **Windows Admin Center** Virtual Machines tool, or we can pivot to the Azure portal and enable a replicated item from within the **Recovery Services Vault**, as shown in *Figure 11.5*:

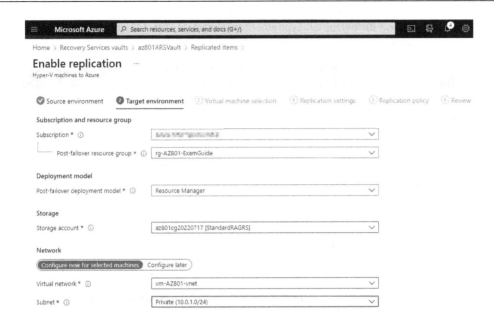

Figure 11.5 – Enabling replication by establishing the Target environment configuration

3. We can then select some or all of the virtual machines from the protected host, as shown in *Figure 11.6*, to continue the replication configuration:

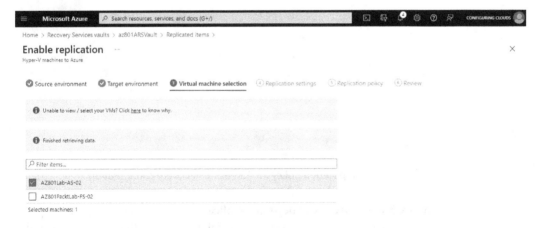

Figure 11.6 – Selecting virtual machines to replicate to Azure Site Recovery

4.   Next, we must select the default OS type and then identify the OS disk and any additional disks to be replicated, as shown in *Figure 11.7*:

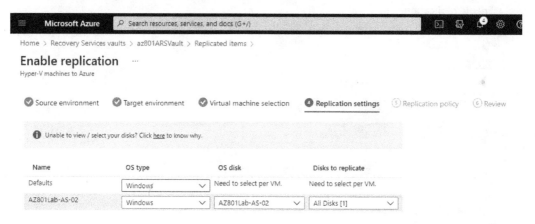

Figure 11.7 – Configuring the replication settings for the OS type and replicated disks

5.   Next, we will be presented with a choice of **Replication policy** to be used, indicating how often the data is copied, what the recovery point retention is, how frequently snapshots are taken, the level of encryption, and any additional retention-specific settings, as shown in *Figure 11.8*:

Figure 11.8 – Choosing a replication policy for Azure Site Recovery configuration

6.   At this point, **Azure Site Recovery** will begin the replication setup. While this protection is established for the virtual machine, we can see additional details on both the replication and failover health, including additional operations for the monitoring, validation, and ongoing configuration management of the replicated item, as shown in *Figure 11.9*:

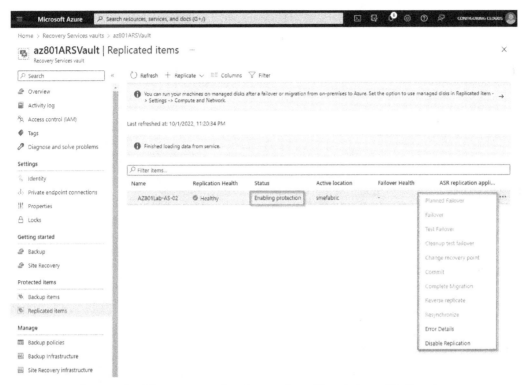

Figure 11.9 – Advanced operations for a replicated item in Azure Site Recovery

Now that the ASR protection is being completed (depending on our network speed, size of the virtual machine, and other dependencies), we have established protection for an on-premises Hyper-V virtual machine.

With that, we have completed this section on configuring ASR for on-premises virtual machines. Next, we will learn about recovery plans and how they are used with ASR.

Configuring a recovery plan

A **recovery plan** is a feature that lets you identify and define a collection of virtual machines that all deliver an application. This recovery plan can then be used to fail over all components of the application at the same time or selectively bring them back up in a failover or failover test. It also allows for additional automation steps to be completed as part of the orchestrated recovery process.

Things to consider when creating a recovery plan primarily based on the hosted virtual machines within the intended plan are shown in *Table 11.1*:

Failover Options	Source Location	Target/Destination Location
Hyper-V to Azure	Select the Hyper-V site name	Select Azure
Vmware to Azure	Select the configuration server	Select Azure
Azure to Azure	Select the Azure region	Select the Azure region
Physical machines to Azure	Select the configuration server	Select Azure
Hyper-V (managed by Virtual Machine Manager) to Azure	Select the **Virtual Machine Manager** (VMM) server	Select Azure

Table 11.1 – Recovery plan failover options and source/target workload selections

General recovery plan recommendations and best practices are as follows:

- A recovery plan can contain machines with the same source and target location.
- The source location must have machines that are enabled for failover.
- A recovery plan can be used for both failover and failback from Azure.
- All VMs in the same recovery plan must replicate into a single subscription. However, VMs can be replicated to different subscriptions and require the use of multiple recovery plans for each target subscription.
- VMware VMs, physical servers, and Hyper-V VMs managed by Virtual Machine Manager can be collocated within the same recovery plan.
- Physical servers can only fail back as VMware virtual machines and will require a VMware infrastructure for failback and reprotection (review the following URL for additional details on this process, as well as best practices: `https://learn.microsoft.com/azure/site-recovery/physical-to-azure-failover-failback#prepare-for-reprotection-and-failback`).

To establish a new recovery plan, let's walk through the following steps:

1. In the **Recovery Services** vault, select **Recovery Plans (Site Recovery)** and then **+ Recovery Plan**.
2. In the **Create recovery plan** dialog box, provide a name for the recovery plan.

3. Choose a source and target based on the machines in the plan.

4. Select **Resource Manager** for the deployment model.

5. Ensure that the source location has machines that are enabled for failover and recovery.

6. In the **Select items virtual machines** dialog box, select the machines or replication group that you want to add to the plan and click **OK**.

7. Once the recovery plan has been created, click the **Customize** button. Note that from the recovery plan, you can complete additional failover and failback operations, as shown in *Figure 11.10*:

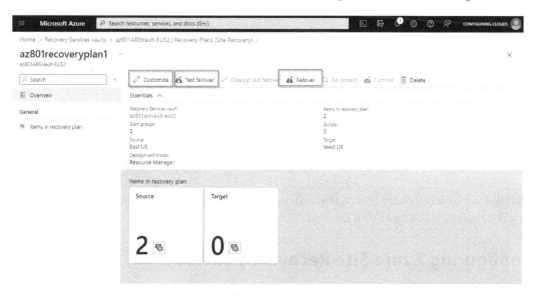

Figure 11.10 – Customizing the recovery plan for Azure Site Recovery

8. Review the recovery plan by expanding the groups, noting that the author has created a second group to allow specific actions to be completed per virtual machine group, as shown in *Figure 11.11* (either an automated script or a manual action with instructions; these steps can be completed via a pre-step or post-step action):

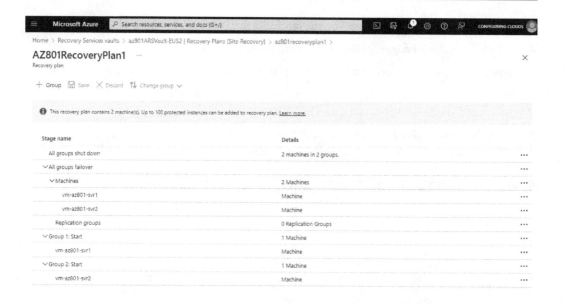

Figure 11.11 – Reviewing a customized recovery plan for Azure Site Recovery

With that, we have completed this section on configuring recovery plans for use within ASR. Next, we will learn how to configure ASR for Azure virtual machines.

Configuring Azure Site Recovery policies

While policies can be created for any of the ASR infrastructure types individually and manually, this doesn't help in managing workloads and virtual machines at scale. As such, it is recommended to utilize **Azure Policy** to apply your disaster recovery policies directly to all resources within subscriptions or resource groups and ensure compliance and remediation so that all virtual machines are protected based on the business requirements and recovery objectives.

Let's walk through how to utilize Azure Policy to complete this configuration, basing the steps on the full quick start located at `https://learn.microsoft.com/azure/site-recovery/azure-to-azure-how-to-enable-policy`:

1. Starting with **Azure Policy** assignments, we will assign a new policy and apply that to the scope of a subscription. Then, we will select the existing policy definition of **Configure disaster recovery on virtual machines by enabling replication via Azure Site Recovery**, as shown in *Figure 11.12*:

Figure 11.12 – Selecting an available policy definition for Azure Site Recovery

2. Next, we can review the **Parameters** options on the **Assign policy** screen to ensure that proper source and target regions are selected, in addition to placement of resource groups and Recovery Services vaults, as shown in *Figure 11.13*:

Figure 11.13 – Updating the parameters for the source, target, and resource group locations

3. Next, we can review the remediation details and **Create a remediation task** to ensure that all virtual machines within the scope of this policy are compliant, as shown in *Figure 11.14*. Note that this remediation task approach automatically creates a **System assigned managed service identity** for automation purposes. In addition, any custom messages can be set under the **Non-compliance messages** page:

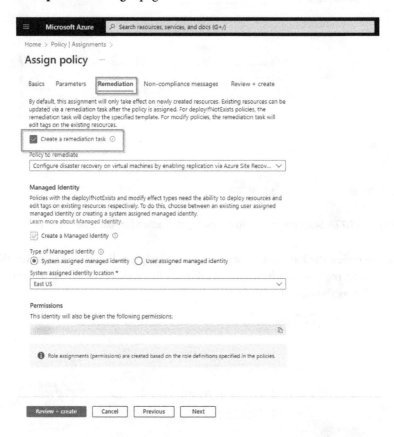

Figure 11.14 – Establishing a remediation task for the Azure policy

4. We can review all the configured settings for the new policy and click **Create** to begin the policy creation and the initial policy assignment to resources within the scope, as shown in *Figure 11.15*:

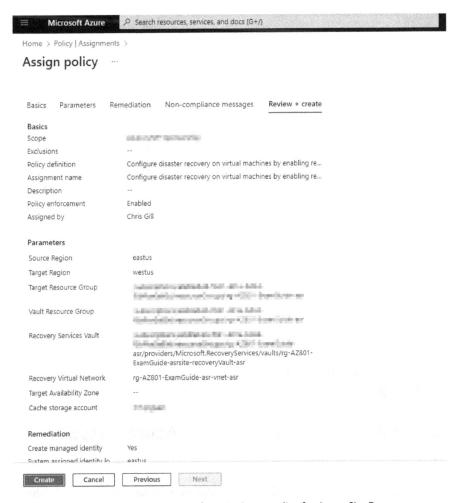

Figure 11.15 – Review + create the new Azure policy for Azure Site Recovery

5. Finally, we can navigate to the **Policy | Compliance** section of Azure Policy and review the effectiveness of the newly created and applied policy, drawing attention to the **Resource Compliance** section, which indicates that the two intended virtual machines are indeed **Compliant**, as shown in *Figure 11.16*:

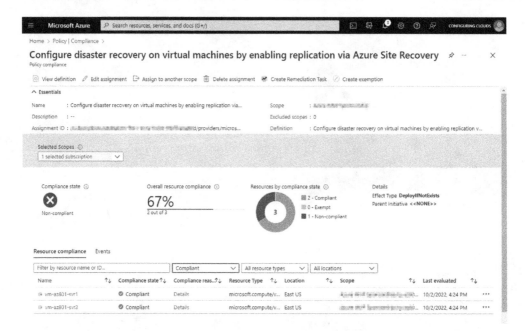

Figure 11.16 – Reviewing the Azure Policy Compliance report for Azure Site Recovery

With that, we have completed this section on configuring ASR policies using Azure Policy. Next, we will learn about configuring ASR for Azure Virtual Machines.

Configuring Azure Site Recovery for Azure Virtual Machines

Throughout this chapter, we have been building up to this point, where we have established all the requirements for configuring ASR for virtual machine protection within Azure. Given a scenario where we have two existing Azure virtual machines already configured and ready for replication, let's walk through the process of ASR with Azure VMs, including monitoring and failover testing.

Let's begin with the process of ASR for Azure Virtual Machines:

1. Within the **Recovery Services Vault** and **Site Recovery Infrastructure** sections for Azure VMs, we will begin by establishing the appropriate network mapping for our source and target locations, as shown in *Figure 11.17*:

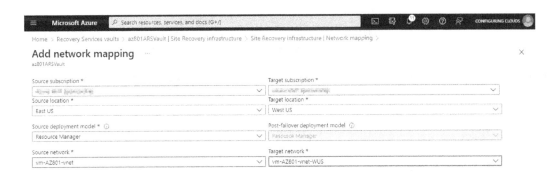

Figure 11.17 – Adding the network mapping

2. We will be presented with a selection of virtual machines that are in scope, as well as the replication settings, including resource groups, network configuration, storage, availability locations, encryption settings, and capacity reservations (if/when required). We can select **Enable replication** to continue our configuration, as shown in *Figure 11.18*:

Figure 11.18 – Replication settings for Site Recovery for Azure VMs

3. After selecting **Enable replication**, ASR will go through a series of tasks and tests to ensure that protection for the virtual machines can be enabled and that resources can be automatically created in the target locations and resource groups, as shown in *Figure 11.19*:

Figure 11.19 – Enabling protection for the replicated items for Azure Site Recovery VMs

4. A deeper look into progress can be achieved by reviewing the **Site Recovery jobs** area, which provides detailed views on each of the tasks, including their timing and status, as shown in *Figure 11.20*:

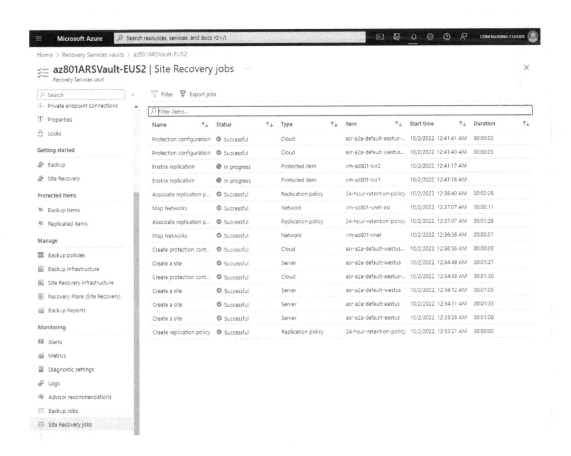

Figure 11.20 – Reviewing the status of the Site Recovery jobs

5. Here, we can see that the virtual machine is currently protected but has a warning, stating that the virtual machine has not yet performed a test failover to validate that all the configuration settings are complete and intact, as shown in *Figure 11.21*. Click the **Test Failover** button to begin the validation steps:

Figure 11.21 – Virtual machine currently protected and replicated, but not tested for failover

6.   Here, we will determine the failover direction and the Azure Virtual Network for our failover test, as shown in *Figure 11.22*:

Figure 11.22 – Setting a failover direction, recovery point, and Azure virtual network for the failover test

7.   Next, we will monitor the failover test's progress to ensure all the steps from our recovery plan have been completed, as shown in *Figure 11.23*:

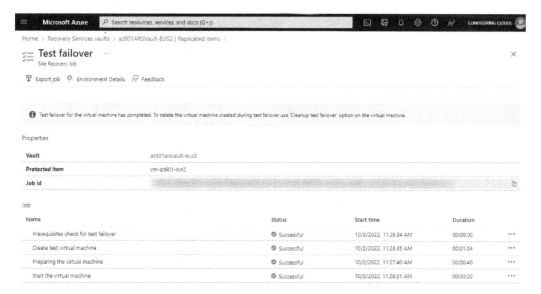

Figure 11.23 – Monitoring the test failover tasks and status updates for our recovery plan

8. Finally, we can confirm that the test failover for our single virtual machine recovery plan has succeeded, indicating a new status that is reminding us to clean up the resources, as well as indicating the recovery point objective in measurable time, as shown in *Figure 11.24*:

Figure 11.24 – Confirmed test failover and failover cleanup now recommended

These same steps can be completed for individual virtual machines, individual applications, or entire recovery plans for both failover testing, failback testing, and real-time failover events. Also, note that once VMs are failed over from one region to another, they must be re-protected to ensure that they can be failed back.

With that, we have completed this section on configuring and managing ASR for Azure Virtual Machines. Next, we will learn how to implement VM replication to a secondary data center or Azure region.

Implementing VM replication to a secondary data center or Azure region

While we previously demonstrated how to implement VM replication from one Azure region to another, there are additional advanced options that allow for zone-to-zone disaster recovery, but this does not protect against regional outages. There are some good reasons for choosing this option, and the following article from Microsoft does a great job explaining certain circumstances where you might want to use this feature, or avoid it for particular reasons: `https://learn.microsoft.com/azure/site-recovery/azure-to-azure-how-to-enable-zone-to-zone-disaster-recovery`.

With that, we have completed this section on implementing VM replication to a secondary data center or Azure region while covering some strategies and high-level best practices consistent with the AZ-801 exam objectives.

Summary

In this chapter, we learned how ASR can be used as a business continuity and disaster recovery option to replicate on-premises and Azure virtual machines to a secondary data center location, whether it's on-premises or in the cloud. We learned how to configure ASR, how to create and implement a recovery plan, how to failover and complete failover testing and monitoring, and how to configure recovery policies consistent with the AZ-801 exam objectives.

In the next chapter, we will learn how to configure Hyper-V hosts for replication, including for managing the replica servers. We will then discuss how to configure VM replication between replica hosts and ultimately perform a failover to learn about the failover orchestration process.

12

Protecting Virtual Machines by Using Hyper-V Replicas

In this chapter, we will learn how to configure Hyper-V hosts for replication, including the management of the replica servers. We will then discuss how to configure VM replication between replica hosts, and finally, perform a failover to learn about the failover orchestration process.

This chapter will include walk-throughs on how to implement and manage **Hyper-V replica** components and workloads on-premises, with hybrid components and services, and in the cloud, consistent with the AZ-801 exam objectives.

In this chapter, we will cover the following topics:

- Technical requirements and lab setup
- Introduction to Hyper-V Replica
- Configuring Hyper-V hosts for replication
- Configuring VM replication
- Managing Hyper-V replica servers
- Performing a VM failover

Technical requirements and lab setup

To successfully follow along and complete the tasks and exercises throughout this chapter and the following chapters in this book, we will need to ensure that the technical requirements from *Chapter 1, Exam Overview and the Current State of Cloud Workflows*, have been completed in full. Many aspects of Hyper-V Replica require a rigorous amount of detail and require infrastructure planning and configuration time for both initial protection as well as ongoing replication and protection. We will focus on the requirements, best practices, and high-level concepts in this chapter, and will be primarily reviewing areas of the Azure portal environment to follow the walk-throughs throughout this chapter, aligning with the AZ-801 exam objectives.

Let's begin with a review of what Hyper-V Replica provides to assist with business continuity and disaster recovery services.

Introduction to Hyper-V Replica

For some time now, **Hyper-V** has provided built-in tools to complete VM replication from one Hyper-V host to another host. Under the hood, Hyper-V Replica creates a copy of a live/running VM to a replica offline VM on another host server to have yet another way to keep your workloads available in the event of a disaster.

While Hyper-V Replica tends to be more frequently configured on Hyper-V servers that are joined to an Active Directory domain and inside of a Hyper-V cluster, it is entirely possible to configure Hyper-V replication for servers that are running in a workgroup. The Hyper-V hosts have no restrictions between the servers – they can be part of a cluster, standalone, or a mix of both. Additionally, the primary replication servers and the secondary replication servers can be physically located in the same data center or in separate geo-locations with replication available over a **wide area network** (**WAN**) link. Additionally, the replication can be completed over **HTTP** (**Kerberos**) or **HTTPS** (**certificate-based authentication**).

Once Hyper-V Replica is enabled for a VM, an initial replication takes place, and any changes after that are identified by change tracking and written to the **UndoLog configuration**. This logging is then replayed in reverse order to the replica virtual hard disk, applying the latest changes asynchronously back to the replicated VM.

In addition, we have **extended replication** (or chained replication) available for use, allowing the secondary host to replicate the VM to a third chained host. Note that you cannot directly replicate from the primary host to a tertiary or downstream host. This is considered configuring a primary and extended replica and is useful when and if the primary and secondary locations are unavailable for failover.

For failover purposes, if a primary or secondary location were to encounter a failure, all failover operations must be completed manually, whether it is a test, planned, or unplanned failover operation. Note that with test and planned failovers, the operations and duration of the planned outage are relatively predictable. With any unplanned failovers, there could be the potential for unexpected data loss or higher durations of unplanned outages.

Finally, as with all disaster recovery products, we have the ability to configure recovery points, allowing for configurable snapshot timing and the retention of snapshots. Note that the maximum time for accessing a recovery point is up to 24 hours in the past.

Now that we have covered an introduction to Hyper-V Replica and what features this technology brings to business continuity and disaster recovery, let's step into how Hyper-V Replica can be established for two standalone hosts that are not part of a domain.

Configuring Hyper-V hosts for replication

As with any of the technologies we have discussed in this book, there are best practices and deployment prerequisites that you must consider to have a successful deployment. Hyper-V Replica is no exception, so let's review some of the talking points before deployment:

- **What workloads will you replicate?** Hyper-V Replica standard replication only maintains the consistency of the VM state during a failover and does not contain the application state. App-consistent recovery points can be created, but this is not available on extended recovery replica sites.

- **How will you complete the initial replication of the VM data?** There are many options available, from completing the initial replication over the existing network immediately or at a later date, to completing the transmission using an already existing restored VM, to even using offline media to deliver the preexisting VM and all snapshots to the replica site.

- **Which virtual hard disks will need to be replicated?** Determine which disks are of utmost importance to be included and identify disks that have rapid change rates (e.g., page file disks) and ensure that they are excluded from replication.

- **How often will you synchronize the data to the replica server(s)?** The available options are currently 30 seconds, 5 minutes, or 15 minutes for replication frequency. For machines that are considered critical, a higher frequency of replication will be warranted.

- **How will you ultimately recover data?** The default configuration of Hyper-V Replica only stores a single recovery point (the latest recovery point), sent from the primary replica to the secondary replica. Additional recovery points can be configured (to a maximum of 24 hourly recovery points), but this will require additional bandwidth and storage overhead in your architectural design.

- **Are the Hyper-V hosts domain-joined or standalone?** If the Hyper-V hosts are domain-joined, **Kerberos authentication** can be used. Alternatively, if the hosts are standalone or a mix of servers, it's recommended to utilize **certificate-based authentication** between Hyper-V Replica hosts. Ensure that the correct Windows firewall rules are enabled, as these two rules are disabled by default: **Hyper-V Replica HTTP Listener (TCP-In)** for **HTTP** usage and **Hyper-V Replica HTTPS Listener (TCP-In)** for **HTTPS** usage.

Now that we have completed a review of the deployment prerequisites and talking points, let's walk through setting up two Hyper-V hosts with one preexisting VM, establishing one primary and one secondary replica server, while using the certificate-based authentication (HTTPS) approach for replication communications between replica hosts:

1.   Starting with the primary Hyper-V replica host, the `New-SelfSignedCertificate` PowerShell cmdlet can be utilized to create a certificate for the primary replica, the secondary replica, and the Root **certificate authority (CA)** certificate. The secondary replica certificate and Root CA certificate must be exported, copied, and imported into the secondary Hyper-V Replica host.

> **Certificates, CA, and certificate-based authentication**
>
> Note that the certificates used for Hyper-V Replica certificate-based authentication must have intended uses of both **Client Authentication** and **Server Authentication**, as provided by the CA (whether local or Certificate Services-based). The root certificate must be exported along with the host certificate and placed in the Root CA and Personal/My certificate stores on the Hyper-V hosts, respectively.

2.   On the secondary Hyper-V host, selecting **Hyper-V Settings** and then selecting **Replication Configuration** will result in the screen shown in *Figure 12.1*, indicating the use of **certificate-based authentication** and requiring the selection of an appropriate certificate for use:

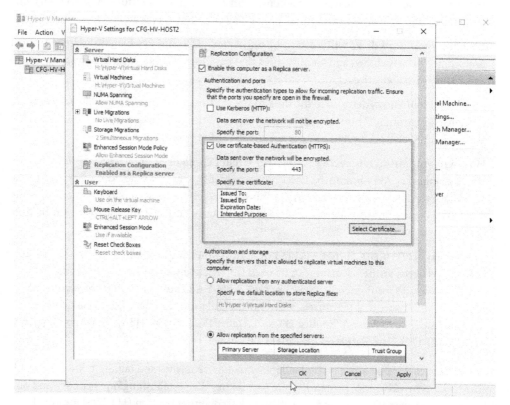

Figure 12.1 – Enabling Replication Configuration for the primary Hyper-V replica

3. Here, we will select the already imported certificate on the secondary replica host to complete the configuration, as shown in *Figure 12.2*:

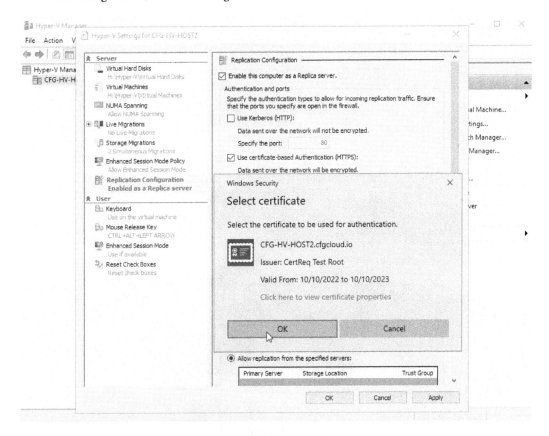

Figure 12.2 – Selecting the appropriate certificate for use on the secondary Hyper-V replica host

4. The very next step will be to allow replication from specified servers by specifying the primary replica server first, followed by naming the replication trust group (otherwise known as **authorization entry**), as shown in *Figure 12.3*:

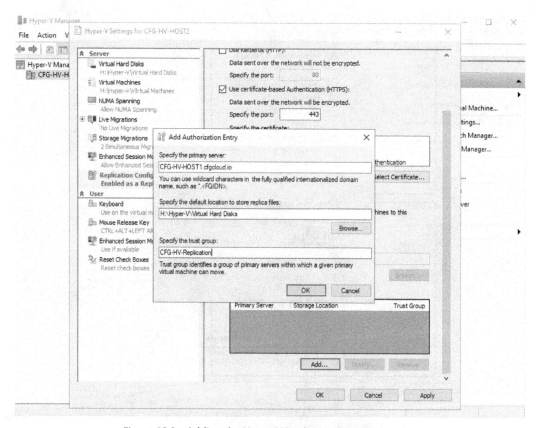

Figure 12.3 – Adding the Hyper-V Replica authorization entry

5. After selecting **OK** in *Step 4*, we are reminded to complete the Windows firewall rule configuration, as shown in *Figure 12.4*, which can be completed by using the following PowerShell command for reference:

```
Enable-Netfirewallrule -displayname "Hyper-V Replica
HTTPS Listener (TCP-In)"
```

The configuration of a new firewall rule for inbound traffic is then successfully created:

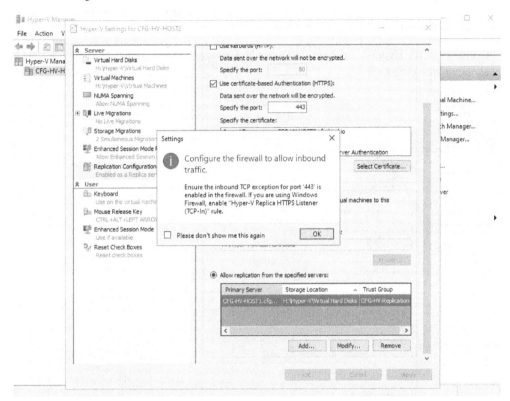

Figure 12.4 – Reminder to enable the Windows firewall Hyper-V Replica HTTPS Listener rule

This completes all steps for the walk-through for establishing a secondary Hyper-V replica for a Hyper-V host using certificate-based authentication.

We have completed this section on configuring Hyper-V hosts for replication and setting up the secondary replica server while using the certificate-based authentication (HTTPS) approach for replication communications between replica hosts. Next, we will learn about configuring the primary Hyper-V replica and configuring VM replication between Hyper-V Replica hosts.

Configuring VM replication

No matter the VM workload, Hyper-V Replica can be utilized for enabling and managing replication between Hyper-V hosts, whether in the same location or physically located in another geographical location. These VM replication features can be enabled using **Hyper-V Manager** or **System Center Virtual Machine Manager** (**VMM**).

Let's walk through setting up the primary Hyper-V Replica host with one preexisting VM, establishing the primary and one secondary replica server hierarchy, while using the certificate-based authentication (HTTPS) approach for replication communications between replica hosts:

1. On the primary Hyper-V host within Hyper-V Manager, we select the VM to be replicated, as shown in *Figure 12.5*:

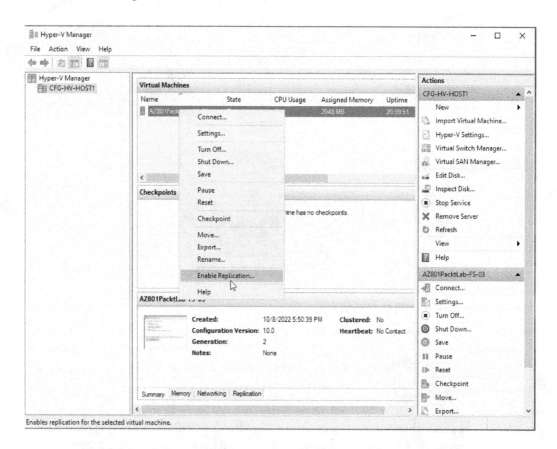

Figure 12.5 – Enabling replication for the intended VM from the primary Hyper-V replica

2. In the next step, we supply the Replica server name for the secondary Hyper-V replica, as shown in *Figure 12.6*:

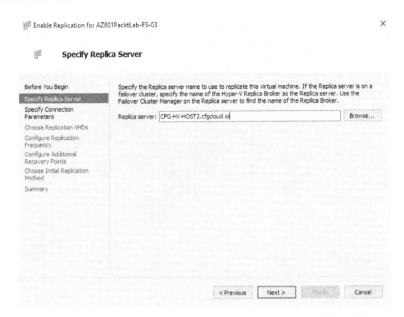

Figure 12.6 – Supplying the secondary Replica server to be used for replication

3. On the **Specify Connection Parameters screen**, we select the certificate issued for the primary Hyper-V Replica host, as shown in *Figure 12.7*:

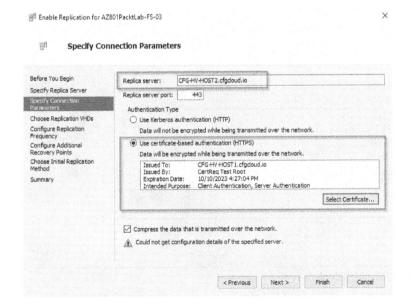

Figure 12.7 – Specifying the connection parameters and certificate for authentication

4. Under **Choose Replication VHDs**, we will select the replication **virtual hard disks (VHDs)** for each of the replicated VMs, as shown in *Figure 12.8*:

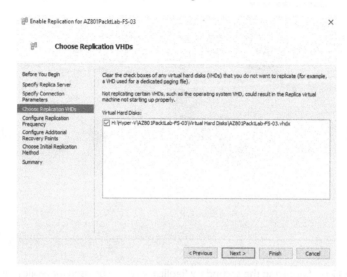

Figure 12.8 – Choosing the replication VHDs for replication configuration

5. For the **Configure Replication Frequency** option, we will select **5 minutes** for replication frequency to the Replica server, as shown in *Figure 12.9*:

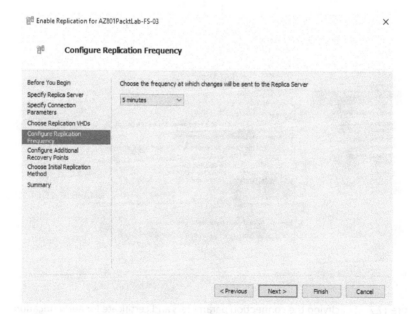

Figure 12.9 – Configuring the replication frequency for the Replica server

6. For **Configure Additional Recovery Points**, we will create additional hourly recovery points for **24** hours of total coverage, with a volume shadow copy snapshot frequency of **4** hours, as shown in *Figure 12.10*, to meet our recovery time and point objectives:

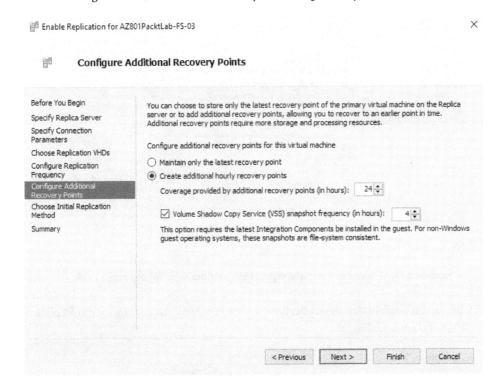

Figure 12.10 – Configuration of additional recovery points for our RTO/RPO

7. For **Choose Initial Replication Method**, we will select **Send initial copy over the network** and **Start replication immediately**. Note that we could also consider sending the copy via external media and can also schedule the initial replication for a date and time in the future, as shown in *Figure 12.11*:

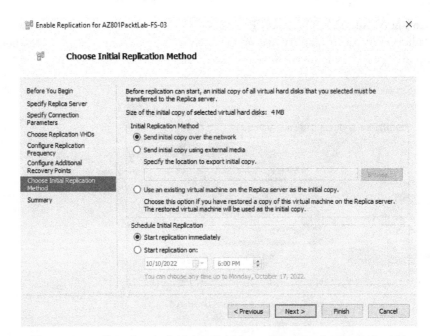

Figure 12.11 – Choosing the initial replication configuration for our replica VM

8. Finally, we can confirm that the replication has been enabled successfully for our Replica host, as shown in *Figure 12.12*:

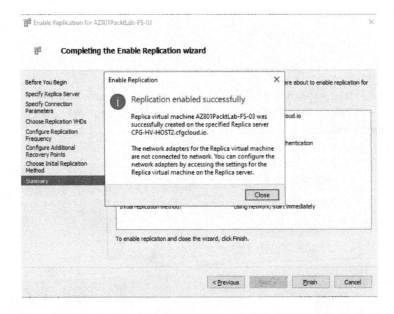

Figure 12.12 – Confirmation that the replication of our VM has succeeded

This completes all steps of the walk-through for establishing Hyper-V primary host and replica VM replication for a Hyper-V host using certificate-based authentication.

We have completed this section on configuring Hyper-V hosts for replication and setting up the primary and secondary Replica servers while using the certificate-based authentication (HTTPS) approach for replication communications between replica hosts. Next, we will learn about managing Hyper-V Replica servers.

Managing Hyper-V replica servers

While we have primarily reviewed a two-Hyper-V-host replica setup, we can also establish replica clusters for Hyper-V failover clusters. With the **Hyper-V Replica Broker Role** configured within the cluster, the cluster can be configured as either a primary cluster or a replica cluster. A replica cluster can use either the Kerberos (HTTP) or certificate-based (HTTPS) authentication type, and setting up allowed replication from a wildcard server configuration (e.g., `*.domain.com`) with the use of the cluster shared storage as a storage location gives high availability to this Hyper-V replica role.

Hyper-V Replica settings are modified using the **Hyper-V Manager** interface for standalone Hyper-V servers. Conversely, Hyper-V failover clusters are configured using the **Failover Cluster Manager** interface. Hyper-V replica health can then be seen per VM by using the **Replication** > **View Replication Health…** option for the replicated VM, as shown in *Figure 12.13*:

Figure 12.13 – Viewing the replication health per VM

Firewall rules are not automatically enabled for each Replica server node in the cluster, and certificate-based authentication configuration for a cluster must be completed by an administrator for each node in the cluster. When utilizing certificate-based authentication for a failover cluster, appropriate machine certificates must be configured for each of the cluster nodes and any **Client Access Points (CAPs)**.

Performing a VM failover

Now that we have completed a review of managing **Hyper-V Replica** servers, let's cover the final topic of this chapter by learning about failover options for our replicated VM:

1. First, we ensure that the VM has been shut down/stopped on the primary Hyper-V replica host before proceeding.

2. From the secondary replica host, we can select the replicated VM, and by selecting the **Replication** menu, we can see all available options for failover, as shown in *Figure 12.14*. Here, we will select the **Test Failover...** option to continue:

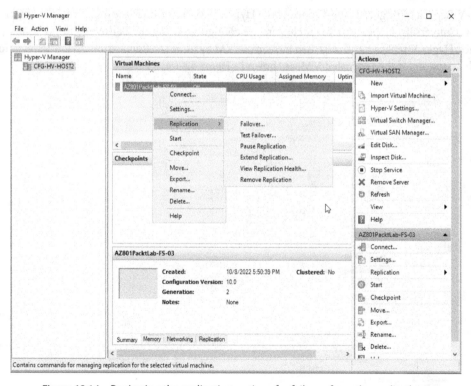

Figure 12.14 – Reviewing the replication options for failover from the replica host

3. On the **Test Failover** screen, we have the option of choosing any of the recovery points, including the latest recovery point, any planned failover recovery points, and any standard recovery points (or snapshots), as shown in *Figure 12.15*. We will select the **Planned Failover Recovery Point** option for this exercise:

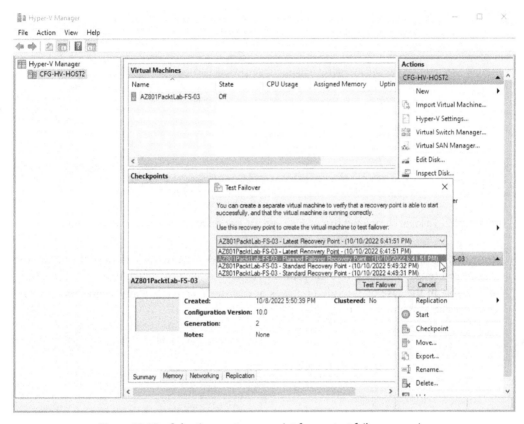

Figure 12.15 – Selecting a recovery point for our test failover exercise

4. To complete the failover validation, we then select **Fail Over** and can confirm that the failover process completed successfully and the VM has been started on the secondary Hyper-V replica, as shown in *Figure 12.16*:

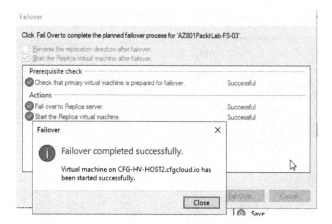

Figure 12.16 – Failover completed successfully and VM started on the secondary Hyper-V replica

5. Now that the VM is running on the secondary Hyper-V replica host, we have the option to complete a reverse replication to the primary Hyper-V replica, as shown in *Figure 12.17*. This is valuable in the case of extended outages or disaster scenarios, and was not selected for this exercise:

Figure 12.17 – Drawing attention to the optional Reverse Replication option

6.   At this point in the exercise, we can stop the test failover to clean up the secondary copy of the VM, as shown in *Figure 12.18*:

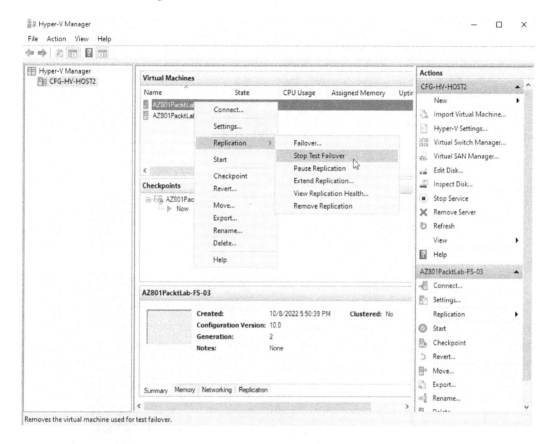

Figure 12.18 – Stopping the test failover to clean up resources on the secondary Hyper-V replica

This completes all steps for the walk-through for establishing Hyper-V primary host and replica VM replication for a Hyper-V host using certificate-based authentication.

Note that when selecting a planned failover from the primary Hyper-V host, you must manually choose the failover task on the Replica server VM to complete the failover process, as shown in *Figure 12.19*:

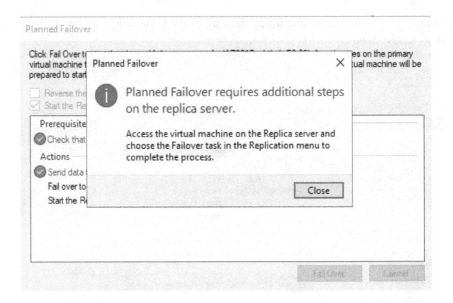

Figure 12.19 – Planned failover requires additional failover task steps on the Replica server

We have completed this section on performing a VM failover within the Hyper-V Replica configuration, covering some strategies, walk-throughs, and high-level best practices consistent with the AZ-801 exam objectives. Next, we will complete a review of all we have learned in this chapter and set the stage for the next chapter of this book.

Summary

In this chapter, we learned how to configure Hyper-V hosts for replication, including the configuration and management of the Replica servers. We then discussed how to configure VM replication between replica hosts, and finally, performed a failover to learn about the failover orchestration process consistent with the AZ-801 exam objectives.

In the next chapter, we will learn how to effectively migrate existing workloads from on-premises infrastructure to Microsoft Azure using existing and new server management and migration tools.

Part 5: Migrate Servers and Workloads

In this section, we will learn how to effectively migrate existing workloads from on-premises infrastructure to Microsoft Azure using existing and new server management and migration tools.

This part of the book comprises the following chapters:

- *Chapter 13, Migrating On-Premises Storage to On-Premises Servers or Azure*

- *Chapter 14, Migrating On-Premises Servers to Azure*

- *Chapter 15, Migrating Workloads from Previous Versions to Windows Server 2022*

- *Chapter 16, Migrating IIS Workloads to Azure*

- *Chapter 17, Migrating an Active Directory Domain Services (AD DS) Infrastructure to Windows Server 2022 AD DS*

13

Migrating On-Premises Storage to On-Premises Servers or Azure

In this chapter, we will identify how to successfully transfer data and share configurations from on-premises Windows servers to other available Windows servers running on-premises or in Microsoft Azure. We will learn how to use **Windows Admin Center** and **Storage Migration Service** to migrate services from one server to another.

We will then learn how to use Storage Migration Service to successfully integrate with and migrate to **Microsoft Azure** and **Azure Virtual Machines**. Finally, we will learn what **Azure File shares** bring to the table, and how we can migrate existing shares to Azure File shares quickly and securely.

In this chapter, we will cover the following topics:

- Technical requirements and lab setup
- Introduction to Storage Migration Service
- Transferring data and sharing configuration
- Cutting over to a new server by using Storage Migration Service
- Using Storage Migration Service to migrate to Azure Virtual Machines
- Migrating to Azure file shares

Technical requirements and lab setup

To successfully follow along and complete the tasks and exercises in this chapter and the following chapters in this book, you will need to ensure that the technical requirements from *Chapter 1, Exam Overview and the Current State of Cloud Workflows*, have been met in full. We will focus on requirements, best practices, and high-level concepts in this chapter and will be primarily reviewing areas of Windows Admin Center and the Azure portal environment in the walk-throughs in this chapter, aligning with the AZ-801 exam objectives.

Let's begin with an introduction to Storage Migration Service within Windows Admin Center and how it can be used to migrate files and shares from existing Windows servers to new Windows servers.

Introduction to Storage Migration Service

Storage Migration Service (**SMS**) allows for the migration of unstructured data such as files and shares from any Windows Server version (even including Windows Server 2003) to other physical machines, virtual machines, Azure Files, or Azure **Infrastructure-as-a-Service** (**IaaS**). The service is part of Windows Admin Center and can be used to automate the creation of a VM within Microsoft Azure when selecting an Azure VM as the destination for migration of storage. Note, however, that multiple file shares cannot be consolidated into a single-destination file share when using Storage Migration Service.

One of the benefits of **SMS** is that it can automatically assign the source server identity to the target server, allowing clients to automatically begin using the migrated data on the destination servers. Orchestration behind the scenes (via PowerShell) automatically configures both the server name and the server IP address(es) to lessen the administrative burden, and any local user accounts for applications or administrative purposes can also be migrated via SMS.

Let's quickly review the acceptable source and destination object types and scenarios for Storage Migration Service.

Here is what Storage Migration Service can consume and migrate as a source:

- SMB shares running on Windows Server 2003, 2008, 2008R2, 2012, 2012R2, 2016, 2019, and 2022 (more information can be found at the following URL: `https://learn.microsoft.com/windows-server/storage/storage-migration-service/overview#requirements-for-source-servers`)

- Linux servers configured with Samba, including versions 3.6, 4.2, 4.3, 4.7, and 4.8 across multiple Linux distributions

- NetApp **Fabric Attached Storage** (**FAS**) arrays

Here is what Storage Migration Service can migrate data and configurations to as a destination (with additional details found at `https://learn.microsoft.com/windows-server/storage/storage-migration-service/overview#requirements-for-destination-servers`):

- SMB via **Azure File Sync** and **Azure Files**

- SMB via Windows Server physical machines, virtual machines, and **Azure Stack HCI** IaaS

- SMB via Microsoft Azure IaaS virtual machines

The prerequisites for Storage Migration Service are as follows:

- An identified **source server**

- A **destination server** configured with 2 CPU cores and at least 2 GB of memory, noting that both Windows Server 2019 and Windows Server 2022 will have roughly double the transfer performance due to the SMS proxy server built into the operating system

- An **orchestrator server** that is either Windows Server 2019 or Windows Server 2022 configured with 2 CPU cores and at least 2 GB of memory and is not the source server, noting that the orchestrator components can be installed on the destination server prior to migration

- The source and destination servers must reside in the same domain and forest

- Either a Windows workstation or Windows Server running the **Windows Admin Center (WAC)** version 2103 or higher

- Installation of the SMS **Proxy**, noting that this will automatically open and configure the following necessary four firewall rules and ports on the device:

  - File and printer sharing (SMB-In)

  - Netlogon service (NP-In)

  - Windows Management Instrumentation (DCOM-In)

  - Windows Management Instrumentation (WMI-In)

- An **administrator account** that is the same on the source, destination, and orchestrator servers *or* a source migration account and a destination migration account, where the source migration account is an administrator on the source and orchestrator servers and the destination migration account is an administrator on the destination and orchestrator servers

Post-migration details and caveats

Please note that part of the orchestration during the cutover process for Storage Migration Service includes changes to the source server, such that the server remains functional but is no longer accessible to users and services under the original server name and IP address(es). This will require eventual cleanup and decommissioning of the source server once all validation has been completed for the new destination/target server. As this could include a potential name change for the server, please note that any certificates used for the device will most likely need to be reissued or re-enrolled for proper authentication, communication, and service encryption (where used).

The high-level process when using SMS involves the following three phases:

- Phase 1: Complete an **inventory devices** process on the source servers that includes creating a scan devices job

- Phase 2: Complete a **transfer data** process between the source and destination devices

- Phase 3: **Cut over to the new servers** by migrating identities and IP address(es)

For additional details and insights into Storage Migration Service, be sure to check out the following URL: `https://learn.microsoft.com/windows-server/storage/storage-migration-service/faq`.

Now that we have covered an introduction to **Storage Migration Service** and what features this technology brings to the migration of files and shares for existing Windows servers, let's step into how Storage Migration Service can be used to migrate files and shares from one on-premises server to another.

Transferring data and sharing configuration

Now that we have completed a review of the deployment pre-requisites and introduction to the service, let's walk through setting up a storage migration Phase 1 and Phase 2 example with one preexisting virtual machine in our lab environment, a new destination/orchestrator server, and an on-premises server to server migration using Storage Migration Service:

1.   Starting with Windows Admin Center, we will locate the destination server, locate **Roles & features**, select **Storage Migration Service Proxy**, then select **+ Install** as shown in *Figure 13.1*:

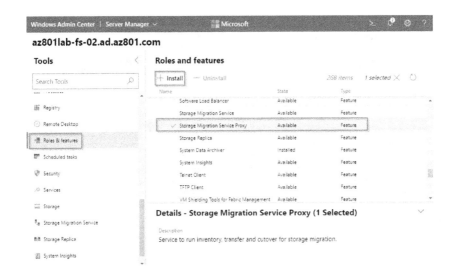

Figure 13.1 – Installing the SMS proxy role on our destination server

2. We then choose the orchestrator server, select **Storage Migration Service** from the left navigation blade, then select **Install** to install the Storage Migration Service feature on this server, as shown in *Figure 13.2*:

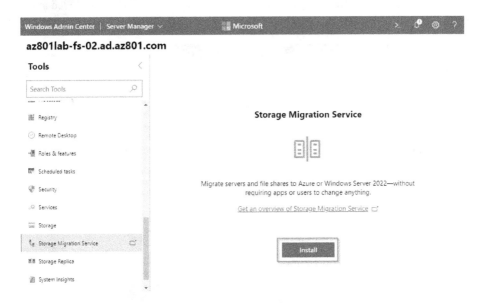

Figure 13.2 – Install Storage Migration Service on our orchestrator server

3. Notice that we are prompted with an overview of how we can migrate the data from an existing server to a new destination, as shown in *Figure 13.3*:

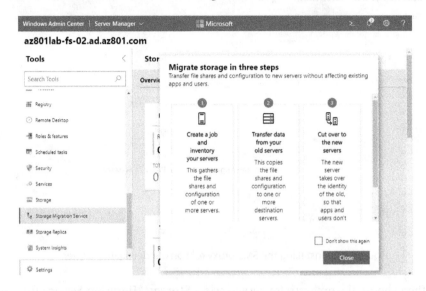

Figure 13.3 – Overview of ways to migrate data using Storage Migration Service

4. We can now scroll down the page after Storage Migration Service has been installed to create a new migration job by selecting + **New job** and then supplying a job name, as shown in *Figure 13.4*:

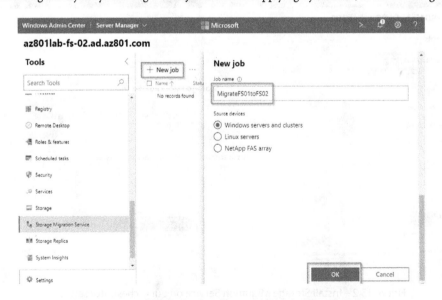

Figure 13.4 – Creating a new Storage Migration Service job

5. We then receive a prerequisite check page that can be reviewed, as shown in *Figure 13.5*, and then select **Next** to continue:

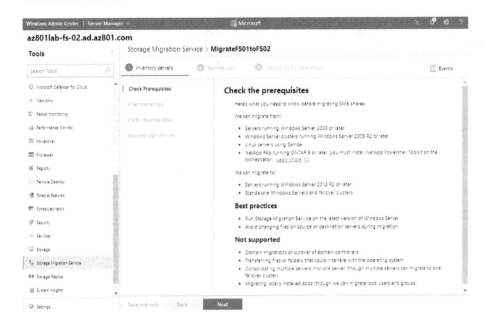

Figure 13.5 – Prerequisite overview for the inventory servers step in SMS

6. We are then prompted to provide appropriate administrative credentials for the source device, as shown in *Figure 13.6*:

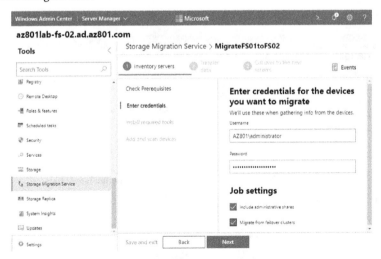

Figure 13.6 – Prompt for credentials for the source server

7. At this point, we use **+ Add a device** and once the source server is added, we utilize the **Start scan** menu item to complete a migration assessment, as shown in *Figure 13.7*:

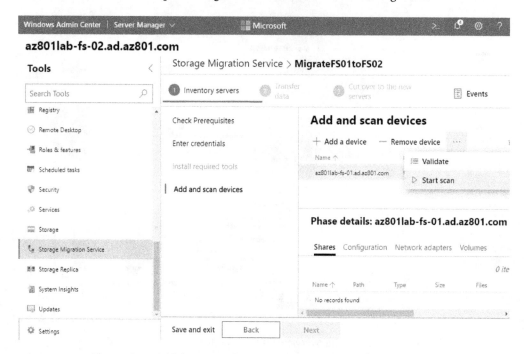

Figure 13.7 – Add the source device and start the scan before migration

8. Once the scan has been completed, we can see that there are four shares that can be migrated and two that we are targeting, as shown in *Figure 13.8*:

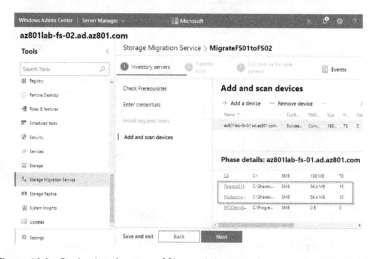

Figure 13.8 – Reviewing the scan of files and shares in Storage Migration Service

9. After selecting **Next**, we begin phase 2 by entering the administrative credentials for the destination server, as shown in *Figure 13.9*:

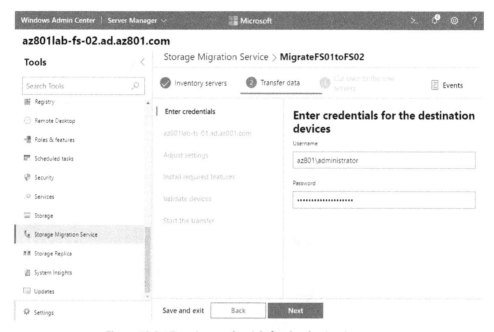

Figure 13.9 – Entering credentials for the destination server

10. Here, we will identify the source server by searching for and selecting the server name, as shown in *Figure 13.10*:

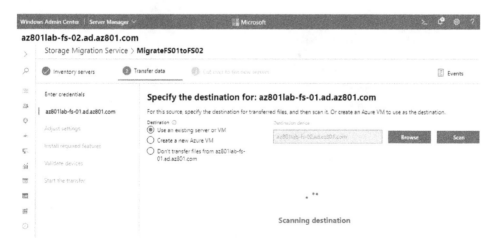

Figure 13.10 – Specifying the destination server for migration

11. Here, we will select the shares intended for transfer to the destination server, as shown in *Figure 13.11*:

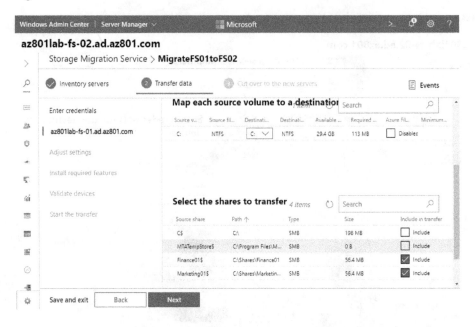

Figure 13.11 – Selecting the shares and files intended for migration

12. Reviewing the **Adjust transfer settings** page, we can select transfer settings and retry intervals to ensure that any migration efforts complete successfully, as shown in *Figure 13.12*:

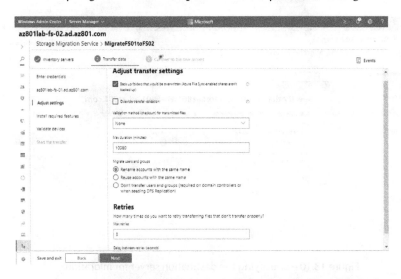

Figure 13.12 – Reviewing the settings for the Storage Migration Service job

13. Next, the system will install any required features automatically, as shown in *Figure 13.13*:

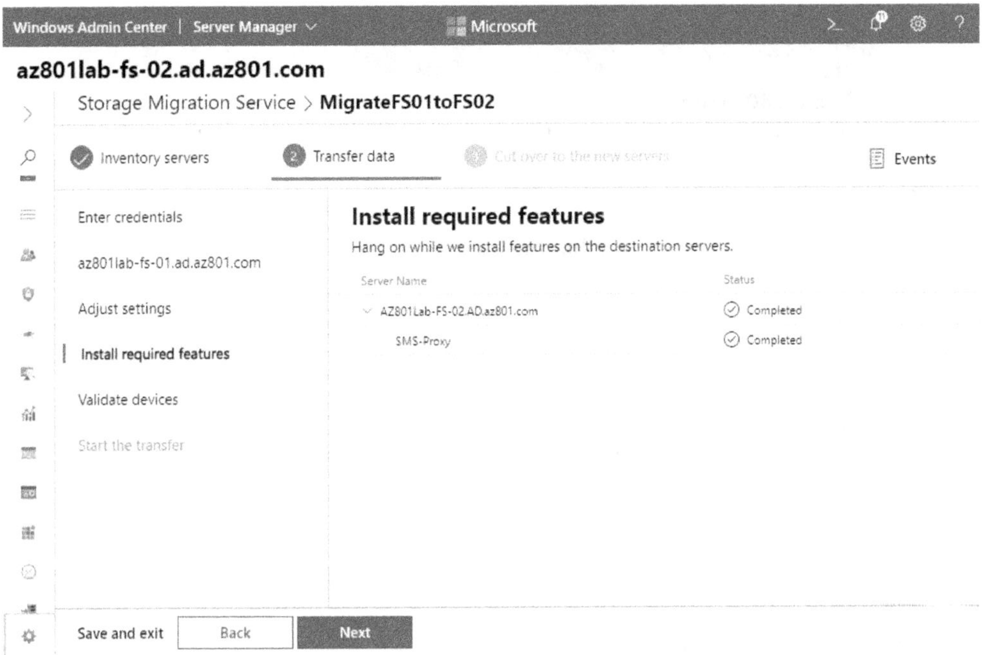

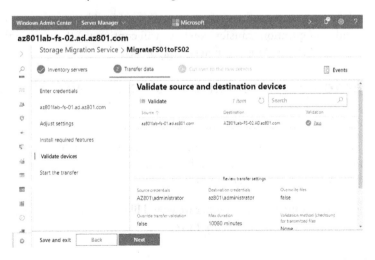

Figure 13.13 – Prerequisite installation checker automatically installs required features

14. A validation pass will now be completed against both servers to determine whether the migration can proceed successfully, as shown in *Figure 13.14*:

Figure 13.14 – Validation of both source and destination servers completed

15. Here, the transfer has been started and successfully completed the transfer of files, shares, and share configuration, as shown in *Figure 13.15*:

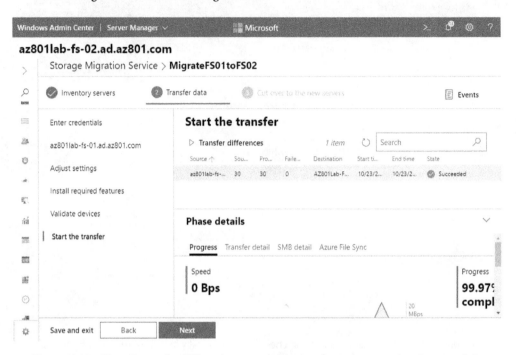

Figure 13.15 – Phase 2 transfer of files, shares, and share configuration completed successfully

This completes phase 2 of the Storage Migration Service migration approach.

We established a job between a source and a destination server and identified all necessary requirements for the initial data and configuration transfer. Next, we will continue with an example of cutover migration as phase 3 of the **Storage Migration Service** approach.

Cutting over to a new server by using Storage Migration Service

Now that we have reviewed the approach for setting up Storage Migration Service for both phase 1 and phase 2 of the migration approach, let's walk through the steps and considerations for phase 3, which involves cutting over to a new server:

1. The first step involves validation of whether we would like to reuse the existing credentials or create new credentials, as shown in *Figure 13.16*:

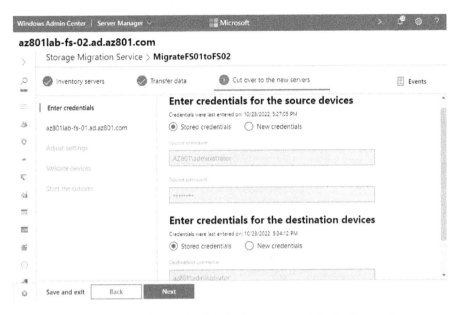

Figure 13.16 – Entering credentials for the source and destination devices

2. In this series of configuration steps, we are asked to make out the network settings from the source server to the destination server as shown in *Figure 13.17*:

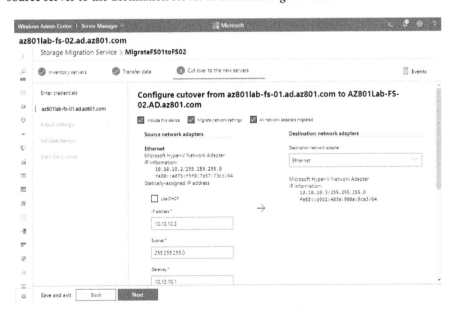

Figure 13.17 – Review and configure the source and destination network adapters

3. Note that as part of this process, we can also rename the source device after the cutover has been completed, as shown in *Figure 13.18*:

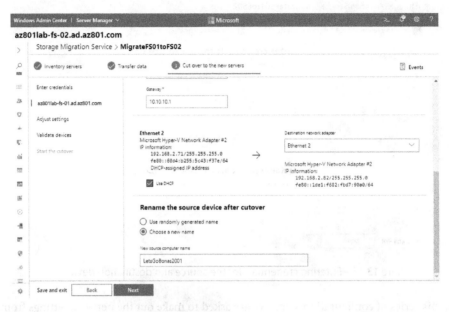

Figure 13.18 – Rename the source device after the cutover has been completed

4. We are prompted to adjust timeout settings and credentials for security and firewall rules, as shown in *Figure 13.19*:

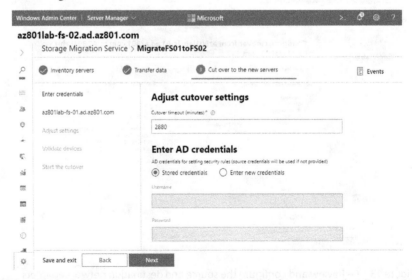

Figure 13.19 – Adjust cutover settings and AD credentials for security rule adjustments

5. We are then presented with one last validation pass to ensure that all configurations are consistent and can be applied to both the source and destination servers, as shown in *Figure 13.20*:

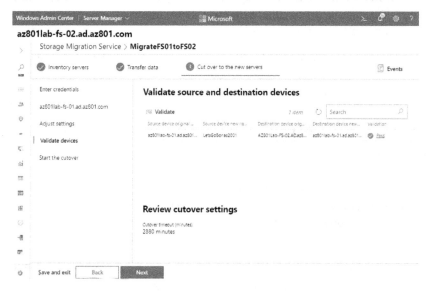

Figure 13.20 – Validate the source and destination device configuration before migration

6. Finally, we are prompted to begin the cutover to a new server with the option to stop the cutover if, during the process, monitoring has determined that there are any anomalies, as shown in *Figure 13.21*:

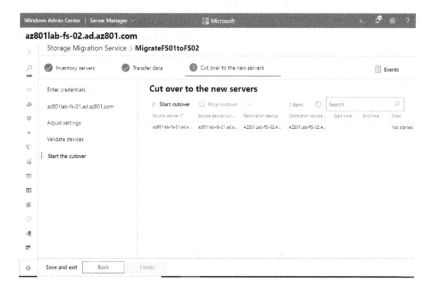

Figure 13.21 – Start the cutover to complete the migration, or stop the cutover if needed

This completes phase 2 of the **Storage Migration Service** migration approach.

For a deep dive into how cutover works in Storage Migration Service, be sure to check out the following URL: `https://learn.microsoft.com/windows-server/storage/storage-migration-service/cutover`.

We established a job between a source and a destination server and ultimately completed a cutover to the new destination server. Next, we will continue with an example of using Storage Migration Service to utilize an Azure VM as part of a migration approach.

Using Storage Migration Service to migrate to Azure Virtual Machines

As we have learned throughout this book, Windows Admin Center tends to make a lot of administrative tasks more straightforward and easily configurable for any environment. Many of the steps from the on-premises source to on-premises destination server migration are identical. Let's focus on some alternative steps to utilize an Azure VM instead of an on-premises server:

1. During this approach, instead of selecting **Use an existing server or VM** as part of phase 2, we select **Create a new Azure VM**, as shown in *Figure 13.22*:

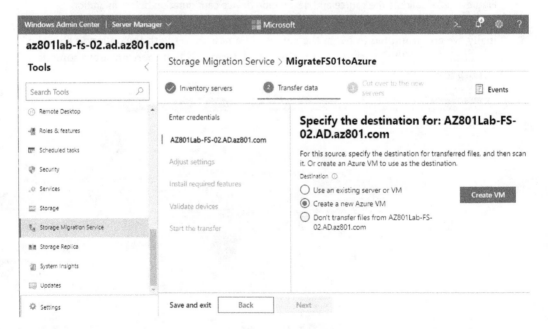

Figure 13.22 – Utilizing the Create a new Azure VM option to automatically create a VM

2. Here, we are presented with the details pane to create an Azure VM on the fly, as shown in *Figure 13.23*:

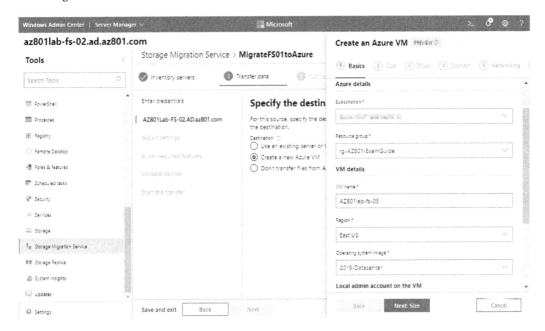

Figure 13.23 – Selecting subscription and Azure resource details for our new Azure VM creation

3. Next, we are prompted to select a virtual machine size, as shown in *Figure 13.24*:

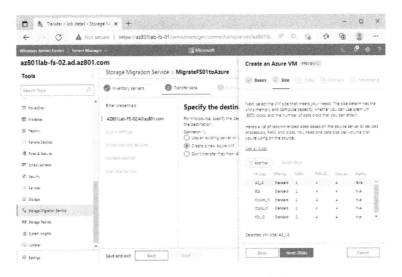

Figure 13.24 – Selecting the appropriate VM size for our new Azure VM

4. We are then prompted to select the operating system disk type and can select from multiple choices, as shown in *Figure 13.25*:

Figure 13.25 – Selecting the appropriate operating system disk type for the Azure VM

5. We are then presented with domain join information and credentials, as shown in *Figure 13.26*:

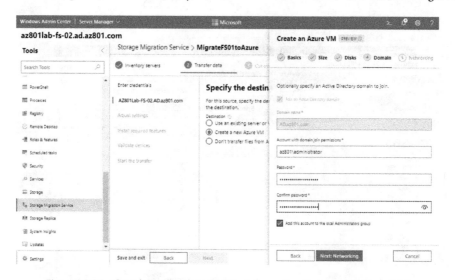

Figure 13.26 – Supplying the domain join information for our new Azure VM

6. Next, we are prompted for the Azure networking configuration that we intend to use, as shown in *Figure 13.27*:

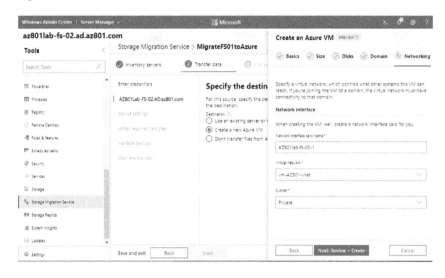

Figure 13.27 – Configure the Azure VM networking interface and virtual network

7. Next, we can begin the Azure VM configuration directly through Windows Admin Center, as shown in *Figure 13.28*:

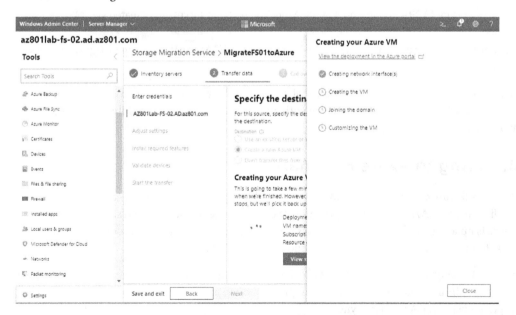

Figure 13.28 – Beginning the Azure VM creation process from Windows Admin Center

8. Finally, we can confirm that the Azure virtual machine has been successfully created and configured for use, as shown in *Figure 13.29*. Steps for the migration can then be completed, as we have learned earlier in this chapter:

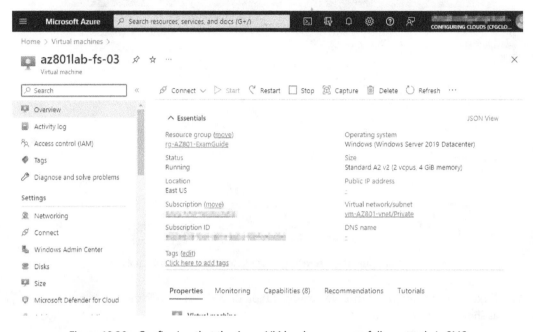

Figure 13.29 – Confirming that the Azure VM has been successfully created via SMS

This completes the section covering the use of Azure VMs for the Storage Migration Service migration approach.

This section covered the use of an Azure VM as the destination server for Server Migration Service source migration. Next, we will continue with an example of using **Azure File shares** as part of a migration approach.

Migrating to Azure File shares

Utilizing a feature called **Azure File Sync**, you can install a storage sync agent that helps to centralize your files inside of Azure, ultimately providing fast local access to your files and shares. Once the sync service is up and running, your on-premises server and Azure Files are continuously in sync, giving access to multiple sites and locations via cloud tiering and a fast local cache.

In the past, configuring and integrating Azure File Sync was quite a lengthy process and, to no surprise, Windows Admin Center helps to make that process much easier. Let's begin with a quick review of the prerequisites for Azure File Sync:

- **Windows Admin Center** 1904 or later (newest version recommended with necessary extension updates installed for the best experience)

- The source file server where **Azure File Sync** will be enabled should be running Windows Server 2012 R2 or newer

- An Azure subscription with existing **Storage Sync Service**, **Azure storage account**, and **Azure File Share** resources created

- A **Windows Admin Center** gateway connection registered to Azure

Once Azure File Sync has been installed and configured, the following tasks will be automatically handled as a welcome addition to many administrators:

- Installation of the Azure File Sync agent and any necessary PowerShell modules

- The update of the Azure File Sync agent to the latest version once available

- Registration of the file server to **Storage Sync Service** in Azure

- The addition of deep links so that advanced configuration of **sync groups** can be completed directly from the Azure portal

Let's complete the following walk-through using Windows Admin Center to establish Azure File Sync for a pre-existing file server virtual machine:

1. Whether you select **Azure hybrid center** or **Azure File Sync** from the left navigation blade in Windows Admin Center, you will be directed to the page shown in *Figure 13.30*, where you can select **Set up** to begin:

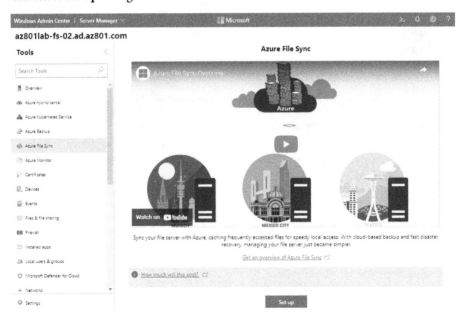

Figure 13.30 – Begin the initial setup steps for Azure File Sync in Windows Admin Center

2. Now we can select an Azure region for the setup of the Storage Sync Service, adjust any necessary settings for Azure (including **Subscription**, **Resource Group**, and **Storage Sync Service**), and adjust any of the Azure File Sync agent settings, as shown in *Figure 13.31*:

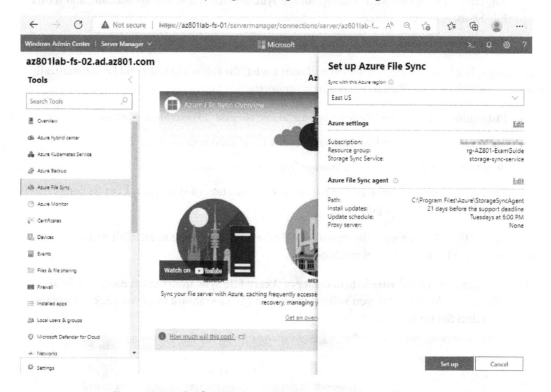

Figure 13.31 – Configuring Azure File Sync settings for the file server

3. After selecting **Set up**, Windows Admin Center will handle the preparation of the resource group, the Storage Sync Service, the download and installation of the Azure File Sync agent to the file server, and server registration with Azure Storage Sync Service, as shown in *Figure 13.32*. Note that this setup can take a considerable amount of time to complete, depending on the speed of the connection to Azure:

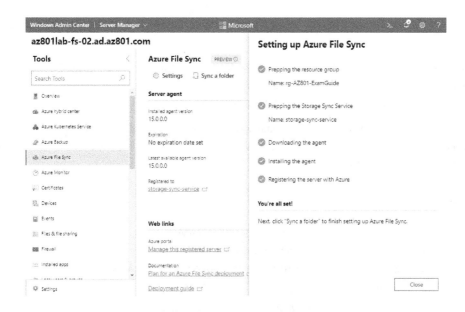

Figure 13.32 – Setting up all of the requirements for Azure File Sync in WAC

4. Finally, we can establish sync folders and sync groups from within Windows Admin Center by selecting **Sync a folder**, then supplying the local folder name, a sync group name, and the Azure file share location with which to sync, as shown in *Figure 13.33*:

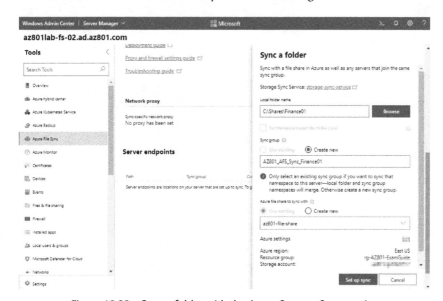

Figure 13.33 – Sync a folder with the Azure Storage Sync service

This completes the section covering the use of Azure File Sync to provide fast, centralized, local access to your files.

For additional aspects of migrations from additional types of storage to Azure File shares, be sure to check out the additional reading at the following URL: `https://learn.microsoft.com/azure/storage/files/storage-files-migration-overview`.

We have completed this section on how to install and configure Azure File Sync and how to establish sync folders and sync groups. Next, we will complete a review of all we have learned in this chapter and set the stage for the next chapter of this book.

Summary

In this chapter, we learned how to use Storage Migration Service to successfully integrate with and migrate to Microsoft Azure and Azure Virtual Machines. We also learned what Azure File Shares is and how we can migrate existing shares to Azure File Shares quickly and securely, consistent with the AZ-801 exam objectives.

In the next section, we will learn how to deploy and configure the **Azure Migrate** appliance. Working with the Azure Migrate appliance, we will then migrate VM workloads to **Microsoft Azure IaaS**, migrate physical workloads to Azure IaaS, and finally, identify additional tools that can be used within Azure Migrate to achieve your migration objectives.

14

Migrating On-Premises Servers to Azure

In this chapter, we will learn how to deploy and configure the Azure Migrate appliance. Working with the Azure Migrate appliance, we will then migrate VM workloads to Microsoft Azure IaaS, migrate physical workloads to Azure IaaS, and finally, identify additional tools that can be used within Azure Migrate to achieve your migration objectives.

In this chapter, we will cover the following topics:

- Technical requirements and lab setup

- Introduction to Azure Migrate

- Migrating by using Azure Migrate tools

- Deploying and configure the Azure Migrate appliance

- Migrating VM workloads to Azure IaaS

- Migrating physical workloads to Azure IaaS

Technical requirements and lab setup

To successfully follow along and complete the tasks and exercises throughout this chapter and the following chapters in this book, we will need to ensure that the technical requirements from *Chapter 1, Exam Overview and the Current State of Cloud Workflows*, have been completed in full. We will focus on the requirements, best practices, and high-level concepts in this chapter, and will be primarily reviewing areas of the **Hyper-V Manager**, the **Azure Migrate** appliance, and the **Azure portal** environment to follow the walk-throughs throughout this chapter, aligning with the AZ-801 exam objectives. This is primarily due to the Azure Migrate appliance performance requirements, which are beyond the scope of a standard lab machine configuration.

Let's begin with an introduction to the Azure Migrate service and how it can be used to discover, assess and modernize existing on-premises virtual and physical machine workloads and migrate them to Microsoft Azure.

Introduction to Azure Migrate

Azure Migrate is a tool that helps you discover and assess on-premises workloads for migration and modernization into Azure. After configuring the Azure Migrate appliance, the service assesses the readiness of a particular VM or workload, providing performance-based recommendations on size, suggesting a migration tool for a successful migration, completing dependency analysis, and giving in-depth cost estimates for running the identified workloads in Azure. Azure Migrate is quite extensible and, as such, integrates with a large amount of third-party **Independent Software Vendor** (**ISV**) tools from some well-known integrators, as shown in *Figure 14.1* and outlined at this URL: `https://learn.microsoft.com/azure/migrate/migrate-services-overview#isv-integration`.

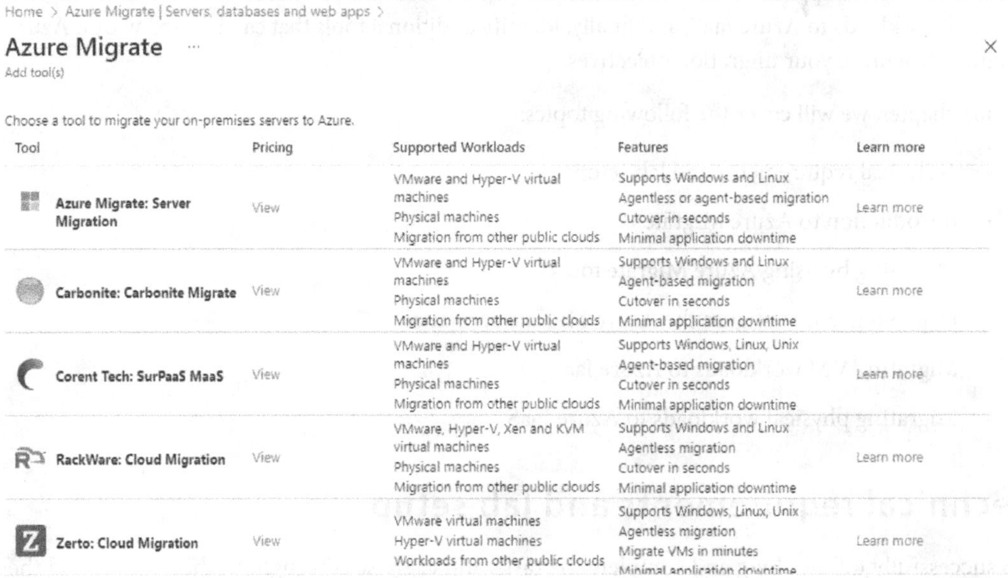

Home > Azure Migrate | Servers, databases and web apps >

Azure Migrate ...
Add tool(s)

Choose a tool to migrate your on-premises servers to Azure.

Tool	Pricing	Supported Workloads	Features	Learn more
Azure Migrate: Server Migration	View	VMware and Hyper-V virtual machines Physical machines Migration from other public clouds	Supports Windows and Linux Agentless or agent-based migration Cutover in seconds Minimal application downtime	Learn more
Carbonite: Carbonite Migrate	View	VMware and Hyper-V virtual machines Physical machines Migration from other public clouds	Supports Windows and Linux Agent-based migration Cutover in seconds Minimal application downtime	Learn more
Corent Tech: SurPaaS MaaS	View	VMware and Hyper-V virtual machines Physical machines Migration from other public clouds	Supports Windows, Linux, Unix Agent-based migration Cutover in seconds Minimal application downtime	Learn more
RackWare: Cloud Migration	View	VMware, Hyper-V, Xen and KVM virtual machines Physical machines Migration from other public clouds	Supports Windows and Linux Agentless migration Cutover in seconds Minimal application downtime	Learn more
Zerto: Cloud Migration	View	VMware virtual machines Hyper-V virtual machines Workloads from other public clouds	Supports Windows, Linux, Unix Agentless migration Migrate VMs in minutes Minimal application downtime	Learn more

Figure 14.1 – Third-party ISVs and integrations with the Azure Migrate service

Azure Migrate supports on-premises **Hyper-V** and **VMware** VMs and physical servers running either Windows or Linux-based operating systems. While Azure Migrate utilizes **Azure Recovery Services** for replication of VM workloads, agentless migration of VMware and Hyper-V VMs to Azure can be completed from a single centralized user interface. This Azure Migrate hub interface also allows you to track and project your migration journey to Microsoft Azure.

The following resources can be assessed, migrated, and modernized via Azure Migrate (additional Azure Migrate details and scenarios can be found at the following URL: `https://learn.microsoft.com/azure/migrate/`):

Resource	Migration options
Servers, databases, and web apps	Existing on-premises servers running SQL Server or web applications can be migrated to new Azure VMs or **Azure VMware Solution (AVS)**
Databases (only)	Existing on-premises SQL Server databases and instances can be migrated to either of the following: • SQL Server running on an Azure VM • **Azure SQL Managed Instance** • An **Azure SQL database**
Virtual Desktop Infrastructure (VDI)	Virtual desktop environments can be assessed and migrated to **Azure Virtual Desktop (AVD)**
Web apps	Existing on-premises web applications can be assessed and migrated to either **Azure App Service** or **Azure Kubernetes Service (AKS)**
Data box	Organizations can migrate larger amounts of data directly to Microsoft Azure using the Azure Data Box feature, which is outlined in more depth at this URL: `https://learn.microsoft.com/azure/databox/data-box-overview`

The Azure Migrate appliance comes in two variations: a preconfigured VM (that is roughly 27 GB once expanded) or a set of PowerShell configuration scripts that can be run on an existing machine, with additional reading on scripted installation available at this URL: `https://learn.microsoft.com/azure/migrate/deploy-appliance-script`

The discovery and assessments within Azure Migrate have no overhead or impact on the performance of the existing on-premises environment, as the appliance can be placed on any server that can remotely communicate with the VM hosts and has a line of sight to identity servers (AD Domain Controller) for domain authentication when needed.

The configuration data collected by the Azure Migrate appliance includes the following:

- VM display name
- IP address(es)
- MAC address(es)
- Operating system
- Number of CPU cores
- Memory (RAM) configuration
- Network interfaces
- Disk configuration, count, and sizes

The performance data collected by the Azure Migrate appliance includes the following:

- CPU utilization
- Memory (RAM) utilization
- Per disk information for the following:
  - Disk read and write throughput
  - Disk read and write operations per second
- Per virtual network adapter information on network utilization, both in and out of the adapter(s)

The Azure Migrate appliance prerequisites are as follows:

- An existing or new **Azure subscription** for the Azure Migrate resources
- Azure AD user privileges for the following:
  - Permission to register the Azure Migrate resource providers for the subscription, including the Contributor role at the subscription level
  - Permission to create Azure AD applications so that Azure Migrate can bridge the authentication and access between the appliance and Azure Migrate, and the second application for the creation of an Azure Key Vault resource (noting that this permission can be revoked once discovery has been established)
- Hardware requirements for a VMware scenario require at least Windows Server 2016 with 32 GB of memory, 8 vCPUs, and a minimum of 80 GB of storage (with additional VMware requirements outlined at this URL: `https://learn.microsoft.com/azure/migrate/migrate-support-matrix-vmware-migration`)

- Hardware requirements for a Hyper-V scenario require at least Windows Server 2016 with 16 GB of Memory, 8 vCPUs, and a minimum of 80 GB of storage (with additional Hyper-V requirements outlined at this URL: `https://learn.microsoft.com/azure/ migrate/migrate-support-matrix-hyper-v-migration`)

- A single user with administrative privileges on the hosts and clusters running the VM workloads you wish to discover (can be a domain or local user)

- One external virtual switch (preferably on a separate or isolated VLAN or subnet) for communication with the internet and Azure, as the appliance will upload data to Azure and needs access only to the Hyper-V host, not to the applications

- Hyper-V integration services enabled so that operating system information can be discovered during the server assessment task

- Download and installation of the Microsoft Azure Site Recovery provider and Azure Recovery Service agent on Hyper-V hosts or cluster nodes

- Allow the listing of the following URLs for internet connection and VM discovery:

  - `*.portal.azure.com`

  - `*.vault.azure.net`

  - `*.microsoftonline.com`

  - `*.microsoftonline-p.com`

  - `*.msauth.net`

  - `management.azure.com`

  - `dc.services.visualstudio.com`

Now that we have covered an introduction to the Azure Migrate service and what features this technology brings to the migration of existing on-premises virtual and physical workloads, let's step into how the Azure Migrate service can be used to discover, assess, and modernize existing on-premises virtual and physical machine workloads by migrating them to **Microsoft Azure**.

Migrating using Azure Migrate tools

Before we begin, I'd like to share some resources that cover two additional scenarios: physical server migration using Azure Migrate at `https://learn.microsoft.com/azure/migrate/ agent-based-migration-architecture`, and VMware server migration located at `https:// learn.microsoft.com/azure/migrate/migrate-support-matrix-physical- migration`.

Let's walk through how we can establish the initial resources needed for the Azure Migrate service as an example:

1. After logging into `https://portal.azure.com` and searching for `Azure Migrate`, we then select **Discover, assess and migrate** from the **Get started** section, as shown in *Figure 14.2*:

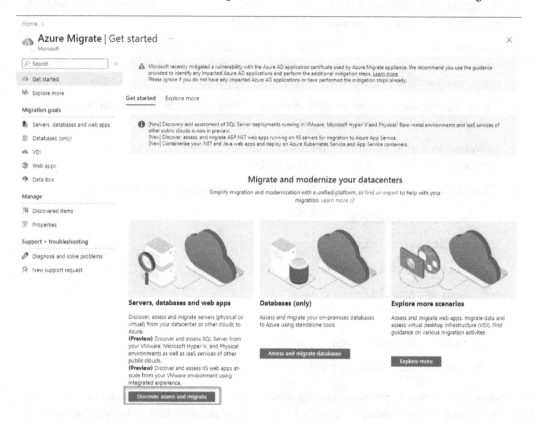

Figure 14.2 – Getting started with the Discover, access and migrate option

2. We then create a new Azure Migrate project by selecting + **Create project** or the **Create project** button, as shown in *Figure 14.3*:

Figure 14.3 – Create a new Azure Migrate project resource

3. We then select an appropriate **Subscription** and **Resource group** option and then supply a **Project** name, and select appropriate **Geography** and **Connectivity method** options for our project, as shown in *Figure 14.4*:

Figure 14.4 – Filling the project details

4.   Once the project has been successfully created, we are presented with a screen showing both **Assessment tools** and **Migration tools**, as shown in *Figure 14.5*. We will then select **Discover** under **Azure Migrate: Discovery and assessment**:

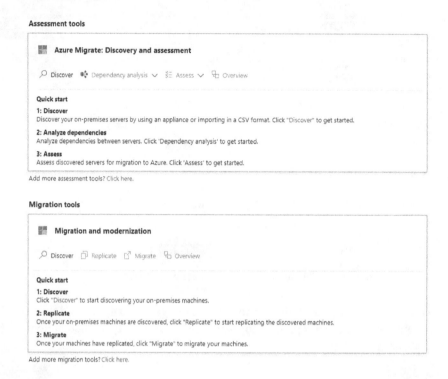

Figure 14.5 – Assessment and Migration tools available from Azure Migrate

5.   As the servers have been virtualized on Hyper-V hosts for this exercise, we will select **Yes, with Hyper-V**, as shown in *Figure 14.6*. Note that we could also select physical or virtual machines running in additional cloud computing environments:

Figure 14.6 – Discover virtualized machines running on Hyper-V

6. Next, we are asked to name our new appliance and press **Generate key**, and download either the VM files or the script installer, as shown in *Figure 14.7*. Here, we have copied the **Project key** details for later use, and we have selected to download the VM image:

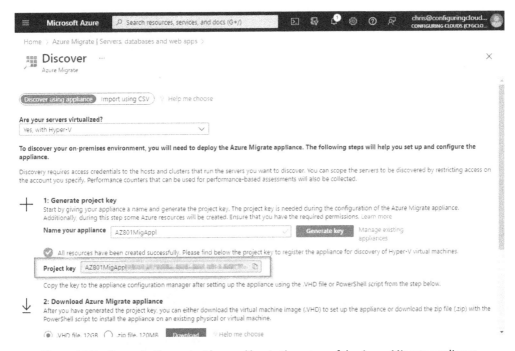

Figure 14.7 – Generate the project key and begin the setup of the Azure Migrate appliance

7. For this exercise, the files were downloaded into the `C:\AZ801PacktLab\Imports` folder and all contents were extracted to `C:\AZ801PacktLab\Imports`.

This completes the initial setup of Azure Migrate tools in the Azure portal.

We established a new Azure Migrate resource in the Azure portal and learned how to complete the initial setup steps. Next, we will continue with an example of how to successfully deploy and configure the **Azure Migrate** appliance to discover and assess an environment.

Deploying and configure the Azure Migrate appliance

Now, let's walk through deploying and configuring the appliance for the Azure Migrate service as an example, to discover and assess virtual workloads with one preexisting Hyper-V host and multiple hosted VMs in a lab environment:

1. We begin the import of the virtual appliance into Hyper-V Manager by right-clicking on the name of the Hyper-V host under **Hyper-V Manager** and selecting **Import Virtual Machine...**, then selecting **Next >** to continue.

2. Here, we have selected the expanded virtual appliance folder location, as shown in *Figure 14.8*:

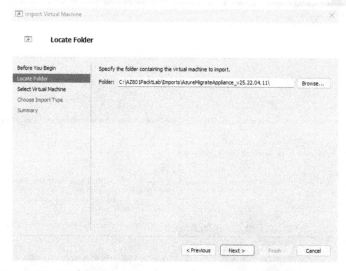

Figure 14.8 – Selecting the folder for the imported Azure Migrate appliance

3. In the **Select Virtual Machine** step, we select the only available VM appliance from the list, as shown in *Figure 14.9*:

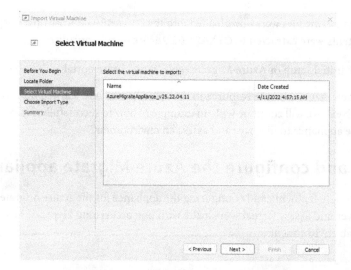

Figure 14.9 – Selecting the Azure Migrate appliance from the list of VMs

4. In the **Choose Import Type** step, we select the **Register the virtual machine in-place (use the existing unique ID)** option, as shown in *Figure 14.10*:

Figure 14.10 – Ensuring that Register the virtual machine in-place is selected

5. For the **Connect Network** step, we will choose the external virtual network to ensure that we have proper internet access for our virtual appliance, as shown in *Figure 14.11*:

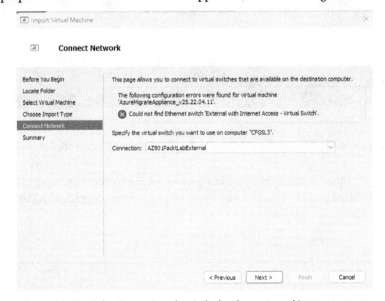

Figure 14.11 – Selecting a virtual switch that has external internet access

6. After completing the virtual appliance import, we then start the virtual appliance, and we are presented with a **License terms** acceptance screen, where we must select **Accept** to continue, as shown in *Figure 14.12*:

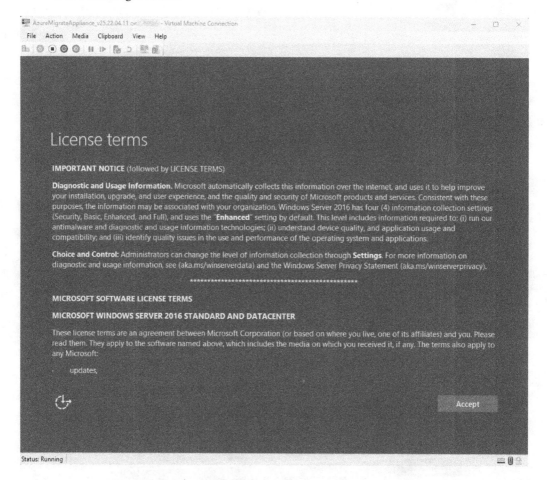

Figure 14.12 – Accepting the license terms for the Azure Migrate appliance

7. We're then prompted to supply an appliance password twice to begin the setup of the appliance; we used our lab password of `Packtaz801guiderocks` to complete this step, as shown in *Figure 14.13*:

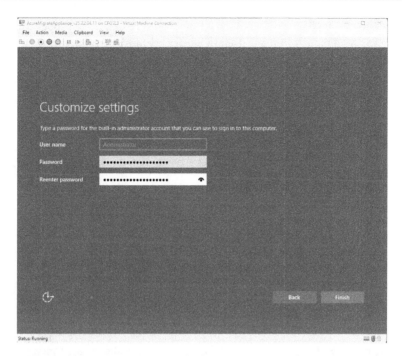

Figure 14.13 – Setting the new administrator password for the appliance

8. Once logged into the appliance, we must wait a few minutes for background setup scripts to complete and fire up the **Azure Migrate Appliance Configuration Manager.** We are then required to select **I agree** on the **Terms of use** screen, as shown in *Figure 14.14*:

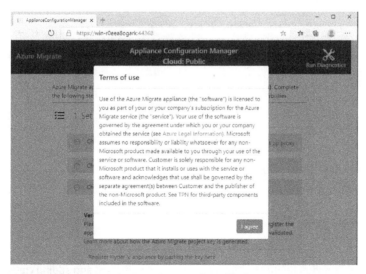

Figure 14.14 – Accepting the EULA/Terms of use for the Azure Migrate appliance

9. The Azure Migrate appliance will take some time to ensure that it has proper connectivity to Azure and will automatically update itself (if needed) before proceeding with the setup. Once ready, we supply the Azure Migrate project key we saved in *Step 6* in the *Migrate using Azure Migrate tools* section, and then select **Verify**, as shown in *Figure 14.15*:

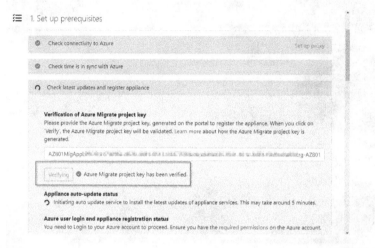

Figure 14.15 – Registering and verifying the new Azure Migrate appliance

10. Next, we must log in to the Azure portal to complete the appliance registration, as shown in *Figure 14.16*. We will then select **Add credentials** to begin adding our already existing Hyper-V host:

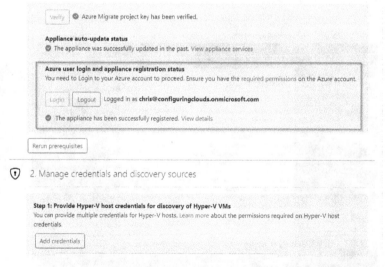

Figure 14.16 – Confirming that the appliance has been registered and adding credentials

11. Here, we key in a **Friendly name** choice for the credential, as well as the **Username** and **Password** details, as shown in *Figure 14.17*:

Figure 14.17 – Adding credentials for our Hyper-V host discovery

12. We then proceed by adding a discovery source with either the host FQDN or IP address and ensure that validation is successful, as shown in *Figure 14.18*:

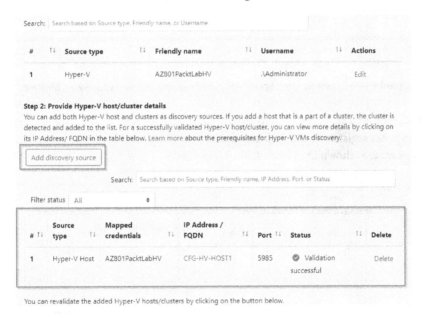

Figure 14.18 – Adding our Hyper-V host as a new discovery source for VMs

13. Here, we opt to disable the dependency analysis for our virtualized lab environment, as shown in *Figure 14.19*. To read more about the benefits of enabling this agentless dependency analysis and discovery in your migration efforts, be sure to read this article: `https://learn.microsoft.com/azure/migrate/concepts-dependency-visualization`.

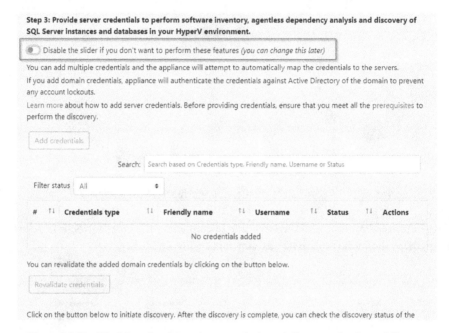

Figure 14.19 – Disabling the dependency analysis and discovery for Azure Migrate

14. At this point, we can start the discovery for our Azure Migrate project, and confirm that the appliance has registered with Azure and that virtualized workloads are being discovered by the appliance, as shown in *Figure 14.20*:

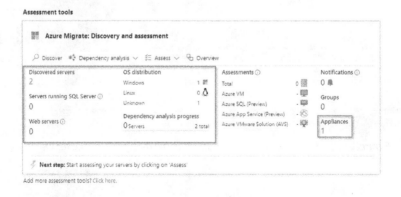

Figure 14.20 – Discovery has begun and has identified an appliance and two VMs

This completes the setup of the Azure Migrate appliance, both on-premises and in the Azure portal, giving us a feel for configuration and ongoing discovery efforts.

We established a new Azure Migrate appliance in our on-premises lab environment as an example, and have learned how to complete the integration with Azure Migrate. Next, we will continue with an example of how to assess and begin the migration of VM workloads to Azure **Infrastructure-as-a-Service** (**IaaS**).

Migrating VM workloads to Azure IaaS

Let's continue learning about migrating VM workloads with a walk-through to assess virtual workloads on one preexisting Hyper-V host and multiple hosted VMs in a lab environment and ensure that replication is configured:

1. Now that we have discovered the lab environment, we will begin a migration assessment for our virtualized workload by selecting **Assess** > **Azure VM**, as shown in *Figure 14.21*:

Figure 14.21 – Beginning a migration assessment for our virtualized workload

2. Here, we review the recommendations by Azure and then select **Next: Select servers to assess** > to continue, as shown in *Figure 14.22*:

Figure 14.22 – Beginning the creation of the assessment resource in Azure Migrate

3. Continuing the creation of our assessment, we supply an assessment name, create a new group, and select the VMs we want to complete the assessment on, then select **Next: Review + create assessment** >, as shown in *Figure 14.23*:

Figure 14.23 – Selecting the assessment name, group name, appliance name, and VMs

4. After the assessment has been created, we notice that a new group is available for review within our project, as shown in *Figure 14.24*, and we can click through to view the assessment details outlining the **Azure readiness** and **Monthly cost estimate (USD)** information:

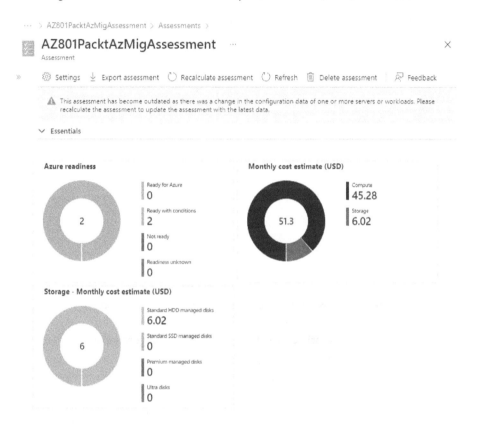

Figure 14.24 – Reviewing the migration assessment completed for Azure Migrate

---

Performance assessment recommendations

It is highly recommended that you allow at least 24 hours to complete an initial performance assessment recommendation from the Azure Migrate appliance. This will allow ample time for performance insights to be fed into your Azure Migrate project, giving the best and most insightful recommendations on sizing, cost, and utilization estimates.

5. Scrolling to the **Migration tools** section, we can now select **Discover** to begin the setup for the migration of our on-premises resources, as shown in *Figure 14.25*:

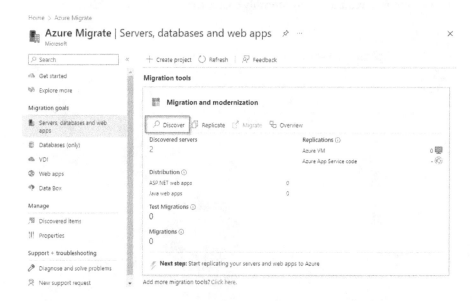

Figure 14.25 – Beginning the Discover phase of the migration

6. Here, we will select the appropriate virtualization platform configuration and our target region for migration, as shown in *Figure 14.26*:

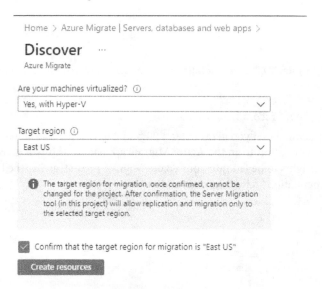

Figure 14.26 – Selecting the appropriate settings for the discovery phase of our migration

7. Next, we begin the setup and configuration of any Hyper-V hosts with the Azure Site Recovery provider and Azure Recovery Services agent by selecting **Download** to gather the proper installation files and download the registration key, as shown in *Figure 14.27*:

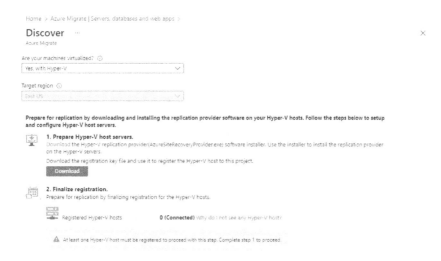

Figure 14.27 – Downloading the Recovery Services provider, agent, and registration key

8. On the physical Hyper-V host, we complete the installation of the **Microsoft Azure Site Recovery** (**MASR**) agent by supplying the necessary configuration options, as shown in *Figure 14.28*:

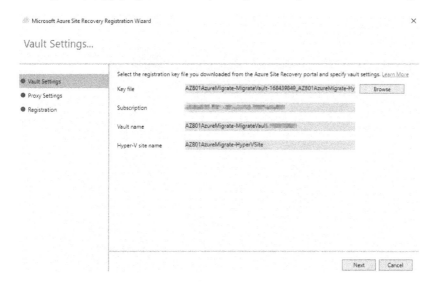

Figure 14.28 – Configuring the MASR agent on the Hyper-V host

9.   Registration of the Hyper-V host is now finalized, as shown in *Figure 14.29*:

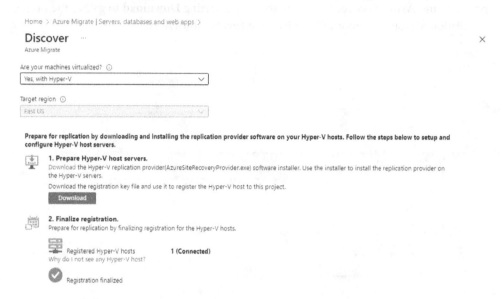

Figure 14.29 – Confirming that the registration of the Hyper-V host is finalized in Azure

10. Next, we begin establishing replication for our Azure Migrate project by identifying what we are migrating and where we are migrating the workloads to, as shown in *Figure 14.30*:

Figure 14.30 – Specifying the intent of our replication setup for Azure Migrate

11. We then continue by selecting the migration settings from an existing assessment, selecting both the group and the assessment, and finally, including the intended VMs for replication and eventual migration, as shown in *Figure 14.31*:

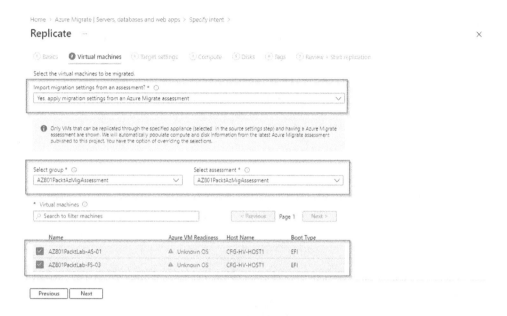

Figure 14.31 – Establishing the VM replication settings for Azure Migrate

12. For **Target settings**, we then select an appropriate subscription and resource group, storage account, virtual network, subnet, and any availability options, as shown in *Figure 14.32*:

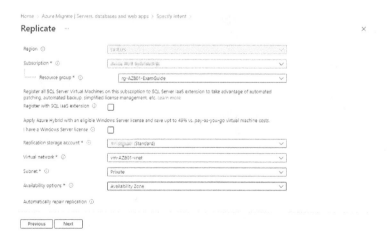

Figure 14.32 – Configuring the target replication settings for Azure Migrate

13. We now have the option to configure and make changes to the **Compute** configuration of our VMs to be placed into Azure, as shown in *Figure 14.33*:

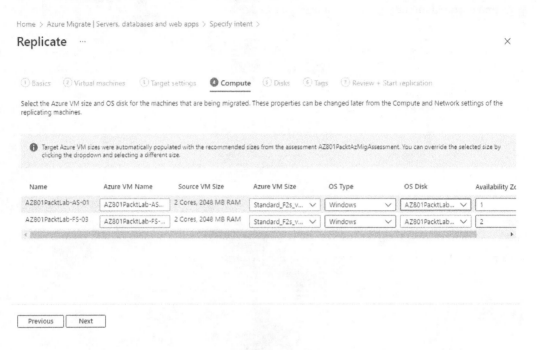

Figure 14.33 – Review and configuration of compute for replicated VMs

14. We can review and make any needed changes to the **Disks** configuration, as shown in *Figure 14.34*, and then select **Create** to confirm and begin the replication:

Figure 14.34 – Review and configure the disk settings for Azure Migrate

15. Reviewing replication progress can be daunting and can take quite a while, and once finished, there will be a visual indicator beneath **Replications** to identify how many VMs have successfully been replicated, as shown in *Figure 14.35*:

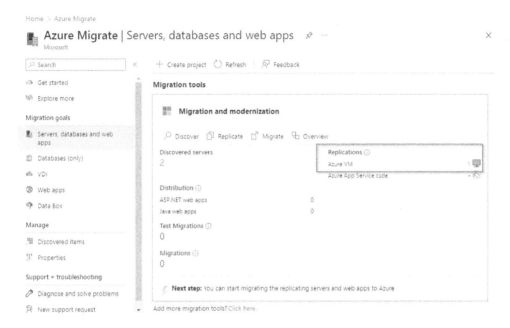

Figure 14.35 – Reviewing active replications within Azure Migrate

16. At this point in the configuration steps, we can opt to select to test a migration or simply complete a migration of a VM, as shown in *Figure 14.36*:

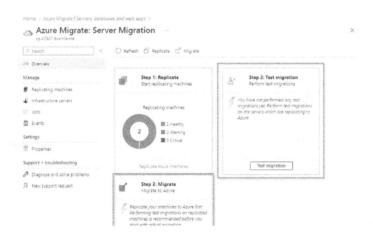

Figure 14.36 – Selecting either Test migration or Migrate within Azure Migrate

17. Similar to what we learned in *Chapter 11, Implementing Disaster Recovery Using Azure Site Recovery, in the Configuring Azure Site Recovery for Azure Virtual Machines section*, we now have the ability to test, cancel, or migrate a VM, as shown in *Figure 14.37*:

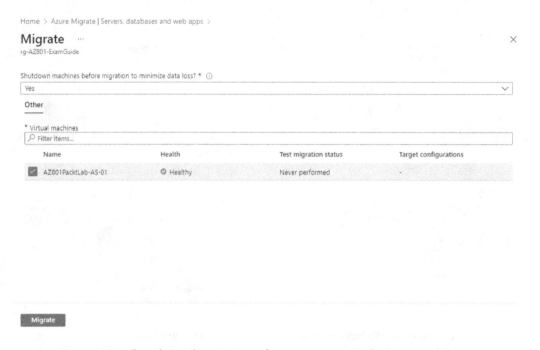

Figure 14.37 – Completing the migration of a server to Azure IaaS using Azure Migrate

This completes the setup of the Azure Migrate appliance for discovery efforts, replicating, testing, and eventually, migrating virtual workloads to Azure IaaS.

> **Replication of Hyper-V VMs to Azure has some limitations**
>
> Do note that you can replicate up to 10 machines together in a group. If you need to replicate more than 10 machines at a time, you must consider doing the replications in batches of 10 at a time to achieve the best results for your migration efforts.

Next, we will continue with an example of how to migrate physical workloads to **Azure Infrastructure-as-a-Service (IaaS)**.

Migrating physical workloads to Azure IaaS

Now that we have learned how to configure the Azure Migrate appliance and how to discover, assess, and replicate VM workloads, we should cover what additional steps need to be completed to achieve the same for physical server workloads. Understanding all the requirements laid out earlier in this chapter in the *Introduction to Azure Migrate* section, the only additional configuration that needs to be completed outside of proper migration planning for the physical server is to include the discovery source as a new host within the Azure Migrate appliance, as shown in *Figure 14.38*:

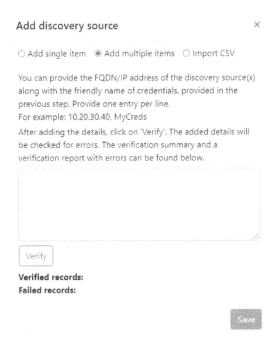

Figure 14.38 – Configuring additional discovery sources with physical hosts

We have completed this section on how to configure the Azure Migrate appliance for additional physical workload migration to Azure IaaS. Next, we will complete a review of all we have learned in this chapter and set the stage for the next chapter of this book.

Summary

In this chapter, we learned about the features of Azure Migrate and how to deploy and configure the Azure Migrate appliance. Working with the Azure Migrate appliance, we then migrated VM workloads to Microsoft Azure IaaS, migrated physical workloads to Azure IaaS, and finally, identified additional tools that can be used within Azure Migrate to achieve migration objectives, consistent with the AZ-801 exam objectives.

In the next chapter, we will learn about available tools for migrating various legacy Windows Server workloads to **Windows Server 2022**. We will dive into how to migrate IIS workloads, Hyper-V hosts, **Remote Desktop Services (RDS)** host servers, **Dynamic Host Configuration Protocol (DHCP)**, and print servers from an older Windows Server version to Windows Server 2022.

15

Migrating Workloads from Previous Versions to Windows Server 2022

In this chapter, we will learn about the available tools for migrating various legacy Windows Server workloads to Windows Server 2022. We will dive into how to migrate **Internet Information Services** (**IIS**) workloads, Hyper-V hosts, **Remote Desktop Services** (**RDS**) host servers, **Dynamic Host Configuration Protocol** (**DHCP**), and print servers from an older Windows Server version to Windows Server 2022 to achieve the AZ-801 exam's preparation objectives.

In this chapter, we will cover the following topics:

- Technical requirements and lab setup
- Migrating IIS
- Migrating Hyper-V hosts
- Migrate RDS host servers
- Migrate DHCP servers
- Migrate print servers

Technical requirements and lab setup

To successfully follow along and complete the tasks and exercises in this chapter and the following chapters in this book, we will need to ensure that the technical requirements from *Chapter 1, Exam Overview and the Current State of Cloud Workflows*, have been completed in full. We will focus on the requirements, best practices, and high-level concepts in this chapter and will be primarily reviewing areas of the legacy Windows Server environment (Windows Server 2016) so that we can follow the walkthroughs throughout this chapter that align with the AZ-801 exam objectives.

Let's begin with some general insights and recommendations on planning migrations for general application servers running on older versions of Windows Server.

General recommendations for migrating Windows Server workloads

Over the years, Windows Server versions have been released on a much more frequent cadence and for a good handful of organizations, aging server deployments have become a commonality. These server versions not only fall into the unsupported range quickly, but they also tend to have no security fixes available, no extended support, and become quite the red flag on environmental audits.

Sure, there may be a few incredibly complex and environmentally specific applications that require a specific version of the OS, specific drivers, hardware, or other individualized configurations. However, for most applications running on these outdated or soon-to-be-outdated OS versions, there is a tried-and-true plan to migrate or modernize existing Windows Server instances to a newer version of the OS. Those steps are outlined as follows:

1. **Discovery** by utilizing enterprise management tools combined with human investigation and inventory analysis to determine the OS version and server architecture, allocated resources, and performance data. This includes doing the following:

    - Determining how many virtual and physical machines are present

    - Determining where the resources are located, how they are being used, who is using them, and who is the subject matter expert

    - Determining the server role or workloads involved with each server

    - Determining what versions of the software are running on the servers

2. **Justification** of whether the migration of a workload not only fulfills the technical direction and requirements of the organization but also answers the following questions:

    - Can the criticality and resiliency of the workload be determined by a human subject matter expert?

    - Does the migration of the workload fit the vision and technical roadmap for the organization and can existing tools be utilized?

    - Are there any features and capabilities in the new OS that could potentially replace any existing third-party solutions in the organization, reducing the overall management footprint?

    - Are there any steps of the migration that can be completed seamlessly without user or service interruption?

3. **Planning** for the steps that need to be taken for a certain technological or requirements-driven direction based on the following considerations:

   - Can a new feature or functionality be achieved by migrating to a newer set of Windows OS capabilities?

   - How large is the current footprint and is there an added benefit in reducing the footprint size or is additional performance gained?

   - Are any additional architectural redesigns needed to ensure that the design of the environment still meets the requirements (old or new)?

   - Does the current support and/or licensing agreement cover the workload migration you are designing and planning? Will additional costs be incurred during the migration?

4. **Migration** from the existing Server OS to a newer or newly installed OS:

   - Can an in-place upgrade achieve the desired results? Typically, this is not the case for any services installed on a server, thus requiring a migration to a new Server OS.

   - Are there any existing servers with a newer OS that could be suitable hosts for this migrated workload?

   - Can the workload (including both role and application configuration state) be backed up and then restored to a new or replacement server?

   - Will the workload that's intended for migration work on a new architecture (and support 64-bit, if migrating from a legacy server OS)?

5. **Continuous improvement** in the overall operational design and architecture of the environment to include earlier evaluation of migration efforts moving forward:

   - Always evaluate new technologies and server OS releases for feature improvements, added functionality, and easier management

   - Ensure that the organizational requirements are kept up to date and reviewed to ensure parity with current workloads

   - Evaluate new capabilities against the existing architecture to simplify and centralize operational efforts where appropriate

   - Complete design and architectural reviews of existing workloads with additional eyes – you may find additional ways to streamline workloads

> **Migration assessments and tools available**
>
> Once known as the Microsoft Assessment and Planning Toolkit (or MAP for short), this tool allowed for an agentless approach to discovering workloads and completing assessments of your environment to help you plan workload migrations. While MAP is still available as a download at `https://www.microsoft.com/download/details.aspx?id=7826`, it is recommended that you utilize **Azure Migrate** to complete the discovery, assessment, and migration steps of your migration journey.

Now, let's learn what tools are available to administrators to migrate **Internet Information Services (IIS)** websites from previous versions of an OS to **Windows Server 2022**.

Migrate Internet Information Services (IIS)

For many organizations, there is a mix of basic web pages running either **HTML**, **Active Server Pages (ASP/ASP)**, or other code, and much larger applications that are incredibly more complex with application server configurations. When tasked with migrating either of these larger applications, multiple tools can be utilized to complete the planning and migration activities.

For example, let's say we have a Windows Server 2016 application server that is hosting three basic IIS web applications, as follows:

Site Name	Application Notes
AZ801-App1	Small website, basic HTML files
AZ801-App2	Medium website, basic ASP files
AZ801-App3	Large application, complex configuration

Table 15.1 – List of IIS example applications

We can begin by installing a tool called **Web Deploy** (or **msdeploy**, when used via the command line) that will not only allow us to publish websites from development tools but also allow us to migrate websites and applications from an older OS to a newer replacement OS. This migration can be done via a copy/paste deployment of the export package or automatically via an **HTTPS** sync mechanism using the **Remote Agent Service**. The great part about using the Remote Agent Service is that publishing can be enabled for a non-administrator user, requiring just enough of an administrative approach to deployment and migration. In addition, Web Deploy can handle security, publish databases, and apply transforms during deployment for situations where application settings or connection strings need to be updated.

Once the Web Deploy application has been downloaded, the installation should be completed on both the source and destination host servers by selecting the `complete` installation if the Remote Agent Service is desired for the migration.

> **Automating the Web Deploy setup and management**
>
> Note that Web Deploy can be set up and configured using PowerShell scripts for a consistent approach to any scale migration efforts. Details about this can be found here: `https://learn.microsoft.com/iis/publish/using-web-deploy/powershell-scripts-for-automating-web-deploy-setup`. In addition, Web Deploy has its own PowerShell cmdlets to assist in managing published files, IIS, sites, web applications, databases, and migration activities, as outlined in the following article: `https://learn.microsoft.com/iis/publish/using-web-deploy/web-deploy-powershell-cmdlets`.

As shown in *Figure 15.1*, the **Web Deploy** tool can be accessed via the **Internet Information Services (IIS) Manager** user interface simply by selecting the website that's been chosen for migration, then right-clicking to select **Deploy > Export Application…**:

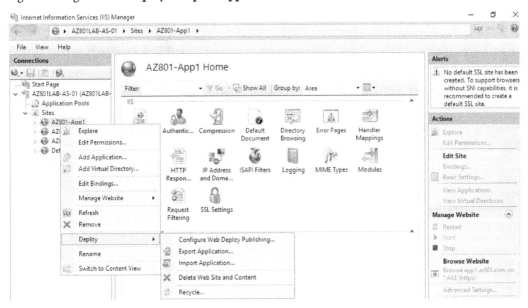

Figure 15.1 – Selecting an application for migration via Web Deploy in IIS Manager

After doing this, we are presented with insights from **Web Deploy**, including what contents will be included with the export package, including the application creation manifest, any filesystem content, system parameters, and any folder permissions, as shown in *Figure 15.2*:

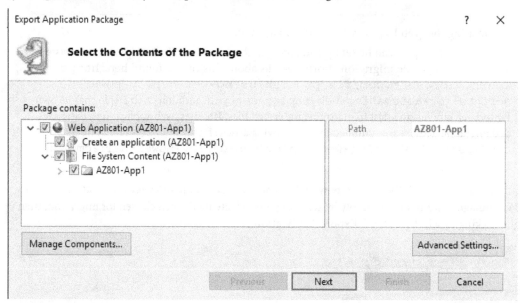

Figure 15.2 – Contents of the export application package via Web Deploy

Ultimately, we are presented with a `.zip` file for our exported application package that contains all the necessary contents and instructions for the destination host, as shown in *Figure 15.3*. This package can then be copied to the destination server or published via the Remote Agent Service, automatically deploying anywhere from one to thousands of web applications from an aging source server to a newer destination server:

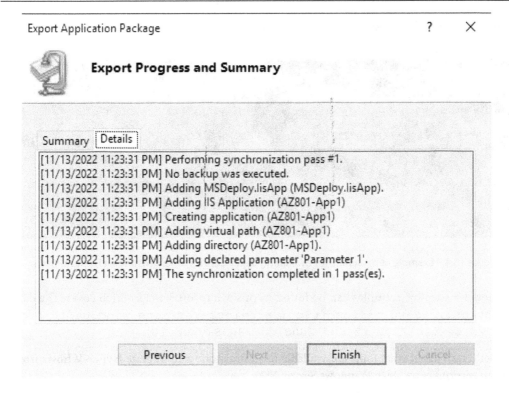

Figure 15.3 – Example Web Deploy synchronization pass for an exported web application

From a command-line perspective, **msdeploy** can be utilized to identify dependencies, configure the destination server, and complete the migration all from a scripted command-line approach. For instance, to determine the dependencies for our AZ801-App1 application within IIS, we could issue the following command:

```
msdeploy -verb:getDependencies -source:apphostConfig="AZ801-
App1"
```

The results are then displayed, as shown in *Figure 15.4*:

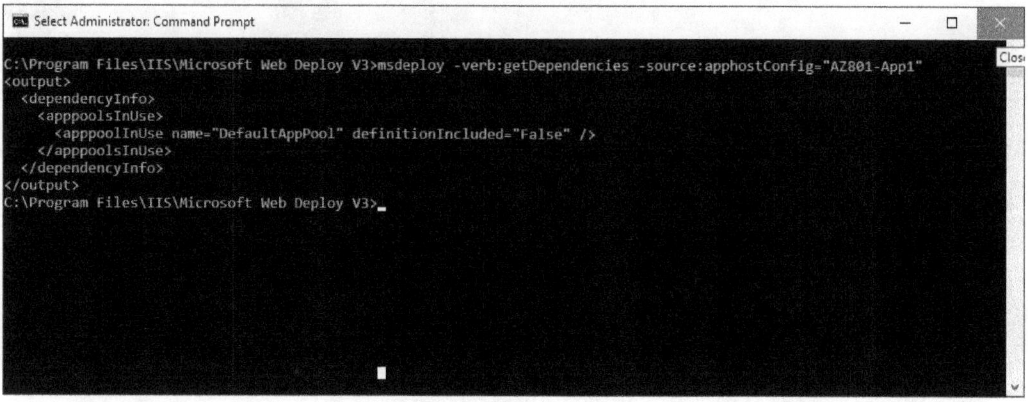

```
C:\Program Files\IIS\Microsoft Web Deploy V3>msdeploy -verb:getDependencies -source:apphostConfig="AZ801-App1"
<output>
  <dependencyInfo>
    <apppoolsInUse>
      <apppoolInUse name="DefaultAppPool" definitionIncluded="False" />
    </apppoolsInUse>
  </dependencyInfo>
</output>
C:\Program Files\IIS\Microsoft Web Deploy V3>
```

Figure 15.4 – Example of msdeploy for evaluating dependencies for an existing web application

Additional msdeploy examples can be found in this Microsoft article, which covers the core tasks: `https://learn.microsoft.com/previous-versions/windows/it-pro/windows-server-2008-r2-and-2008/dd569098(v=ws.10)`.

Now, let's learn about what tools are available to administrators to migrate **Hyper-V hosts** from previous versions of an OS to **Windows Server 2022**.

Migrating Hyper-V hosts

There are several ways that Hyper-V migration can be achieved when migrating from a legacy Windows Server OS version to Windows Server 2022. These migration options can be mixed and matched to achieve the **service-level agreements** (**SLAs**) and requirements of the business regarding the migration plan.

Some thoughts to take into consideration when planning a Hyper-V migration are as follows:

- For critical workloads that require high availability and short downtimes or maintenance periods, cross-version live migration can be considered as it requires no downtime to complete the migration

- When using **Hyper-V Replica** servers, the Replica server must be upgraded first so that it can receive replication from down-level servers or at the same Windows Server OS version

- When moving virtual machines to a new server running an updated Windows Server OS version, the replication settings must also be updated to indicate the new **Hyper-V** server or **Hyper-V Replica Broker** so that replication can resume post-migration

- If the environment is using certificate-based authentication for **Hyper-V Replica** server(s), after the primary virtual machine has been migrated to a new server, the certificate thumbprint must also be updated for the virtual machine

Let's review the various Hyper-V migration options available for use when planning migration and upgrading to newer Windows Server host OS versions:

Migration Selection	Overview	Considerations
In-place upgrade	This migration approach can be completed simply by inserting media and installing the new OS	<ul><li>All virtual machines hosted on the server being upgraded must be powered off during the upgrade process</li><li>No additional hardware is required to complete this migration approach</li></ul>
Cross-version live migration	This migration approach requires no downtime for the hosted virtual machines and moves the virtual machine files from one Hyper-V cluster to another	<ul><li>If the virtual hard disk files are stored on a Scale-Out File Server or SMB 3.0 share that is accessible to both source and destination servers, only the memory and VM configuration files need to be moved</li><li>Will require additional hardware and configuration to expand the existing cluster and create the destination cluster</li><li>May take additional time for migration, depending on where the VM files are stored</li><li>The virtual machine will not be protected by the cluster services during the move/copy operations</li><li>The virtual machine will need to be removed from the existing cluster before being added to the new cluster; only after successful migration can high availability be re-added to the migrated virtual machine</li></ul>

Migration Selection	Overview	Considerations
Export and Import	This migration approach allows for full control of the migration of virtual machines, one at a time	• Will require additional hardware and configuration to expand the existing cluster and create the destination cluster • The virtual machine is completely shut down during the export and import processes and will incur additional migration time • The import process for the virtual machine will remove any existing **Hyper-V Replica** configuration settings; these will need to be reestablished post-import via the **Use an existing virtual machine on the Replica server as the initial copy** option
Copy Cluster Roles Wizard	This migration approach is a wizard-based process that completes the role, configuration, and VM copy process while automatically testing and rolling back if any issues are encountered	• Will require additional hardware and configuration to expand the existing cluster and create the destination cluster • Roles and configuration are automatically copied from test or existing clusters to production or upgraded clusters • VMs on the same **Clustered Shared Volume** are migrated all at the same time, and VMs must be shut down for a short period during this process • Hyper-V replication configuration settings will need to be reestablished post-copy via the **Use an existing virtual machine on the Replica server as the initial copy** option

Table 15.2 – Migration selections and considerations matrix

Additional Hyper-V cluster migration details

For additional optional reading on Hyper-V **Scale-Out File Server (SOFS)** or **Cluster Shared Volumes (CSV)** migrations, be sure to review `https://learn.microsoft.com/previous-versions/windows/it-pro/windows-server-2012-r2-and-2012/dn486784(v=ws.11)` and `https://learn.microsoft.com/previous-versions/windows/it-pro/windows-server-2012-r2-and-2012/dn486822(v=ws.11)`.

Now that we have covered how to design and plan the migration of Hyper-V hosts to a new Windows Server OS version, let's learn about what tools are available to administrators to migrate RDS from previous versions of the OS to Windows Server 2022.

Migrating RDS host servers

As a rule of thumb for RDS deployments, all **Remote Desktop (RD) Connection Broker** servers must be migrated first, unless your architecture does not contain any RD Connection Broker servers.

The order of migration and recommendations for multiple role servers for RDS are as follows:

1. **RD Connection Broker(s):**

    A. If the RD Connection Brokers are deployed in a high-availability cluster, one host with the Connection Broker role can be left in the cluster while in-place upgrades are completed on the other hosts. Then, they can be added back to the cluster once upgraded.

    B. Ensure that all certificates for the RDS configuration have been exported (preferably as a PFX with a password) and imported into the new RDS configuration or environment.

2. **RD Session Host(s):**

    A. Farms can contain Session Hosts with different Windows OS versions – it is highly recommended to use hosts with the same Windows Server version within the same RDS collection.

    B. Ensure that the RDS Licensing Host is installed on the latest Windows Server version and be sure to successfully activate the RDS CALs for licensing.

    C. Add any new hosts to the session collection and sign out of all sessions on the old host servers. Then, remove the servers that require migration.

3. **RD Virtualization Host(s).**

4. **RD Web Access Host(s):**

    A. The RDS-Web/Access role must be installed, and certificates should be imported.

    B. Microsoft Web Deploy can be used to copy the RD website settings from IIS Manager between RDS servers with the same OS version.

    C. Offline transfer of the IIS settings can be completed with the following commands:

    • Run the following command on the source host:

```
msdeploy -verb:sync -source:apphostconfig="Default Web
Site" -dest:archivedir=c:\RDSMigration\rdweb
```

- Run the following command on the target host:

```
msdeploy -verb:sync -source:archivedir=c:\RDSMigration\
rdweb -dest:appHostConfig="Default Web Site"
```

- If the RD web access URL has changed, ensure that the URL has been updated within the RDS Single Sign-On policy.

5. **RD Licensing Host** and **RD Gateway Host(s)** (can be completed at any time during the migration).

Microsoft detailed process and recommendations on RDS migration

As with any migration, there may be additional configurations that are outside the scope of general documentation and practices as everyone has different workloads and environmental requirements. As such, here are two detailed articles that go into greater depth surrounding both migration and update considerations: `https://learn.microsoft.com/windows-server/remote/remote-desktop-services/migrate-rds-role-services` and `https://learn.microsoft.com/windows-server/remote/remote-desktop-services/upgrade-to-rds`.

Now, let's look at what tools are available to administrators to migrate DHCP servers from previous versions of an OS to Windows Server 2022.

Migrating DHCP servers

Migrating DHCP servers requires that we carefully plan and collect data. This can be implemented using the following four-phase process:

1. **Pre-migration** efforts include identifying prerequisites for the migration and preparing both the source and destination servers.

2. **Migration** efforts consist of confirming the destination server both before and post-migration, exporting the source configuration using cmdlets, copying the export, and importing the source configuration to the destination server.

3. **Post-migration** efforts consist of verifying the migration. If the migration succeeds, the process has been completed, leaving us with only having to clean up the source server. If the process fails, additional troubleshooting steps need to be completed.

4. **Troubleshooting** efforts could consist of restoring the DHCP service on the source server or doing additional troubleshooting using the Windows Server Migration Tools `ServerMigration.log` file, with additional details located at `https://learn.microsoft.com/previous-versions/windows/it-pro/windows-server-2012-r2-and-2012/dn495424(v=ws.11)`.

The overall requirements for completing a successful DHCP server migration are as follows:

- Domain Administrator rights to authorize or deauthorize DHCP servers
- Local Administrator rights to install and manage the DHCP server on the host
- Write permissions to the migration store location
- Server Manager module installed and imported into your PowerShell session
- DHCP Windows Feature installed on the destination server
- Identify all IP address configurations on the source server

Now that we have established the planning and requirements for a DHCP migration, let's review the general tasks for completing a DHCP migration:

1. On the destination server, if the DHCP service is running after installation, use the following command before proceeding:

   ```
   Stop-Service DHCPServer
   ```

2. On the source server, launch PowerShell as an administrator and then use the following command before proceeding:

   ```
   Add-PSSnapin Microsoft.Windows.ServerManager.Migration
   ```

3. Begin to collect data from the source server by running the following PowerShell cmdlet as an example, where path can be an empty or nonempty directory or UNC share for storing the migration data:

   ```
   Export-SmigServerSetting -featureID DHCP -User All -Group
   -IPConfig -path <storepath> -Verbose
   ```

4. Provide a password to encrypt the migration store data.

5. On the source server, run the following command to delete the DHCP authorization:

   ```
   Netsh DHCP delete server <Server FQDN> <Server IPAddress>
   ```

6. On the destination server, update the following command in the following instances:

   A. SourcePhysicalAddress is equal to the source IP address (or addresses, separated by commas).

   B. TargetPhysicalAddress is equal to the destination IP address (or addresses, separated by commas).

C.   path points to a directory or UNC share where the migration data exists:

```
Import-SmigServerSetting -featureid DHCP -User
All -Group -IPConfig All -SourcePhysicalAddress
<SourcePhysicalAddress> -TargetPhysicalAddress
<TargetPhysicalAddress> -Force -path <storepath> -Verbose
```

7.   To begin the DHCP service on the destination server, use the following command before proceeding:

```
Start-Service DHCPServer
```

8.   Finally, authorize the DHCP server using the following command, noting that the command is case-sensitive, replacing the following:

A.   Server is the fully-qualified domain name of the server.

B.   Server IPAddress is the migrated IP address of the destination server:

```
netsh DHCP add server <Server FQDN> <Server IPAddress>
```

Now, let's look at what tools are available to administrators to migrate **print servers** from previous versions of an OS to Windows Server 2022.

Migrating print servers

Migration of printer configurations and shares must be completed from the source server to the target server, just like DHCP and file services migrations. One of the main concerns is the need for new printer drivers that are 64-bit and compatible with newer versions of the Windows Server OS, as well as with modern clients.

Microsoft has a print migration wizard and a command-line tool you can use for migrating printer services that effectively backs up and exports printer settings, queues, and drivers. This file can then be utilized as a restore file on the destination server to migrate the print services.

For this walkthrough, we will utilize the **Print Management** MMC or snap-in, which can be run locally or remotely (to run it remotely, you will need to connect to the remote server first).

The **Print Management** tool can be accessed via the Start menu by either selecting **Windows Administrative Tools** > **Print Management** or typing `printmanagement.msc` and selecting it when it's returned in the search results. Once opened, selecting **Print Management** and right-clicking to select **Migrate Printers…** opens the **Printer Migration** wizard, as shown in *Figure 15.5*:

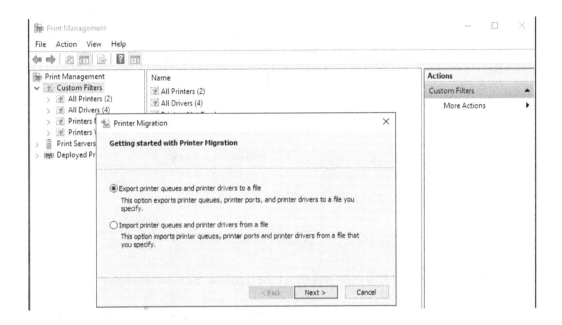

Figure 15.5 – Opening the Printer Migration wizard from the Print Management tool

After selecting a print server and then reviewing an identified list of objects to be placed into the `.printerExport` file, the wizard will export drivers, queue configuration, ports, and additional configuration information and will direct you to an event log review, as shown in *Figure 15.6*:

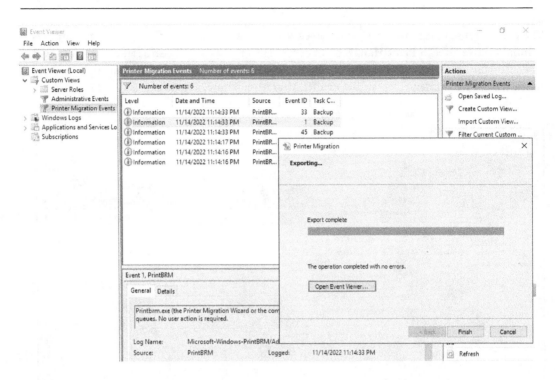

Figure 15.6 – Completing the printer export and reviewing the event log entries

To finish up on the destination server, we can launch the **Print Management** tool once more, this time selecting **Import printer queues and printer drivers from a file**, then locating the .printerExport backup file and selecting **Next** to complete the import and migration. The final cutover steps include renaming the source host to SERVERNAME_OLD, giving a different IP address to the source host, renaming the destination host to the previous source hostname while also updating the IP address to the source IP, and rebooting the destination host to finalize the migration.

The printbrm.exe command-line tool and additional migration scenarios

Details on the printbrm.exe command-line tool can be found at the following URL: https://learn.microsoft.com/previous-versions/windows/it-pro/windows-server-2012-R2-and-2012/jj134237(v=ws.11)?redirectedfrom=MSDN. You are encouraged to review this documentation.

Additional details on the migration process, including unsupported scenarios, can be found at the following URL: https://learn.microsoft.com/previous-versions/windows/it-pro/windows-server-2012-r2-and-2012/jj134150(v=ws.11)#unsupported-scenarios.

With that, we have learned how to migrate print servers from previous versions of the OS to Windows Server 2022. Next, we will review what we have learned in this chapter and set the stage for the next chapter of this book.

Summary

In this chapter, we learned about the available tools for migrating various legacy Windows Server workloads to Windows Server 2022. We examined how to migrate IIS workloads, Hyper-V hosts, RDS host servers, DHCP, and print servers from an older Windows Server version to Windows Server 2022, consistent with the AZ-801 exam objectives.

In the next chapter, we will learn about the available tools for migrating IIS workloads to **Microsoft Azure**. We will dive into how to migrate IIS workloads to **Azure Web Apps**, and how to migrate IIS workloads to Windows containers by using the **Dockerfile** technology.

16

Migrating IIS Workloads to Azure

In this chapter, we will learn about the available tools for migrating **Internet Information Services** (**IIS**) workloads to **Microsoft Azure**. We will dive into how to migrate IIS workloads to **Azure App Service** and **Azure Web Apps**, and how to migrate IIS workloads to Windows containers using the **Dockerfile** technology to achieve the AZ-801 exam's preparation objectives.

In this chapter, we will cover the following topics:

- Technical requirements and lab setup
- Migrating IIS workloads to Azure Web Apps
- Migrating IIS workloads to containers

Technical requirements and lab setup

To successfully follow along and complete the tasks and exercises throughout this chapter and the following chapters in this book, we will need to ensure that the technical requirements from *Chapter 1, Exam Overview and the Current State of Cloud Workflows*, have been completed in full.

We will be primarily reviewing our **Hyper-V** virtual machine environment and **Microsoft Azure** to complete the exercises and tasks throughout this chapter to align with the AZ-801 exam objectives. Please ensure that both the AZ801PacktLab-DC-01 and AZ801PacktLab-FS-01 virtual machines have been powered on and are showing a Running state in **Hyper-V Manager**. We will not be utilizing the failover cluster VMs during this chapter, so they can remain in a powered-off state. We will focus on the requirements, best practices, and hands-on walkthroughs in this chapter that align with the AZ-801 exam objectives.

Let's begin by establishing an IIS installation and two basic IIS applications on our `AZ801PacktLab-FS-01` Windows Server:

Begin a **Virtual Machine** (**VM**) connection to `AZ801PacktLab-FS-01` and select **View > Enhanced Session**. Then, click **Connect** to connect to the VM so that we can copy and paste commands from the host into the VM.

1. Log in to the VM using the administrator credentials for the VM.

2. To automatically install IIS on our server, launch an administrative **PowerShell** session, paste the following command into PowerShell, and press *Enter*. Note that you may need to open an instance of `Notepad.exe` and copy it there first before copying the full command into PowerShell:

    ```
    Enable-WindowsOptionalFeature -Online
    -FeatureName IIS-WebServerRole,IIS-WebServer,IIS-
    CommonHttpFeatures,IIS-ManagementConsole,IIS-
    HttpErrors,IIS-HttpRedirect,IIS-WindowsAuthentication,IIS-
    StaticContent,IIS-DefaultDocument,IIS-
    HttpCompressionStatic,IIS-DirectoryBrowsing
    ```

3. Confirm that IIS has been installed by reviewing the output screen, noting that there are no errors, as shown in *Figure 16.1*:

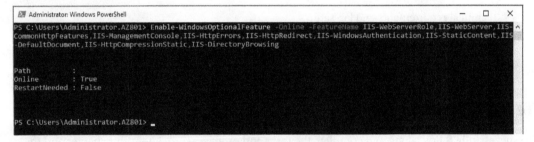

Figure 16.1 – Reviewing that IIS has been installed

4. Next, let's also install **Docker** into our VM by running the following two PowerShell commands by pressing *Enter* and *Y* after each command to trust the installation of **NuGet** and the installation of the **Docker** package:

    ```
    Install-Module -Name DockerMsftProvider -Repository
    PSGallery -Force
    Install-Package -Name docker -ProviderName
    DockerMsftProvider
    Restart-Computer
    ```

5. Confirm that the Docker components have been installed by reviewing the output screen, making sure that there are no errors (red text), as shown in *Figure 16.2*:

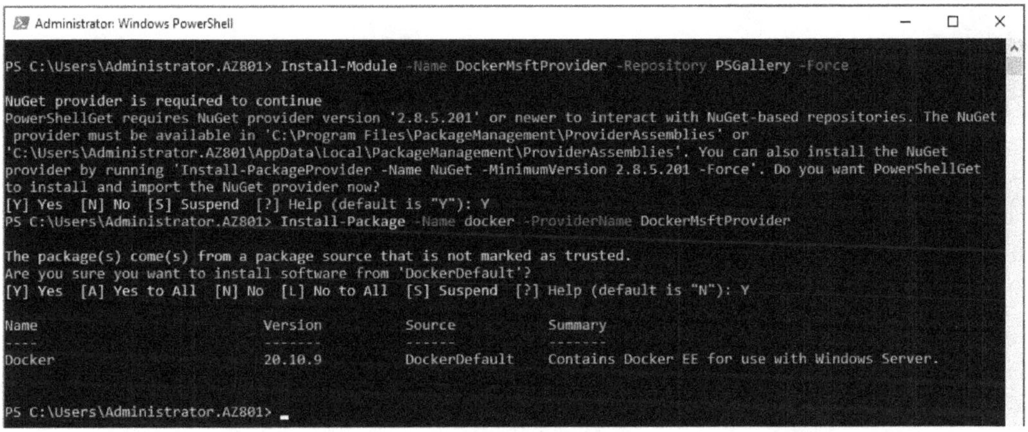

```
Administrator: Windows PowerShell                                                                    —    □    ×

PS C:\Users\Administrator.AZ801> Install-Module -Name DockerMsftProvider -Repository PSGallery -Force

NuGet provider is required to continue
PowerShellGet requires NuGet provider version '2.8.5.201' or newer to interact with NuGet-based repositories. The NuGet
 provider must be available in 'C:\Program Files\PackageManagement\ProviderAssemblies' or
'C:\Users\Administrator.AZ801\AppData\Local\PackageManagement\ProviderAssemblies'. You can also install the NuGet
provider by running 'Install-PackageProvider -Name NuGet -MinimumVersion 2.8.5.201 -Force'. Do you want PowerShellGet
to install and import the NuGet provider now?
[Y] Yes  [N] No  [S] Suspend  [?] Help (default is "Y"): Y
PS C:\Users\Administrator.AZ801> Install-Package -Name docker -ProviderName DockerMsftProvider

The package(s) come(s) from a package source that is not marked as trusted.
Are you sure you want to install software from 'DockerDefault'?
[Y] Yes  [A] Yes to All  [N] No  [L] No to All  [S] Suspend  [?] Help (default is "N"): Y

Name                    Version         Source          Summary
----                    -------         ------          -------
Docker                  20.10.9         DockerDefault   Contains Docker EE for use with Windows Server.

PS C:\Users\Administrator.AZ801> _
```

Figure 16.2 – Reviewing that Docker has been installed

6. Still in the open PowerShell terminal, let's run the following script, which will create new directories, place some basic **HTML** files into the directories, and create two new IIS websites automatically:

```
cd c:\windows\system32\inetsrv
New-Item -Path c:\Web\AZ801-App1 -ItemType Directory
New-Item -Path c:\Web\AZ801-App2 -ItemType Directory
New-Item -Path c:\Web\AZ801-App1 -Name index.html
-ItemType File -Value "<h1>Hello Packt AZ801 Readers!</
h1><p>This is our simple page for AZ801 running on
site:</p><h2>AZ801-App1</h2>"
New-Item -Path c:\Web\AZ801-App2 -Name index.html
-ItemType File -Value "<h1>Hello Packt AZ801 Readers!</
h1><p>This is our simple page for AZ801 running on
site:</p><h2>AZ801-App1</h2>"
.\appcmd add site /name:AZ801-App1 /id:2 /
physicalPath:c:\Web\AZ801-App1 /bindings:http/*:82:az801-
app1.az801.com
.\appcmd add site /name:AZ801-App2 /id:3 /
physicalPath:c:\Web\AZ801-App2 /bindings:http/*:83:az801-
app2.az801.com
```

7. This completes the technical requirements and lab setup for this chapter.

Let's continue by learning how to migrate existing IIS workloads to Azure App Service and Azure Web Apps.

Migrating IIS workloads to Azure Web Apps

As we have learned throughout this guide, dependencies are key for all workloads, and IIS applications are no different. Admins, architects, and engineers alike must take into consideration different application requirements, ports used, identities used for hosting the applications, and authentication type(s) used, among other application and environmental-specific dependencies. The Microsoft Azure team has provided the following Wiki documenting the various readiness checks for any IIS to Azure migration efforts: https://github.com/Azure/App-Service-Migration-Assistant/wiki/Readiness-Checks.

In addition to the Wiki, the **Azure App Service Migration Assistant** can be utilized as a set of tools and applications to help transform and migrate applications to the cloud. This assistant acts as a set of assessments on specific web applications and determines details and dependencies to complete a smooth transition to the cloud, including any warnings or errors that could prevent migration. This tool also allows you to complete the migration from within the tool, allowing you to customize the destination within Microsoft Azure and export an **Azure Resource Manager** (**ARM**) template for repeatable use for automation efforts.

Similarly, **Azure App Service** provides many features over traditional server environments, helping you cut hardware costs, software and hardware updates, configuration changes, and overall setup efforts. Not only does Azure App Service allow you to closely monitor performance and insights, but you can quickly detect where in the application hosting chain you may be experiencing failures. Here, you can scale up or scale out when needed to give flexibility for hosting needs, and you can use deployment slots to establish live staging and testing environments for agility in promoting or swapping test and production versions of applications.

For more information on the Azure App Service Migration Assistant, be sure to visit the following URL: https://appmigration.microsoft.com. As an example, I have used my blog URL of https://cfgcloud.io to run the free compatibility report located on the Azure App Service migration tools page. The results confirm that my site is supported for migration to Microsoft Azure, as shown in *Figure 16.3*:

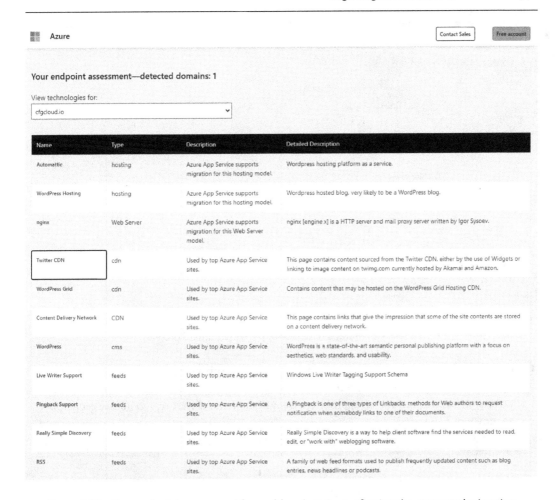

Figure 16.3 – Free endpoint assessment for my blog domain, confirming the supported migration

Now, let's begin our hands-on exercise with the Azure App Service Migration Assistant!

1. While connected to and logged into `AZ801PacktLab-FS-01`, navigate to `https://portal.azure.com` and log in while utilizing the Global Administrator account you created in *Chapter 1*.

2. Select **Azure Migrate** from the list of Azure services or simply search for `Azure Migrate` and select it to continue.

3. Within **Azure Migrate**, select the **Web apps** blade under **Migration goals**, and then select **+ Create project**, as shown in *Figure 16.4*:

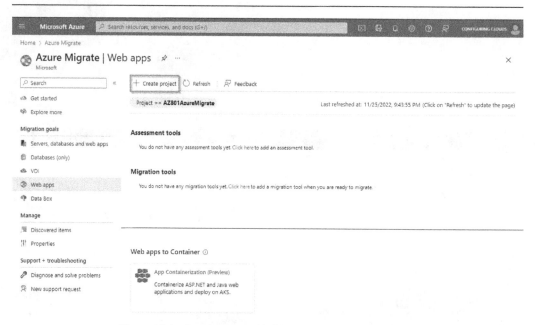

Figure 16.4 – Creating a new Web apps Azure Migrate project

4.   On the **Create project** screen, select an appropriate **Subscription**, the **Resource group** option we have been using for our AZ801 resources, a suggested **Project** name of AZ801-WebAppMigration, and an appropriate **Geography** (your closest representative geography), as shown in *Figure 16.5*:

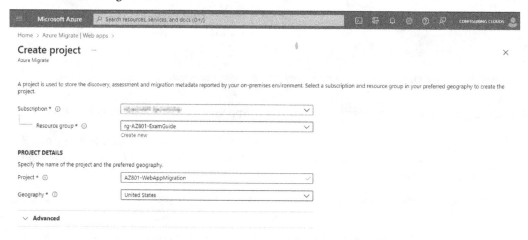

Figure 16.5 – Selecting the appropriate project details based on your geography and resources

5.   From the **Azure Migrate | Web Apps** page, under **Assessment tools**, select **1: Download App Service Migration Assistant**, as shown in *Figure 16.6*, then accept the **Microsoft Software License Terms** shown:

Figure 16.6 – Download App Service Migration Assistant

6. Once the MSI has finished downloading, double-click it to install it on our IIS server VM.

7. From the Start menu, locate **Azure App Service Migration Assistant** and run it. Then, choose **AZ801-App1** from the list and select **Next**, as shown in *Figure 16.7*:

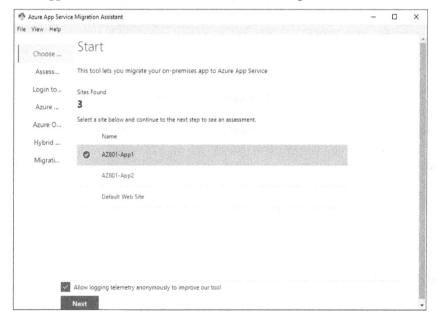

Figure 16.7 – Choosing an app to assess via Azure App Service Migration Assistant

8. Notice that the **Assessment** screen indicates that there is an error preventing automatic migration, stating that the port binding for our application is not currently supported, as shown in *Figure 16.8*. No worries – let's work to resolve this blocker!

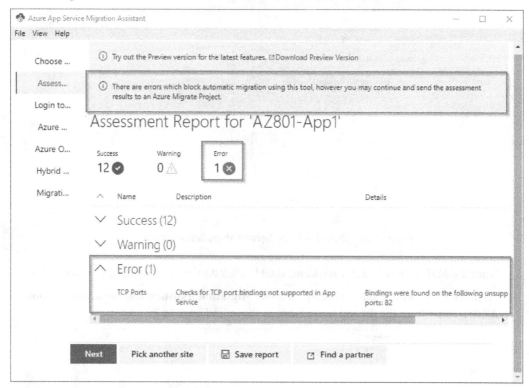

Figure 16.8 – Troubleshooting an error picked up by the Migration Assistant tool

9. From the Start menu, locate **Windows Administrative Tools** > **Internet Information Services (IIS) Manager** and select it to launch it. Expand our server's name and then **Sites**, then locate **Default Web Site** in the list and select **Stop** from the right-hand menu under **Manage Website**.

10. Locate **AZ801-App1** in the list and select it, then select **Bindings…** from the right-hand menu under **Edit site**. Select the lone entry in the list and select the **Edit** button, then change **Port:** from 82 to 80 and select **OK**. Finally, click **Close**.

11. Back in **Azure App Service Migration Assistant**, return to **Choose a site** and rerun the assessment for AZ801-App1. Note that this time, it results in 13 successful checks. Click **Next** to continue.

12. On the **Login to Azure** page, select the **Copy Code & Open Browser** button, as shown in *Figure 16.9*. Note that you will receive two to three prompts, asking you if you trust or are trying to sign into **Azure App Service Migration Assistant** – select **Yes** or **Continue** where appropriate, then select **Next** to continue:

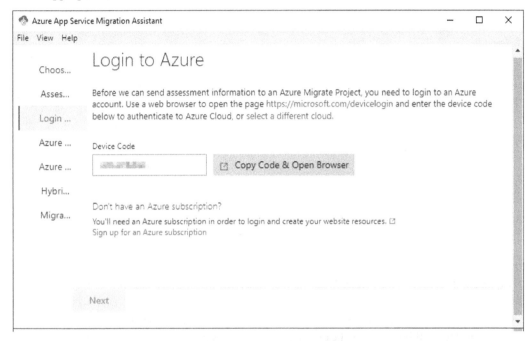

Figure 16.9 – Logging into Azure to tie our Migration Assistant results to our Azure Migrate project

13. Back in the **Azure App Service Migration Assistant** tool and **Azure Migrate Hub**, select the project we created earlier under **Azure Migrate Project**, as shown in *Figure 16.10*, and click **Next** to continue:

Figure 16.10 – Selecting the correct Azure Migrate Project for the Azure Migrate Hub connection

14. You will be shown the **Azure Options screen**, so ensure you select the **Subscription** property you created for this guide, use the existing **Resource group** of `rg-AZ801-ExamGuide`, and select a unique **Destination Site Name** (we recommend the name of the app, your first and last initial, and the date). Then, select **Create new** under **App Service Plan** and select an appropriate nearby **Region**. Then, click the **Migrate** button, as shown in *Figure 16.11*:

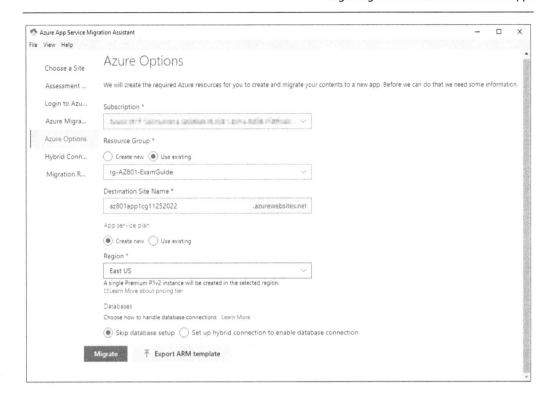

Figure 16.11 – Configuring Azure Options for the Migrate project

15. While the application automatically migrates to Azure, we can click **Export ARM Template to export our template** for later use, click to **Migrate another site**, or choose **Go to your website** to visit the newly migrated site. Select **Go to your website**, as shown in *Figure 16.12*:

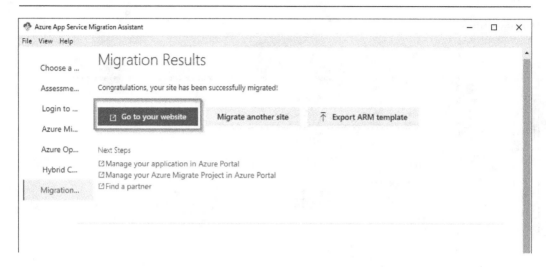

Figure 16.12 – Completing the site migration and visiting our migrated website

16. With some luck and careful configuration, you should now be able to access a previously on-premises application hosted within **Microsoft Azure**, as shown in *Figure 16.13*:

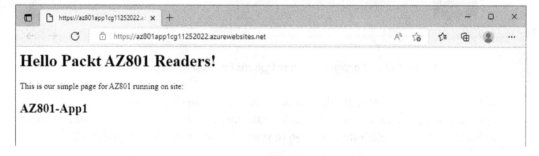

Figure 16.13 – Reviewing our newly migrated website running in Azure App Service

While this was an incredibly basic application that we migrated, imagine the possibilities of completing this type of migration for enterprise applications at scale! With that, we have completed this section on how to migrate IIS workloads to Azure Web Apps. Next, we will learn how to migrate existing IIS workloads to containers using container technology.

Migrating IIS workloads to containers

At a high level, containers are considered isolated environments where an application or service can run without external interference. The container is typically defined in a file, with the application and dependencies built with configuration pieces of the host resources (think OS, CPU, RAM, and network) and then logically separated from other containers running on the same host or cluster.

These containers can then run on a variety of platforms and operating systems, such as **Linux**, **macOS**, **Apache Tomcat**, **Windows**, and **Azure Kubernetes Service** (**AKS**). In addition, Windows containers specifically can be created and managed with a variety of tools, from simple file editors to **Visual Studio** or **VSCode**, **Docker**, or **Windows Admin Center**. To simplify our efforts, we will focus on the use of **Docker** in this chapter. So, let's begin by learning how to create a **Dockerfile** and container image for migration:

1. Ensure that `AZ801PacktLab-FS-01` has been powered off (i.e., it shows an **Off State** in **Hyper-V Manager**). Then, open **PowerShell** as an administrator on your host device and enter the following commands to install the necessary tools:

    ```
    New-Item -Path c:\AZ801PacktLab\Containers -ItemType
    Directory
    Install-Module Image2Docker
    Import-Module Image2Docker
    ```

2. While in the open PowerShell terminal on your host device, enter the following command to assess our VM and create the Dockerfile:

    ```
    ConvertTo-Dockerfile -ImagePath C:\AZ801PacktLab\VMs\
    AZ801PacktLab-FS-01.vhdx -Artifact IIS -OutputPath C:\
    AZ801PacktLab\Containers
    ```

3. After a few minutes, a new configuration export will be completed and our **Dockerfile** will be created, as shown in *Figure 16.14*:

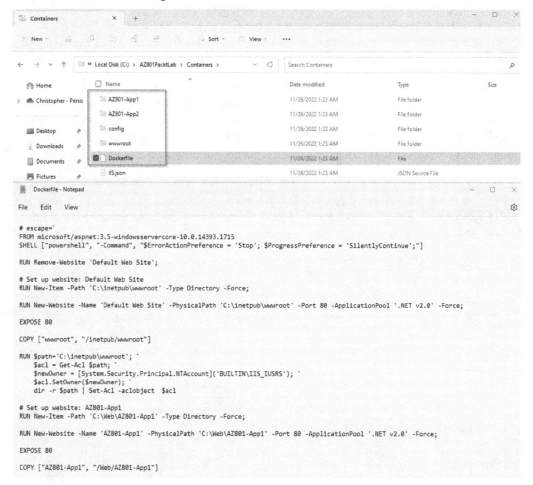

Figure 16.14 – Creating and reviewing a Dockerfile based on the AZ801PacktLab-FS-01 IIS configuration

4. At this point, we only need to build our Dockerfile to create the necessary image to migrate the websites from the VM to another container host. To complete the image build, we can run the following command:

```
Docker build -t webserver .
```

5. Note that this build process will take quite some time to complete. This process can be monitored from within the PowerShell terminal, as shown in *Figure 16.15*:

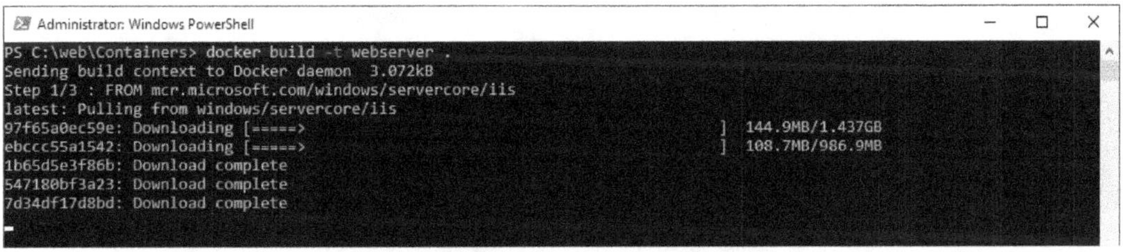

Figure 16.15 – Building our Dockerfile into a container image

---

Container management and deployment information (optional)

At this point, the container image that has been created can be pushed into a private container registry using **Azure Container Registry**. Once the application has been published, the image can then be deployed to a **Kubernetes cluster** running within **Azure Kubernetes Service**. While this may not be a requirement in the AZ-801 exam, it's valuable knowledge as more and more applications transition to a cloud-native and serverless architecture.

Creating a private container registry using Azure Container Registry can be reviewed here: `https://learn.microsoft.com/azure/container-registry/container-registry-get-started-powershell`; creating a Kubernetes cluster on Azure Kubernetes Service can be reviewed here: `https://learn.microsoft.com/azure/aks/learn/quick-windows-container-deploy-cli`.

Additional examples of Windows containers provided by Microsoft can be found at the following URL for those who wish to learn more about container solutions: `https://learn.microsoft.com/virtualization/windowscontainers/samples?tabs=Application-frameworks`.

With that, we have learned how to migrate IIS workloads to containers. Next, we will review what we have learned in this chapter and set the stage for the next chapter of this book.

Summary

In this chapter, we learned about the available tools for migrating IIS workloads to Microsoft Azure. We dived into how to migrate IIS workloads to Azure Web Apps, and how to migrate IIS workloads to Windows containers using the Dockerfile technology to achieve the AZ-801 exam preparation objectives, consistent with the AZ-801 exam objectives.

In the next chapter, we will learn how to determine the various approaches for moving **Domain Controllers** to **Windows Server 2022**. We will learn how to use the **Active Directory Migration Tool** to migrate **Active Directory** objects such as users, groups, and Group Policy objects. We will also learn how to migrate Active Directory objects to a new Active Directory forest. Finally, we will learn about the best practices for successfully upgrading an existing forest to Windows Server 2022.

Figure 15.15 – Building a Dockerfile into a container image

- Container management and deployment (optional)

At this point, the container image that has been created can be pushed into a repository using Azure Container Registry. Once the application image is built, the image can then be deployed to a Kubernetes cluster running within Azure AKS. If the deployment has been run out in the AKS cluster, its publicly provisioned access to the application transition to production ready for a production environment.

Creating a private container registry using Azure Container Registry is performed through the `New-AzContainerRegistry` command. A registry can be created in a resource group using the `New-AzContainerRegistry` cmdlet. Images can also be deployed to an Azure Kubernetes Service cluster, which can be created using the `New-AzAksCluster` cmdlet. The application container deployed in the cluster can be created using the `kubectl apply` cmdlet and deployed using the `kubectl rollout status deployment` cmdlet.

Typical examples of Windows containers using the `New-AzContainerGroup` cmdlet and deployment can be done using the `kubectl apply` cmdlet. The examples covered in this chapter have also demonstrated how to build PowerShell workloads in containers. Now we will move on to see what we have learned in this chapter and set the stage for the next chapter of this book.

Summary

In this chapter, we learned about the available tools for building IIS workloads in other such scenarios. We also discussed how to create IIS workloads in Azure with Azure containers in mind. We discussed how Kubernetes components work using the `kubectl` tool to build Kubernetes components that are consistent with the Azure PowerShell cmdlets.

In the next chapter, we will learn how to design different solutions approaches for moving Domain Controllers to Windows Server 2024. We will learn how to build the Active Directory integration and learn to migrate Active Directory objects to the systems, groups and resources. You will also learn how to manage Active Directory with the PowerShell and modules, and we will learn about the best practices with administering and developing Windows Server 2024.

17

Migrating an Active Directory Domain Services (AD DS) Infrastructure to Windows Server 2022 AD DS

In this chapter, we will learn how to determine the various approaches for moving **Domain Controllers** to **Windows Server 2022**. We will learn how to use the **Active Directory Migration Tool** to migrate **Active Directory** objects such as users, groups, and Group Policy objects. We will also learn how to migrate Active Directory objects to a new Active Directory forest. Finally, we will learn best practices for successfully upgrading an existing forest to Windows Server 2022 to achieve the AZ-801 exam's preparation objectives.

In this chapter, we will cover the following topics:

- Technical requirements and lab setup
- Upgrading an existing forest
- Migrating to a new Active Directory forest
- Migrating AD DS objects, users, groups, and Group Policy objects using the Active Directory Migration Tool

Technical requirements and lab setup

To successfully follow along and complete the tasks and exercises throughout this chapter and the following chapters in this book, we will need to ensure that the technical requirements from *Chapter 1, Exam Overview and the Current State of Cloud Workflows*, have been completed in full. We will focus on the requirements, best practices, and walkthroughs in this chapter that align with the AZ-801 exam objectives.

Let's begin with a general overview of **Active Directory Domain Services** (**AD DS**), including what features are new and notable in **Windows Server 2022**.

OSs continually reach the end of their support cycle and many organizations undergo a process to migrate away from these legacy and unsupported OSs via either server life cycle or workload modernization. **Active Directory** migrations from legacy operating systems tend to follow a similar process. With **Windows Server 2008 R2** leaving extended support, and **Windows Server 2012 R2** nearing an extended end date and soon entering extended security support, organizations must consider migration to newer operating systems while completely removing **Domain Controllers** that are older than Windows Server 2012 R2 from the domain.

Those that have been working with Active Directory for some time know that every Windows Server release includes AD DS improvements in the form of both domain and forest functional levels. Examples of some of the features over the years were a change from **File Replication Service** (**FRS**) to **Distributed File Service Replication** (**DFS-R**) for **SYSVOL** folder content replication, **Authentication Policies** and **Policy Silos**, **Privileged Access Management** (**PAM**), and DC-side protections for **Protected Users**, to name a few. For a full detailed list of all the features introduced over the years for both domain and forest functional levels, be sure to check out this Microsoft article: `https://learn.microsoft.com/windows-server/identity/ad-ds/active-directory-functional-levels`.

However, the world changes, and cloud services are used more frequently, identity services tend to be more cloud-managed nowadays, and advancements in strong authentication, security, insights, and a zero-trust approach have taken hold. **Windows Server 2016** was the last operating system to include forest and domain functional level changes, as both **Windows Server 2019** and **Windows Server 2022** took on a security-first modern approach for server roles within an environment.

Let's step into a repeatable approach for upgrading an existing AD DS environment with a migration process for an existing forest.

Upgrading an existing forest

While the process of upgrading forest and domain functional levels and migrating **Flexible Single Master Operation** (**FSMO**) roles to a new server takes only a few minutes to complete, some prerequisites need to be considered for a successful upgrade or migration. The following list outlines the best practices and requirements for an AD DS migration process:

- Do the homework first and create a detailed implementation plan, including current Domain Controllers, server configuration down to software, additional server roles, disk, and network configuration. Consider including an audit of the current AD infrastructure (including Group Policy) to determine baseline health.

- Evaluate and document the business requirements and maintenance windows before proceeding with the Active Directory migration. Include any existing monitoring solutions as these will most likely need to be updated or reconfigured after the upgrade.

- Ensure that you have a system state backup, ensuring that the boot files, registry, SYSVOL folder, Certificate Server, cluster database, and registration databases are included in case of emergency.

- Consider patching the existing environment to the highest level possible before migration to ensure there are no outstanding security issues or known migration blockers.

- Identify the destination resources, whether they are physical or virtual, assign a static IP address, and install the necessary AD DS role (and accompanying roles such as DNS and DFS-R, which are installed automatically).

- Ensure that proper system or network time is synchronized in both the source and destination domains where objects are being migrated.

- Map out the current AD forest design so that it includes any parent-child relationships, tree-root, shortcut, forest, external, and realm trusts. Trust flows must be included to determine transitive versus non-transitive characteristics, the direction of existing trusts, specific authentication mechanisms used, and any SID filtering.

- If the environment was sourced from Windows Server 2008 or 2008 R2 Domain Controllers, the environment is already running DFS-R instead of FRS for SYSVOL folder replication. If the environment was sourced from Windows 2000 Server or Windows Server 2003, there is a chance that FRS is still in use and should undergo SYSVOL migration using the following Microsoft documentation: `https://learn.microsoft.com/windows-server/ storage/dfs-replication/migrate-sysvol-to-dfsr`.

- Promote new servers on newer OSs to DCs while demoting the older DCs as needed during migration to ensure legacy systems are removed from the domain. Ensure that the new DCs are introduced to the forest root level first, then continue to the domain tree levels where appropriate.

- Consider completing a migration test with a validation account to ensure that permissions are intact post-migration. When running a least-privileges model for AD DS, you will need to run multiple tests to validate all identified permission sets.

- Communicate the migration details and any outages to the business and share any changes that the end users will encounter after the migration.

Now that we have covered the requirements and suggested best practices, let's discuss the process for introducing and promoting a new Windows Server 2022 Domain Controller to help with upgrading an existing forest. As a reminder, please do *NOT* complete these tasks in your lab environment as we will utilize this environment throughout the remaining chapters of this book! In addition, we highly recommend that you review the section *Installing a New Windows Server 2022 Domain Controller*, in *Chapter 2*, for additional details on establishing a new server as a Domain Controller:

1. Ensure that you are signed into the new server (intended to be the new Domain Controller) as an Enterprise Administrator.

2. Confirm that a static IP address has been configured using the `ipconfig /all` command.

3. Once the server has been fully prepared with an OS installation that has been fully patched, ensure that the server has been added to the existing domain.

4. Launch PowerShell on any existing DC and then run `dfsrmig /getmigrationstate`. If the environment is already using DFS-R, the command will return a global state of (`'Eliminated'`). If the state does not show eliminated, an FRS to DFS-R migration will need to be completed, as previously discussed.

5. Launch PowerShell as an Administrator to begin server configuration.

6. As a best practice, run the following command to complete the prerequisites check before initiating the full DC promotion process, with the expectation of a `Success` status message encouraging us to proceed:

```
Test-ADDSForestInstallation -DomainName ad.az801.com
-InstallDns
```

7. The following command can be issued to install all necessary DC features and roles:

```
Install-WindowsFeature -Name AD-Domain-Services
-IncludeManagementTools
```

8. Configuration can be completed on the new Windows Server 2022 DC by running the following command:

```
Install-ADDSDomainController -CreateDnsDelegation:$false
-InstallDns:$true -DomainName "ad.az801.com" -SiteName
"SecondADSite" -ReplicationSourceDC "AZ801PacktLab-DC-01.
az801.com" -DatabasePath "C:\Windows\NTDS" -LogPath
"C:\Windows\NTDS" -SysvolPath "C:\Windows\SYSVOL"
-Force:$true
```

The preceding parameters are explained in *Table 17.1* for a detailed reference:

Install-ADDSDomainController parameter	Description
`-CreateDnsDelegation`	This defines whether the DNS delegation that references the AD-integrated DNS should be created.
`-InstallDns`	This defines whether the DNS role should be installed on the new AD Domain Controller. For a new forest, DNS is automatically installed as the Global Catalog server for the forest and domain.
`-DomainName`	This defines the **fully-qualified domain name** (**FQDN**) for the AD domain.

Install-ADDSDomainController parameter	Description
-SiteName	This defines the AD site name, with the default value being 'Default-First-Site-Name'.
-ReplicationSourceDC	This defines the AD replication source. When this parameter is not supplied, the nearest available DC will be used for initial replication.
-DatabasePath	This defines the folder path for the AD database.
-LogPath	This defines the location for storing the domain log files.
-SysvolPath	This defines the location for the **system volume** (**SYSVOL**) folder, with a default location of C:\ Windows.
-Force	This defines whether the command will bypass the best practices and recommendations warnings when installing the Domain Controller.

Table 17.1 – Parameter definitions for the Install-ADDSDomainController PowerShell cmdlet

9. You will then be prompted to supply a SafeModeAdministrator password. This should be a complex password to be used only for **Directory Services Restore Mode** (**DSRM**).

10. Next, we need to identify which Domain Controllers are currently hosting the **FSMO** roles by running Get-ADDomain | Select-Object InfrastructureMaster, RIDMaster, PDCEmulator and then Get-ADForest | Select-Object DomainNamingMaster, SchemaMaster on the Domain Controller, as shown in *Figure 17.1*:

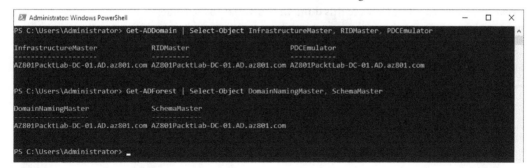

Figure 17.1 – Identifying the current FSMO role holders using PowerShell cmdlets

11. Next, the FSMO roles can all be migrated to the new Windows Server 2022 Domain Controller by running `Move-ADDirectoryServerOperationMasterRole -Identity AZ801PacktLab-DC-22.AD.az801.com -OperationMasterRole DomainNamingMaster, InfrastructureMaster, PDCEmulator, RIDMaster, SchemaMaster` on any Domain Controller, as shown in *Figure 17.2*. Note that there are additional ways to complete these steps using the user interface, as outlined in the following article: `https://activedirectorypro.com/transfer-fsmo-roles/`:

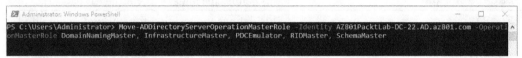

Figure 17.2 – Migrating the FSMO roles to the new Windows Server 2022 Domain Controller

12. At this point, we're ready to decommission any old Domain Controllers from the environment and can begin by logging into the to-be-deprecated DC as an Enterprise Administrator.

13. From the CMD prompt or PowerShell, we can run the `dcpromo` command to begin uninstalling the **Active Directory Domain Services** role from the server(s).

But Chris, the dcpromo command has been deprecated for a while now

While `dcpromo` has been deprecated as of Windows Server 2012, older Windows OSs such as Windows Server 2003, Windows Server 2008, and Windows Server 2008 R2 do not have the luxury of using Server Manager or PowerShell to replace `dcpromo` operations.

To remove Active Directory Domain Services in Windows Server 2012 R2 or later, consider launching **Server Manager** and selecting **Manage** > **Remove Roles and Features**. Then, select the DC host and uncheck the box for **Active Directory Domain Services**. After, select **Remove Features** and then select **Demote this domain controller** to continue.

You can also utilize the `Uninstall-ADDSDomainController -DemoteOperationMasterRole -RemoveApplicationPartition` PowerShell command to remove the **AD DS** role from the DC.

Continuing in Server Manager to remove the AD DS server role, select **Demote this domain controller** to commence the demotion process, as shown in *Figure 17.3*:

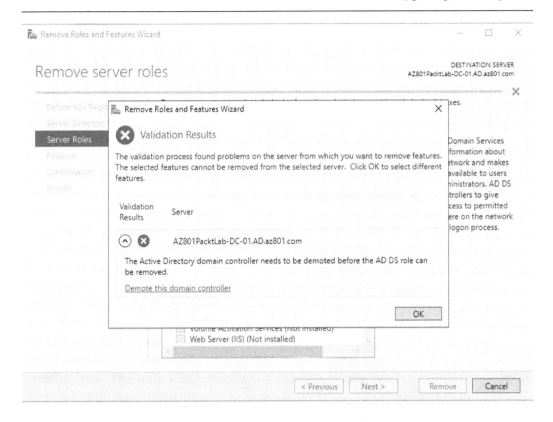

Figure 17.3 – Example of using Server Manager to demote a domain controller

14. As part of the process, we are asked whether this is the last domain controller in the domain; therefore, we will not select the checkbox for **Delete the domain because this server is the last domain controller in the domain**.

15. Moving on, we are then prompted with a screen reading **Delete the DNS delegations pointing to this server**. Here, we will accept the defaults.

16. Next up, we are presented with a prompt to supply credentials for the Domain Administrator. We need to complete this step to confirm that the DC has been removed from the domain.

17. We will then be prompted for a new local administrator password for the server once it has been returned to the domain as a member server. At this point, the server can also be rebooted to complete the demotion process.

18. It is recommended that this server be completely decommissioned from the environment and not reused, or in-place upgraded, ensuring there are no residual artifacts from the prior Domain Controller configuration.

The next step requires raising the domain and forest functional level. To complete these tasks, we can run the following PowerShell commands on the new Windows Server 2022 Domain Controller:

19. To upgrade the domain functional level, run the following command:

```
Set-ADDomainMode -identity ad.az801.com -DomainMode
WinThreshold
```

20. To upgrade the forest functional level, run the following command:

```
Set-ADDomainMode -identity ad.az801.com -ForestMode
Windows2016Forest
```

21. Finally, we can verify that our migration has been completed successfully by running the following two PowerShell commands:

```
Get-ADDomain | fl Name,DomainMode
Get-ADDomain | fl Name,ForestMode
```

This completes the migration from Windows Server 2008 R2 to a newer Windows Server 2022 AD DS environment. Note that these same steps can be completed for migrating Windows Server 2012, Windows Server 2016, or Windows Server 2019 to Windows Server 2022 as per your workload and identity life cycle requirements.

Now, let's discuss how to migrate objects such as users, groups, and **Group Policy** objects using the **Active Directory Migration Tool**.

Migrating AD DS objects, users, groups, and Group Policy objects using the Active Directory Migration Tool

For those that have been in and around Active Directory for years or even those new to this technology, the **Active Directory Migration Tool** (**ADMT**) has been the go-to tool for quite some time. While ADMT may be an incredibly useful migration tool, the code base has been deprecated by Microsoft, meaning that no further work is being done on this tool. A detailed support policy and known issues for ADMT can be found at the following URL: https://learn.microsoft.com/troubleshoot/windows-server/identity/support-policy-and-known-issues-for-admt.

That said, ADMT 3.2 can still be used for migrations and the tool supports the following objects and features for migration:

- User Account Migration
- Group Account Migration
- Computer Account Migration

- Security Translation

- Service Account Migration

- Managed Service Account Migration

- Password Migration (after a password migration filter DLL has been installed and correctly configured for use as part of **Password Export Server** or **PES**)

Along with the features of ADMT 3.2, as shown in *Figure 17.4*, the tool also includes a reporting tool that provides insights into the progress of migrated user accounts, migrated computer accounts, expired accounts, account references, and account name conflicts:

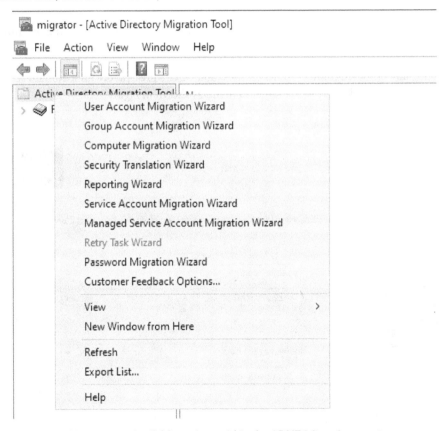

Figure 17.4 – Available options within the ADMT 3.2 environment

The following requirements and best practices are needed for a successful ADMT installation and migration:

- Recommended installation of **Microsoft SQL Server 2012** Express edition for the best performance and least integration issues

- **ADMT** 3.2 must be installed on the target domain controller to avoid unconstrained delegation requirements for ADMT

- Disabling **Windows Defender Credential Guard** on the target domain controller (ADMT server)

- Disable **LSA** protection on the target domain controller (the **ADMT** server) to allow the password export server to run

- Establish a two-way trust relationship, ensuring domain-wide authentication is selected

- Create an **ADMT** administrative user in the destination domain, then set this account up as a Domain Administrator on the new domain and ensure that the source domain has this user set up as part of the `Builtin\Administrators` group

- Set the **ADMT** admin account to have local administrator privileges for all the machines in the source domain for successful computer migration

When selecting any of the migration wizard options, you will be presented with a domain selection screen to help translate the source to the target domain, as shown in *Figure 17.5*:

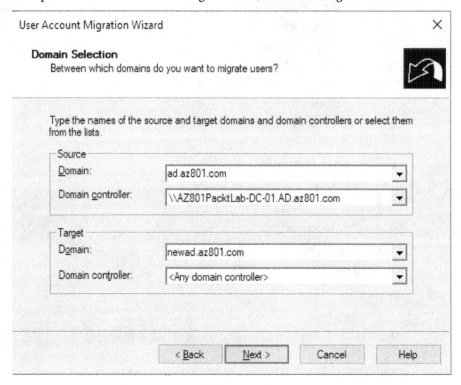

Figure 17.5 – ADMT User Account Migration Wizard – Domain Selection example

Let's step into some of the ways that we can migrate objects from a source domain to a target domain.

For **user account** migrations, the following high-level process can be completed:

1. Open the **ADMT** console.

2. Select **Action > User Account Migration Wizard**.

3. When prompted for a **Domain Selection**, select **Source** and **Target** domains as appropriate.

4. When prompted for **User Selection**, select user accounts individually (or utilize an include file containing the objects you want to migrate).

5. When prompted for **Organizational Unit Selection**, select **target OU**.

6. When reviewing **Password Options**, select **Migrate Password**.

7. On the **Account Transition Options** page, select the appropriate options for your migration.

8. On the **User Account** page, enter the `username`, `password`, and `domain` properties of your **ADMT** administrator account.

9. On the **User Options** page, select the appropriate options for your migration.

10. On the **Object Property Exclusion** page, select the appropriate options for your migration.

11. Finally, on the **Conflict Management** page, select the appropriate options for your migration.

12. Select **Finish** to complete the **User Account Migration Wizard** area. Confirm that the migration has been completed successfully by selecting **View Log**.

For **group account** migrations, the following high-level process can be completed:

1. Open the **ADMT** console.

2. Select **Action > Group Account Migration Wizard**.

3. When prompted for a **Domain Selection**, select **Source** and **Target** domains as appropriate.

4. When prompted for **Group Selection**, select group accounts individually.

5. When prompted for **Organizational Unit Selection**, select **target OU**.

6. On the **User Options** page, select the appropriate options for your migration.

7. On the **Object Property Exclusion** page, select the appropriate options for your migration.

8. Finally, on the **Conflict Management** page, select the appropriate options for your migration.

9. Select **Finish** to complete the **Computer Account Migration Wizard** area. Confirm that the migration has been completed successfully by selecting **View Migration Log** to review the full list of computers, task completion status, and full path to each computer's migration log file.

For **computer account** migrations, the following high-level process can be completed:

1. Open the **ADMT** console.

2. Select **Action > Group Account Migration Wizard**.

3. When prompted for a **Domain Selection**, select **Source** and **Target** domains as appropriate.

4. When prompted for **Computer Selection**, select computer accounts individually (or utilize an include file containing the objects you want to migrate).

5. When prompted for **Organizational Unit Selection**, select **target OU**.

6. On the **Translate Object** page, select the appropriate options for your migration.

7. On the **Security Translation Options** page, uncheck all the available options.

8. On the **Object Property Exclusion** page, select the appropriate options for your migration.

9. On the **Conflict Management** page, select the appropriate options for your migration.

10. Finally, on the **ADMT Agent Dialog** page, select **Run pre-check and agent operation**, and then select **Start**.

11. Select **Finish** to complete the **Group Account Migration Wizard** area. Confirm that the migration has been completed successfully by selecting **View Log**.

For **Group Policy** migrations, the following high-level process can be completed. ADMT does not migrate **Group Policy Objects (GPOs)** between forests and domains:

1. Utilize the **Group Policy Management Console (GPMC)** area to back up any or all the GPOs destined for migration to the target.

2. Copy the backup GPOs and then paste them into a known location on the target domain controller.

3. Within **GPMC**, locate the **Group Policy Objects** node, right-click on it, and select **Open Migration Table**. Select **Tools**, then **Populate From Backup**, and then select the objects destined for migration.

4. The table will populate with values that can be updated based on the target domain properties. The resulting XML file can then be used during import to address any changes to security principals or **UNC** paths.

5. Finally, utilize the GPMC, import any or all the GPOs destined for migration to the target, and utilize the migration table XML when needed to complete the translation.

Now that we have learned how to migrate objects such as users, groups, and Group Policy objects using the Active Directory Migration Tool, let's learn how to migrate AD DS to a new Active Directory forest.

Migrating to a new Active Directory forest

Our review of all the recommendations and best practices in this chapter builds up to this opportunity to discuss and review migrating to a new Active Directory forest (called an interforest migration).

Before we get started, let's quickly address a question that I hear often to this day: What's the difference between interforest and intraforest migrations? For most organizations, a recent merger or acquisition is involved that requires restructuring Active Directory Domains within a forest or multiple forests, though it could just be simply to clean up a dysfunctional Active Directory design. For additional details on restructuring, be sure to review the following article: `https://learn.microsoft.com/previous-versions/windows/it-pro/windows-server-2003/cc755720(v=ws.10)?redirectedfrom=MSDN`.

Let's look at the following table to review the differences between the two options:

Intraforest Migration	Interforest Migration
Relocates objects from source domains in one existing forest to target domains in a single forest. Computer and managed service accounts are copied, and the original accounts remain in the source domain.	Objects are cloned from source domains in one existing forest to target domains in another (existing or new) separate forest.
Objects that are migrated into the target domain no longer exist in the source domain.	Objects that are migrated exist in both the source and the target domains.
Passwords are always copied into the target domain for accounts.	Copying passwords to the target domain is optional.
SID history is maintained for objects, users, and computers.	Migrating SID history to the target domain is optional.
Easier to perform.	More complex to perform and requires extensive planning and documentation.

Table 17.2 – Comparison table between interforest and intraforest migrations

Now that we have covered the basics, let's walk through the high-level steps needed to complete an interforest migration for Active Directory:

1. Install **ADMT** on the target Domain Controller, as well as **SQL Server 2012 Express** edition (or on an ADMT server, so long as additional requirements are met, as outlined here: `https://learn.microsoft.com/troubleshoot/windows-server/identity/support-policy-and-known-issues-for-admt`).

2.   Establish **conditional forwarding** in each domain, cross-configuring the destination domain as a conditional forwarder in the source domain, and the source domain configured in the destination domain, as shown in *Figure 17.6*:

Figure 17.6 – Configuring a Conditional Forwarder in the destination forest

3.   Update the existing **Group Policy** settings so that they include a **DNS** suffix search to look for the existing domain first, then the new domain in this order.

4.   Establish a two-way trust relationship, ensuring forest-wide authentication is selected. Note that this trust should be validated within **Active Directory Domains and Trusts** once configured.

5.   Create an **ADMT** administrative user in the destination domain, then set this account up as a Domain Administrator on the new domain and ensure that the source domain has this user set up as part of the Builtin\Administrators group. In addition, the admin account will need to have local administrator privileges to all the machines in the source domain.

6.   Establish a **Password export server** (PES) to migrate the passwords between the forests running the service under an authenticated account. Note that you will need to create an encryption key on the **ADMT** server within the source domain by running the following command in PowerShell:

```
admt key/option:create /sourcedomain: <DNS or NetBIOS
name of source domain> /keyfile: <path to store the
encrypted key file> /keypassword: {< password >}
```

7.   Disable any SID filtering.

8.   Migrate service account creation and configuration.

9.   Migrate group objects using the ADMT **Group Account Migration Wizard**.

10. Migrate user objects using the ADMT **User Account Migration Wizard**.

11. Reapply any **access control lists** (ACLs) necessary for success and translate security on any member servers using the security translation wizard.

12. Migrate computer accounts and notify users.

13. Complete any resource updates and **File & Print** server migration.

14. Complete **Application Server** migration.

15. Clean up SID history.

16. Decommission the old source servers and domains (after taking a successful backup). Be sure to remove all the trust relationships between the source and target domains, including conditional forwarders, and disable any unnecessary accounts that were migrated.

These generalized steps should help you successfully migrate AD DS objects from any source domains you are managing to new or existing target domains. If you encounter any errors, a rollback process can be completed in most cases. The following article outlines the process and its caveats: `https://learn.microsoft.com/previous-versions/windows/it-pro/windows-server-2008-r2-and-2008/cc974385(v=ws.10)`.

With that, we have learned how to upgrade an existing AD DS forest. Next, we will review what we have learned in this chapter and set the stage for the next chapter of this book.

Summary

In this chapter, we learned how to determine the various approaches for moving Domain Controllers to Windows Server 2022. We reviewed how to use the Active Directory Migration Tool to migrate Active Directory objects such as users, groups, and Group Policy objects. We also reviewed how to migrate Active Directory objects to a new Active Directory forest. Finally, we covered best practices on how to successfully upgrade an existing forest to Windows Server 2022 to achieve skills consistent with the AZ-801 exam objectives.

In the next chapter, we will learn about the various ways we can monitor and be alerted on Windows Server performance and availability events. We will learn how to use **Performance Monitor** and **Data Collector Sets**, learn how to monitor servers and event logs using **Windows Admin Center**, and monitor overall server health using **System Insights**. We will also learn how to deploy and configure **Azure Monitor Agents** to collect performance data and alert on predefined thresholds. Finally, we will discuss how to use the **Azure** diagnostics extension in Azure VMs, as well as how to monitor performance using **VM Insights**.

Part 6: Monitor and Troubleshoot Windows Server Environments

In this section, we will learn about various ways we can successfully monitor and troubleshoot Windows Server environments running in both on-premises and cloud infrastructures.

This part of the book comprises the following chapters:

18

Monitoring Windows Server Using Windows Server Tools and Azure Services

In this chapter, we will learn about the various ways we can monitor and alert ourselves to Windows Server performance and availability events. We will learn how to use performance monitor and Data Collector Sets, learn how to monitor servers and event logs using Windows Admin Center, and monitor overall server health using System Insights.

We will also learn how to deploy and configure Azure Monitor Agents to collect performance data and alert on predefined thresholds. Finally, we will discuss how to use the Azure diagnostics extension in Azure VMs, as well as how to monitor performance using VM Insights to achieve the AZ-801 exam's preparation objectives.

In this chapter, we will cover the following topics:

- Technical requirements and lab setup
- Configuring monitoring and alerting for Windows Server
- Deploying Azure Monitor Agents
- Collecting performance counters to Azure
- Creating Azure Monitor alerts
- Monitoring Azure VMs using the Azure Diagnostics extension
- Monitoring Azure VMs performance using VM Insights

Technical requirements and lab setup

To successfully follow along and complete the tasks and exercises throughout this chapter and the following chapters in this book, we will need to ensure that the technical requirements from both *Chapter 1, Exam Overview and the Current State of Cloud Workflows*, and *Chapter 2, Securing the Windows Server Operating System*, have been completed in full. We will be using both the Domain Controller and the File Server VMs to complete the exercises and tasks throughout this chapter to align with the AZ-801 exam objectives. We will focus on the requirements, best practices, and hands-on walkthroughs in this chapter that align with the AZ-801 exam objectives.

Let's begin with some general insights and recommendations on the various tools that are made available to us for monitoring both the applications and the OS running on Windows Server.

Configuring monitoring and alerting for Windows Server

Over the years, Windows Server has increased the number of built-in tools that can be used for both monitoring and troubleshooting your Windows Server workloads. Monitoring and ease of troubleshooting are arguably just as important as implementing the environment itself.

Consider this: if you are unable to properly identify performance and reliability and are unable to continuously evaluate and optimize the overall performance of your Windows Server workloads both on-premises and in a hybrid architecture, your application and user experience will suffer thus reducing the success of your implementation. Let's begin by learning about two variants of the **Performance Monitor** tool available to a **Windows Server** administrator.

Monitoring Windows Server using Performance Monitor

Performance Monitor has been the go-to tool for quite some time (since circa 1993) and is still a component on every Windows Server OS to date. Performance Monitor can be accessed via the start menu on any system (or by entering `perfmon.msc` from the command line), and provides the following system performance information:

- Overall system diagnostics and system performance monitoring
- A collection of detailed information for troubleshooting and historical tracking
- Analysis of the performance of a system concerning the applications and services hosted on the server

Performance Monitor comprises three main components:

- **Monitoring Tools**, which allow us to create and adjust current counters from the server, including but not limited to memory and processor performance, communication protocol performance, network performance, and additional deeper insights into application workload counters. These details can then be displayed in line, histogram bar, or report (raw data) format.

- **Data Collector Sets**, which include user-defined (such as Server Manager Performance Monitor built-in counter sets), system performance and diagnostics, Event Trace sessions, and Startup Event Trace sessions.

- **Reports**, which include both user and system-defined, covering both Server Manager and system diagnostics and performance default reports for the server.

These Performance Monitor components are displayed in *Figure 18.1* for our **AZ801LAB-FS-01** machine as an example:

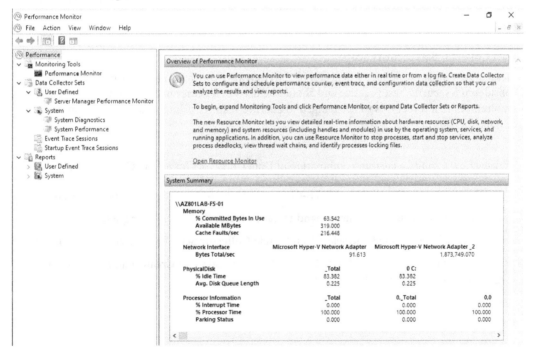

Figure 18.1 – Overview of Performance Monitor characteristics on a server

Now that we have an overview of what Performance Monitor brings to the table for troubleshooting and insights for a Windows Server, let's dig a bit deeper into how to configure and utilize Data Collector Sets for troubleshooting and monitoring.

Creating and configuring Data Collector Sets

As we just learned, Data Collector Sets incorporate a custom set of performance counters, system configuration data, and event tracing to help you collect and review either real-time or historical performance data for your systems. To get started quickly, default templates are provided for monitoring and collecting data and include system diagnostics, system performance, **Windows Data Access Components (WDACs)**, and basics that include simple objects for tracing.

The beauty of Data Collector Sets is that these can also be customized and are meant to be portable collections of counters, event traces, and system configuration data that can be saved as a Data Collector Set and then run in tandem to collect and allow the results to be reviewed.

In general, Data Collector Sets help to consolidate multiple data collection points into a single collection that is portable for use on multiple systems. The Data Collector Set can be used to generate alerts or used to trigger other on-demand or scheduled data collector events. The Data Collector Sets and Performance Monitor data can then be organized collectively for an overall performance or insights review of an application or system and its event logs.

> **Data Collector Sets security scope**
>
> Please note that Data Collector Sets run in the context of the System user by default. This is considered a security best practice and should be used in all cases unless you have a specific requirement that forces this to be changed to another security principle for the environment or implementation.

Let's review how to build a custom combination of Data Collectors for a Data Collector Set:

1. While logged into the **AZ801LAB-FS-01** virtual machine, open **Performance Monitor**.
2. Within the left navigation pane, expand **Data Collector Sets**, then **User Defined**.
3. Right-click **User Defined** and select **New** > **Data Collector Set**. Then, select **Next** to continue.
4. For the name, enter `PacktAZ801-FS-01-Custom` and select **Create manually (Advanced)**, as shown in *Figure 18.2*. Then, select **Next**:

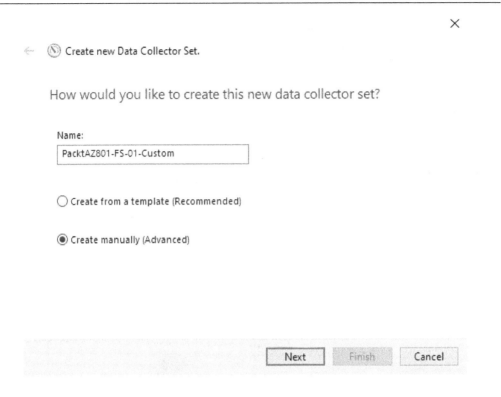

Figure 18.2 – Creating a new Data Collector Set within Performance Monitor

5. Accept the default **Create data logs**, and then select **Performance Counter** and select **Next** to continue.

6. On the **Which performance counters would you like to log?** page, select the **Add…** button:

    A. **Network Interface > Bytes Total/sec**

    B. **Processor > % Processor Time**

    C. **System > Processor Queue Length**

Select the following available counters by scrolling to each, selecting the counter, and selecting **Add >>**, as shown in *Figure 18.3*. Then, select **OK** to continue:

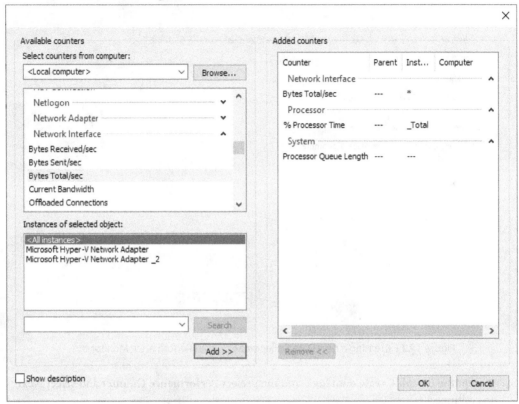

Figure 18.3 – Selecting Available counters for our custom Data Collector Set

7. Leave **Sample interval** set to **15** seconds, then select **Next**.

8. Note where the root directory is currently stored and select **Next**.

9. On the **Create the data collector set** page, take a note of the default security context for the Data Collector Set, then select **Finish**.

10. After being returned to the **User Defined Data Collector Sets** page, select our new PacktAZ801-FS-01-Custom set, right-click, and select **Start**, as shown in *Figure 18.4*:

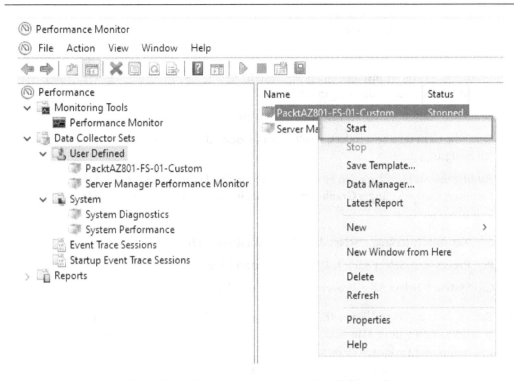

Figure 18.4 – Starting our new custom Data Collector Set

11. Let the data collector run for a few minutes and then right-click on our custom collector and select **Stop**.

12. Locate **Reports** > **User Defined** > **PacktAXZ801-FS-01-Custom** and then the **AZ801LAB-FS-01_XXXXXXX-000001** report to review the collected data to complete this exercise.

In this exercise, we became familiar with Performance Monitor, set up a custom Data Collector Set, and reviewed the output of the collected performance data via Performance Monitor reporting. Next, we will learn how to achieve a similar approach using **Windows Admin Center**.

Monitoring servers and configuring alerts using Windows Admin Center

If there is one thing you have learned from this exam guide, it is that **Windows Admin Center** continues to improve over time and revisit new ways to complete the same tasks you have formerly, with more efficiency. With the introduction of Windows Admin Center version 1910 a few years ago, Performance Monitor was reimagined to have a new design, approach, and overall usability increase for administrators. This includes new reporting features, the ability to search for objects and counters, reviewing definitions about each of the over 1,000 counters, better visualizations, and building monitoring workspaces.

Let's attempt the same exercise for setting up a Data Collector Set, but this time using Windows Admin Center:

1. Open a new browser instance and navigate to `https://az801lab-fs-01/` using your administrator credentials to log in if necessary.

2. Select **az801lab-fs-01.ad.az801.com** from the list of available machines.

3. Once the **General** page renders, scroll down to locate the **Performance Monitor** tool on the left navigation page.

4. Adding the counters here can be done by selecting and searching for **Object**, **Instance**, and **Counter**, then selecting a **Graph type** and clicking the **Add Counter** button for each of the following counters:

    A. **Network Interface > Select All > Bytes Total/sec > Line**

    B. **Processor > Select All > % Processor Time > Line**

    C. **System > Select All > Processor Queue Length > Line**

5. Review the counters and live metrics on the page, then adjust each of the **Graph types** properties to reflect a new type of **Min-max**, as shown in *Figure 18.5*, to help quickly identify anomalies and outliers for performance issues over time:

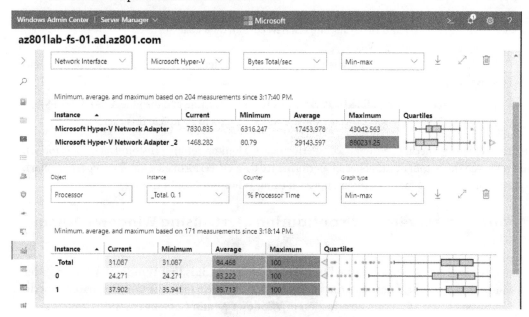

Figure 18.5 – Visualization of min-max graph types

6. Once you're happy with the layout of the report, navigate to the top of the **Performance Monitor** page and select **Save**. On the **Save workspace** prompt, enter `PacktAZ801-FS-01-Custom` as your **Workspace name**, then select **Save workspace** to continue.

7. Select **Performance Monitor** to be returned to the **Performance Monitor and Workspaces** view, noting that you now see the saved **Workspace**, which can be easily downloaded and reused for another server, or selected to return to the real-time monitoring workspace. This step completes this exercise.

As you can see, the Windows Admin Center route gives us a much more streamlined and visual approach to troubleshooting and monitoring systems. In addition, many additional features build upon what Performance Monitor has provided to us for all these years.

Now that we have learned about how Windows Admin Center brings a modern approach to performance monitoring of Windows servers, let's review how **System Insights** can be used to offer predictive monitoring and analytics to our workloads.

Monitoring servers using System Insights

Starting with Windows Server 2019, a new capability called System Insights was introduced to allow for predictive analytics based on machine learning models that were local on the server. This new service helps us quickly analyze system data and overall system performance and events, and provides proactive and prescriptive insights into server functionality, allowing administrators to address issues in advance of any occurrences.

The default capabilities within System Insights are as follows:

- **CPU capacity forecasting**, which gives us insights into CPU forecasts
- **Networking capacity forecasting**, which gives us individual network adapter usage forecasts
- **Total storage consumption forecasting**, which gives us total storage consumption forecasts across all locally attached drives
- **Volume consumption forecasting**, which gives us the total storage consumption forecast for each storage volume on the server

The **System Insights** extension can be installed using PowerShell or via Windows Admin Center, and up to 1 year of historical data can be stored locally for each server. Let's begin by setting up and configuring System Insights on a Windows Server 2022 instance:

1. While remotely connected to the **AZ801LAB-FS-01** virtual machine, select one of two options to proceed:

    I. Using PowerShell, run the `Add-WindowsFeature System-Insights -IncludeManagementTools` command to install System Insights on the server directly.

II.   Conversely, the rest of this exercise will be done using **Windows Admin Center**. To install the **System Insights** extension, navigate to the VM within **Windows Admin Center**, then navigate to **System Insights** in the left navigation and select **Install**, as shown in *Figure 18.6*:

Figure 18.6 – Installing the System Insights extension from within Windows Admin Center

2.   It may take a few minutes for the extension to be installed. Once finished, you will see four default capabilities listed on the screen.

3.   Knowing that our File Server is a little tight on disk space at this point in our lab exercises, let's select **Total storage consumption forecasting** and then select **Invoke** from the menu to force an evaluation now, as shown in *Figure 18.7*:

Figure 18.7 – Invoking the Total storage consumption forecasting System Insights capability

4. After a few seconds, we should see a warning indicating that **Disk usage is forecasted to exceed the available capacity**. If this does not appear, do not fret – we will continue through the exercise with additional settings.

5. Select the **Total storage consumption forecasting** link to open the **Forecast** screen. Then, scroll down to review the **TotalStorage forecast** graph, as shown in *Figure 18.8*:

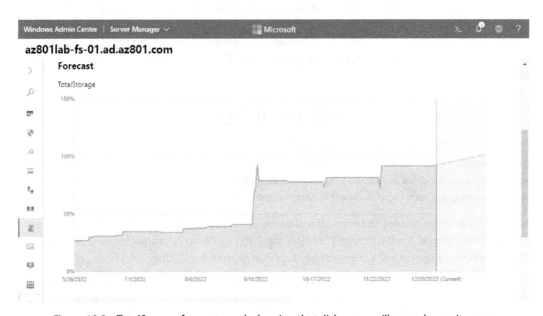

Figure 18.8 – TotalStorage forecast graph showing that disk usage will exceed capacity soon

6. Scroll to the top of the forecast page and select **Settings**.

7. Note that we can now adjust the schedule to fit any business needs.

8. Select **Actions** on the left and take note that **System Insights** is quite flexible and allows for custom scripts to be run for each of the various status levels reported. In addition, different credentials can be utilized to run all the scripts under one shared credential. Alternatively, by using the slider for **Use same set of credentials for all actions**, each script can utilize a different credential.

9. Select the **Close** button at the bottom to return to **System Insights**. This step completes this exercise.

Managing capabilities and remediation actions

By default, System Insights creates all output files as **JSON** files, which provides us with a clear programmatic approach to management and customization. As such, here are two resources that will give more depth surrounding the management of capabilities and the ability to automate remediation actions for System Insights: `https://techcommunity.microsoft.com/ t5/storage-at-microsoft/creating-remediation-actions-for-system- insights/ba-p/428234` and `https://learn.microsoft.com/windows- server/manage/system-insights/managing-capabilities`, respectively.

In addition, a great overview of the capabilities of System Insights is provided by Microsoft at this URL: `https://learn.microsoft.com/windows-server/manage/system-insights/ understanding-capabilities`. Now that we have learned about System Insights and how to incorporate this into any Windows servers running version 2019 and above, let's discuss how to best manage event logs on **Windows Server**.

Managing Windows Server event logs

We have covered a wide range of topics so far. Now, it's finally time to cover **Windows Server** event logs as the topic fits so perfectly with troubleshooting and monitoring. Event logs have proven to give insights into events for Windows OSs in general, including information and warning and error messages covering not only the Windows OS components but any installed applications on the systems.

Event Viewer is typically used to review and filter for events across multiple event logs (including both Windows logs and Application and Services logs) and allows us to customize views to narrow the scope of how we can review and investigate logs during troubleshooting.

Within Event Viewer, scheduled and automated tasks can be created based on individual entries within event logs (using **Attach Task To This Event…**) or attached to custom views (using **Attach Task To This Custom View…**) under the **Actions** > **Administrative Events** menu. This is demonstrated in *Figure 18.9* as an example:

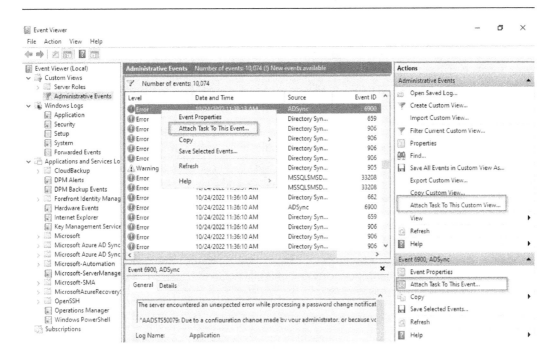

Figure 18.9 – Ability to Attach Task To This Event… or Attach Task
To This Custom View… within Event Viewer

There is also the topic of using Event Viewer to create event subscriptions in a publisher and subscriber relationship. This requires the use of a **Windows Event Collector** (WEC) server and the use of **Windows Event Forwarding** (WEF). For review, we recommend reviewing *Chapter 4, Identifying and Remediating Windows Server Security Issues Using Azure Services*, where we learned about WEF and WEC in the *Introduction to hybrid security using Microsoft Sentinel and Microsoft Defender for Cloud* section. There are two main types of subscriptions, and they are identified as follows:

- **Collector-initiated**, also known as a pull subscription, can identify all servers that the collector will receive (or pull) events from. This approach is typically where you are seeking to pull the same type of events to a centralized location for review and is typically handled via **Group Policy**. This also requires only one subscription definition for all of the members in the subscription group.

- **Source computer-initiated**, also known as a push subscription, where the source computers push events to the centralized collector server. The subscription definition is created on the source computer and the definition can be done manually or via Group Policy. This is because each member in the subscription group can push different events than what other members are sending, or the source computer has different requirements for the push subscription (bandwidth, subscription changes, and so on).

While the setup of WEF could take up an entire chapter on its own, it is valuable to know how to configure WEF in an environment (`https://learn.microsoft.com/defender-for-identity/configure-event-forwarding`) and know about some of the best practices on how to properly establish event forwarding in a larger environment (`https://learn.microsoft.com/troubleshoot/windows-server/admin-development/configure-eventlog-forwarding-performance`).

Event logs for Windows Server in general cover the built-in log types of Application, Security, Setup, System, and Forwarded events. For the Application and Services logs, four main subtypes are covered: Admin, Operational, Analytic, and Debug (where Analytic and Debug are typically only enabled for advanced troubleshooting and tracing as they incur a performance penalty when enabled).

It's also important to know that each event log has properties that determine the location of the log file, the maximum size of the log file, the permissions on the logs, any backup options for the logs, and the behavior for when a log file becomes full. These event log configurations can be fully managed from **Active Directory Group Policy Objects** or additional modern management solutions to enable configuration compliance at scale.

To no surprise, event logs can also be managed from within **Windows Admin Center** simply by navigating to the **Events** node in the left navigation on any device, as demonstrated in *Figure 18.10*:

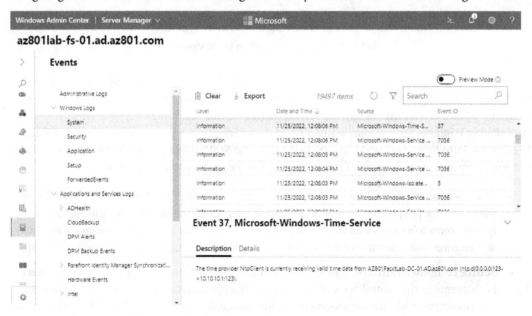

Figure 18.10 – Reviewing event logs in Windows Admin Center

This allows the administrator to quickly review, filter, and search any available events for a selected system, including full event details. We can also clear or export events based on administrative permissions granted to the current administrative account.

Now that we have learned two ways to interact with and utilize Windows Server event logs, let's learn about how to deploy and manage **Azure Monitor Agents** on Windows Server.

Deploying Azure Monitor Agents

In *Chapter 4, Identifying and Remediating Windows Server Security Issues Using Azure Services*, in the *Installing the Azure Arc Connected Machine Agent* section, we learned that once a list of prerequisites is met, setting up **Azure Monitor Agent** (as part of the **Azure Connected Machine agent**) happens behind the scenes. In this section, we will discuss Azure Monitor Agent while reviewing the various ways to install Azure Monitor Agent on a server.

It is important to know that **Log Analytics Agent** will be deprecated by August 2024 (announcement available at the following URL: `https://learn.microsoft.com/azure/azure-monitor/agents/azure-monitor-agent-migration`), and it is recommended to migrate to Azure Monitor Agent before this date to avoid service interruption in any log data ingestion to Azure Monitor and other services. For more details, be sure to check out this article: `https://learn.microsoft.com/azure/azure-monitor/agents/azure-monitor-agent-migration`.

As a refresher, Azure Monitor utilizes Log Analytics Workspaces and other Azure resources to collect performance and telemetry data, group machines, outlines and accounts for various Workspace Data Sources, utilize Insights, respond to alerts and metrics, and visualize the collected data with workbooks and dashboards. Simply put, this monitoring involves collecting and reviewing insightful data surrounding the operation, performance, scale, and behavior of any hosted applications and services. Under no circumstance should installing, upgrading, or uninstalling Azure Monitor Agent require a server to be restarted.

You can install and deploy Azure Monitor Agent in the following ways:

- The **Azure VM extension**, where this can be push-installed via the portal, **PowerShell**, the **Azure CLI**, Azure policies, or via a Resource Manager template for either Windows or Linux. Additional installation and configuration details can be found at `https://learn.microsoft.com/azure/azure-monitor/agents/azure-monitor-agent-manage`. The following prerequisites must be met:

  - **Permissions** require a **Virtual Machine Contributor**, **Azure Connected Machine Resource Administrator**, or any role that includes the `Microsoft.Resources/deployments/*` action.

  - For a non-Azure virtual machine, the **Azure Arc-connected Machine agent** must be installed first.

  - **Authentication** must fall into either a user-assigned managed identity (used for large-scale deployments), a system-assigned managed identity (used for small deployments or a proof of concept), or simply not required as it is enabled with **Azure Arc-enabled servers**.

- **Networking**, which is appropriate for firewalls, ensures that the **Azure Resource Manager** service tag is enabled on the virtual network for a specific virtual machine(s). Access must also be granted to a set of HTTPS endpoints for network traffic and communication:

  - `global.handler.control.monitor.azure.com`

  - `(Your Virtual Machine Region Here).handler.control.monitor.azure.com`

  - `(Your Log Analytics Workspace ID here).ods.opinsights.azure.com`

  - If private links are being used, the **Data Collection Endpoints** (**DCEs**) must also be configured while using the following article as reference: `https://learn.microsoft.com/azure/azure-monitor/essentials/data-collection-endpoint-overview?tabs=portal#components-of-a-data-collection-endpoint`

- The Windows client installer ultimately requires the **Azure extension framework**, which is provided by installing the Azure Arc agent. Additional installation and configuration details can be found at this link: `https://learn.microsoft.com/azure/azure-monitor/agents/azure-monitor-agent-windows-client`. The same prerequisites for the **Azure VM extension** apply, with a few additional changes:

  - **C++ Redistributable version 2015** or higher must be installed

  - Windows MSI must be installed, which can be downloaded directly from the following link: `https://go.microsoft.com/fwlink/?linkid=2192409`

  - For **Authentication**, the Azure AD device token is utilized

  - For **Configuration**, the **Data Collector Rules** are associated with **Monitored Objects** (**MOs**), which are mapped to all **Azure AD** tenant devices

  - For **Networking**, only proxy support is available

- **Windows Admin Center** (via either **Azure hybrid center** or **Azure Monitor**, so long as the **Windows Admin Center** gateway has been registered with your Azure tenant). Additional installation and configuration details can be found at this link: `https://learn.microsoft.com/windows-server/manage/windows-admin-center/azure/azure-monitor`.

Azure Policy and Microsoft best practices

Please note that any Azure Policies for both Virtual Machines and Virtual Machine scale sets require the user-assigned managed identity. For any Azure Arc-enabled servers, policies will rely on the system-assigned managed identity for the resource.

Once you have successfully installed Azure Monitor Agent on the desired resource(s), we can configure data collection and assign appropriate **Data Collection Rules** for the resources.

Now that we have covered the basics on the various ways to get Azure Monitor Agent installed onto servers, let's review how to collect performance counters for Azure.

Collecting performance counters for Azure

In *Chapter 4, Identifying and Remediating Windows Server Security Issues Using Azure Services*, in the *Establishing a Data Collection Rule for the Azure Monitor Agent* section, we learned that **Data Collector Rules** are a prerequisite for sending any alerts to **Microsoft Azure** for event collection, monitoring, troubleshooting, and alerting. We highly recommend that you review this chapter and section to refamiliarize yourself with the overall setup and architecture before proceeding to the *Deploying Azure Monitor Agents* section of *Chapter 18*.

For additional details on collecting data for use with Azure Monitor Agent, be sure to check out this article: `https://learn.microsoft.com/azure/azure-monitor/agents/data-collection-rule-azure-monitor-agent`.

Now that we have reviewed how Data Collector Rules contribute to how performance counters and event log information can be collected for Microsoft Azure, let's learn how to create Azure Monitor alerts.

Creating Azure Monitor alerts

Azure Monitor is a solution for managing, collecting, analyzing, and acting on both events and telemetry from your hybrid environments. Azure Monitor allows you to calculate proper performance baselines for your computing environments by collecting and monitoring real-time information and statistics. It also helps in determining trend analysis of performance over time, helping to identify and resolve resource bottlenecks, and ultimately assisting in overall capacity planning and scalability for all organizations.

There are currently four types of Azure Monitor alerts and they are as follows:

- **Metric alerts** allow you to alert on already computed data that is already available in metric data and requires little to no data transformation

- **Log alerts** allow you to utilize logic operations or queries via the **Kusto Query Language (KQL)** to monitor events stored in logs

- **Activity log alerts** allow you to audit all resource actions and include both Service Health and Resource Health alerts

- **Smart detection alerts** are created automatically within **Application Insights**, are typically related to **App Service** environments, and require at least 24 hours of application learning to determine app behavior baselines

Knowing that we already have an Azure Monitor resource created as part of our work with **Azure Arc**, we can simply dive into the creation of a new Azure Monitor alert rule via the portal:

1. To begin, visit `https://portal.azure.com` while utilizing the Global Administrator account you created in *Chapter 1*.

2. Select **Azure Monitor** from the list of Azure services or simply search for `Azure Monitor` and select it to continue.

3. Select **Alerts** from the left navigation blade, select **+ Create**, and then select **Alert rule**, as shown in *Figure 18.11*:

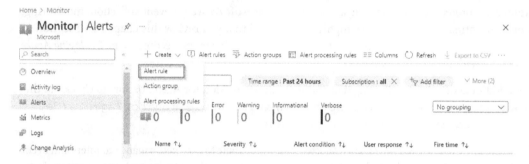

Figure 18.11 – Creating a new Azure Monitor Alert rule

4. While on the **Select a resource** page, change **Filter by resource type** to `Servers - Azure Arc`, then select **AZ801Lab-FS-01** from the list, as shown in *Figure 18.12*. Then, select **Done**:

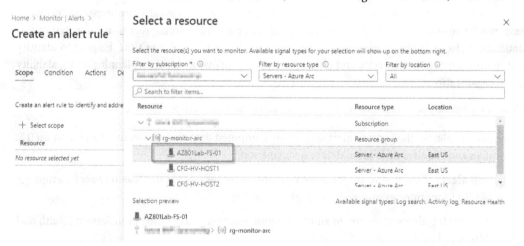

Figure 18.12 – Selecting our on-premises Virtual Machine for alerting purposes

5. Select **Next: Condition >** at the bottom of the page.

6. On the **Select a signal** page, set **Signal name** to `Install patches on Azure Arc machines (Microsoft.HybridCompute/machines)` and then select **Review + create** to continue.

7. The **Details** page will load, asking you to create an **Alert rule name**. Supply an **Alert rule name** of `PacktAZ801-AzMonitor-ArcPatching`. Select **Next: Tags >** to continue.

8. Select **Review + create**, then **Create** to complete the alert configuration. This step completes this exercise.

9. [Optional] To trigger the alert we created, we must visit **Azure Arc**, locate our server, complete a **Check for updates** under **Updates**, and then complete a **One-time update** of the virtual machine to write the appropriate events to the activity log. If you choose to complete this validation activity, you should see the results shown in *Figure 18.13*:

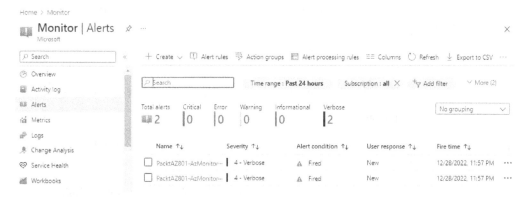

Figure 18.13 – Two alerts triggered during the Azure Arc patch installation for our server

> **Azure Monitor Action groups**
>
> I highly encourage you to review how action groups are created to notify users about an alert and what actions to take to resolve the issue. The available notification types are email, SMS, Azure app push notifications, and Voice. To learn more about the setup and configuration of Action groups, be sure to review the following article: `https://learn.microsoft.com/azure/azure-monitor/alerts/action-groups`.

Now that we have provided an overview of Azure Monitor, including how to create an Azure Monitor alert, let's learn how to monitor Azure VMs using the Azure Diagnostics extension.

Monitoring Azure Virtual Machines using the Azure Diagnostics extension

The **Azure Diagnostics** extension is yet another agent in Azure Monitor that is responsible for collecting monitoring data, including event logs and performance data, from the guest OS of Azure virtual machines. The Azure Diagnostics extension does not require any additional software on the virtual machine and utilizes a diagnostic storage account for sending diagnostic data.

Note that the Azure Diagnostics extension can only be used with Azure resources and while there is no cost associated with using the extension, costs might be incurred based on any data consumed.

The primary uses of the Azure Diagnostics extension include the following:

- Sending limited data to **Azure Monitor Metrics** to take advantage of the Metrics Explorer and real-time metric alerts

- Sending diagnostics data to **Azure Storage** for review with **Azure Storage Explorer** or simply for archiving purposes

- Sending diagnostics data to third-party tools and providers using **Azure Event Hubs**

- Collecting VM boot diagnostics

Let's learn how to install and configure the Azure Diagnostics extension for Windows using the Azure portal:

1. To begin, visit `https://portal.azure.com` while utilizing the Global Administrator account you created in *Chapter 1*.

2. Select **Virtual Machines** from the list of Azure services or simply search for `Virtual Machines` and select it to continue.

3. Ensure that either **vm-az801-svr1** or **vm-az801-svr2** has a status of **Running**. If neither is in the **Running** state, simply select one or both by putting a mark in the checkbox and then select **Start**.

4. Select either of the running virtual machines. Then, locate the **Monitoring** blade section and select **Diagnostic settings** from the left navigation.

5. On the **Overview** tab, select an appropriate diagnostics storage account (that we previously created – the name of your storage resource will be different), and select **Enable guest-level monitoring**, as shown in *Figure 18.14*:

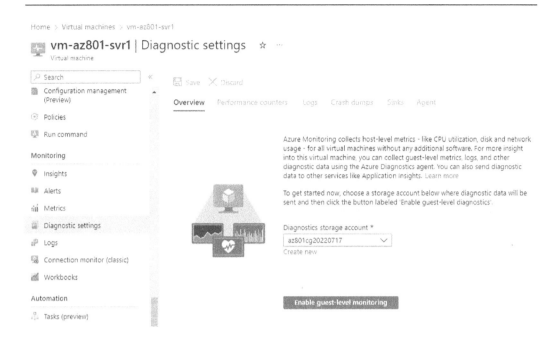

Figure 18.14 – Enabling guest-level monitoring using the Diagnostic settings for the virtual machine

6. Select the **Performance counters** tab or **Configure performance counters**, then review the selected **Basic** predefined performance counters.

7. Select the **Logs** tab, noting that you can adjust the default Application, Security, and System logs gathered, choose a custom selection of logs, and gather IIS logs.

8. Select the **Crash dumps** tab and temporarily select **Enabled** to review the options for configuring a specific process to monitor, including the storage location for any gathered crash dumps and the type of crash dump. Select **Disabled** as we will not be gathering crash dumps in this exercise.

9. Select **Sinks** and note that you will most likely see that **Azure Monitor** requires a managed identity to continue this configuration. Select the **The Azure Monitor sink requires a managed identity entry** banner in yellow to establish a **managed identity**, as shown in *Figure 18.15*:

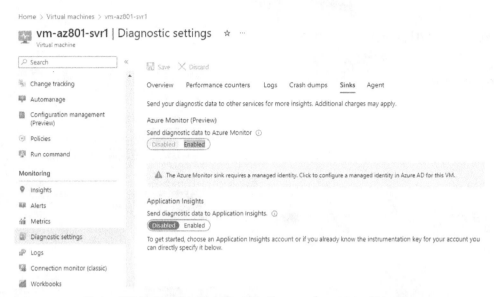

Figure 18.15 – Enabling Sinks requires the use of a managed identity

10. On the **Identity** page that opens, change the status to **On** for **System assigned** and select **Save**, as shown in *Figure 18.16*. Then, select **Yes** regarding the **Enable system assigned managed identity** prompt that appears. Select the **X** icon on the **Identity** page to return to our **Diagnostic settings** configuration:

Figure 18.16 – Enabling system assigned managed identity for our virtual machine

11. If you still cannot change the **Azure Monitor** slider from **Disabled** to **Enabled**, select the **X** icon in the right corner of the **Diagnostics settings** area and repeat *step 6* through *step 9*. Select **Enabled** for **Azure Monitor** and then select **Save**, as shown in *Figure 18.17*:

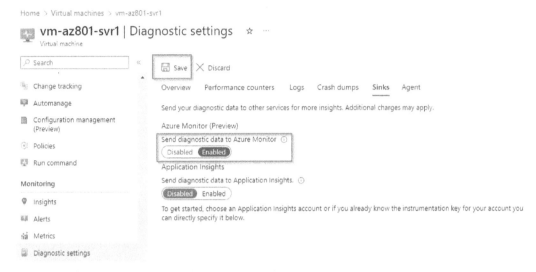

Figure 18.17 – Enabling the Azure Monitor diagnostic data connector

12. On the **Agent** tab, review the settings, noting that you can configure the amount of disk to be used for storing logs, enable or disable the diagnostic troubleshooting logs, adjust the logging level, and **Remove Azure Diagnostics Agent** to reinstall the agent or troubleshoot log collection. While this action removes the agent, the existing diagnostic data stored in the storage account will persist.

13. At this point in this exercise, we can return to **Azure Monitor** and then open the **Metrics** blade. Here, we can select `Available Memory Bytes` as our `Metric` to give us a feel for what diagnostic logging is now being automatically sent to **Azure Monitor**, as shown in *Figure 18.18*. This step completes this exercise:

Figure 18.18 – Reviewing newly written diagnostic data for our Azure VMs in Azure Monitor

Now that we have learned how to monitor Azure VMs using the Azure diagnostics extension and have some insight into the power of these diagnostics logging and monitoring tools, let's learn how to monitor Azure VMs using **VM Insights**.

Monitoring Azure Virtual Machines performance using VM Insights

Azure VM Insights monitors the overall health and performance of your environment's virtual machines and virtual machine scale sets, keeping track of any running processes, as well as their dependencies on other resources throughout the environment. Like **System Insights**, VM Insights provides administrators with predictable availability and performance information for both systems and applications and grants insights into the availability of upstream or downstream dependencies.

VM Insights offers support for the following Windows and Linux workloads:

- Azure virtual machines and virtual machine scale sets.
- On-premises virtual machines.

- Hybrid virtual machines connected and onboarded with **Azure Arc**. Note that you will need to install the dependency agent for Windows and Linux servers, as detailed at the following URL: `https://learn.microsoft.com/azure/azure-monitor/vm/vminsights-enable-hybrid#dependency-agent`.

- Virtual machines that are hosted in other cloud environments.

Note that VM Insights stores all collected data inside Azure Monitor Logs and while there is no cost associated with using VM Insights, costs will be incurred based on any data activity within the Log Analytics workspace.

Let's learn how to identify unmonitored virtual machines and enable and configure VM Insights for a virtual machine using the Azure portal:

1. To begin, visit `https://portal.azure.com` while utilizing the Global Administrator account you created in *Chapter 1*.

2. Select **Monitor** from the list of Azure services or simply search for `Monitor` and select it to continue.

3. Navigate to the **Insights** section on the left navigation pane and select **Virtual Machines**. Then, select **Configure Insights**, as shown in *Figure 18.19*:

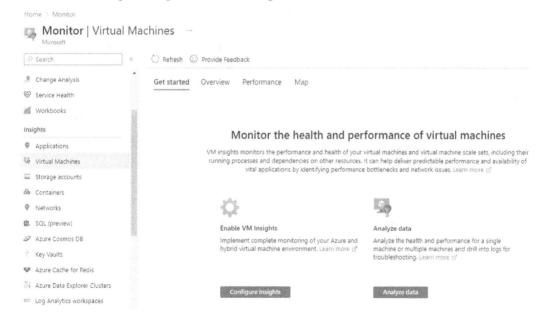

Figure 18.19 – Configuring VM Insights using Azure Monitor and the Azure portal

4. Select the **Not monitored** tab and then locate any of the virtual machines listed under the **rg-az801-examguide** resource group. Then, select **Enable** for **vm-az801-svr1**, as shown in *Figure 18.20*:

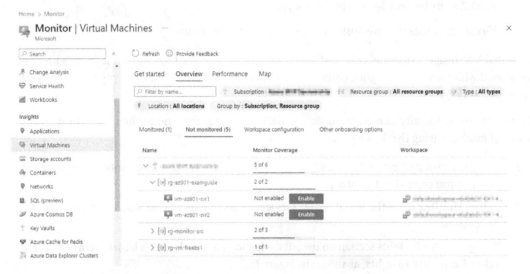

Figure 18.20 – Enabling virtual machines for VM Insights

5. On the **Azure Monitoring Insights Onboarding** page, select the **Enable** button to begin the configuration. This step can take 5-10 minutes to complete, as shown in *Figure 18.21*:

Figure 18.21 – Enabling Azure Monitor Insights Onboarding for VM Insights

6.   Next, accept the defaults on this **Monitoring configuration** page, noting that you can adjust the **Subscription** and **Data Collection Rule** properties used for the monitoring configuration, as shown in *Figure 18.22*. Select the **Configure** button to continue with the configuration:

Figure 18.22 – Establishing Monitoring configuration for VM Insights

7.   Now that the configuration is reporting metrics and VM Insights data to the Log Analytics workspace, we can select the **Performance** tab under **Insights** for the individual **vm-az801-svr1** virtual machine to review and evaluate overall system performance with monitored historical **VM Insights**, as shown in *Figure 18.23*. This step completes this exercise:

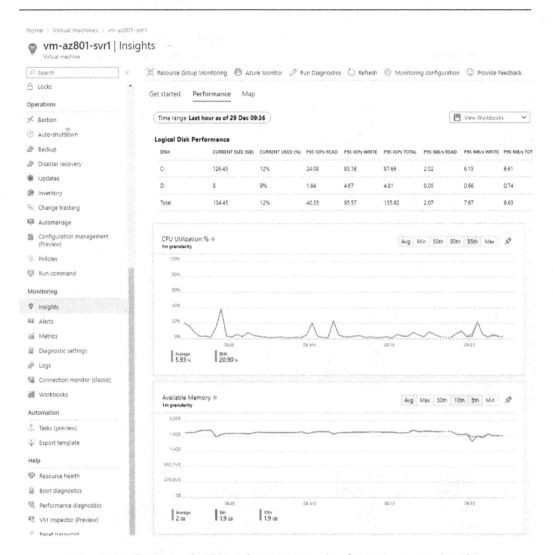

Figure 18.23 – Reviewing the VM Insights monitoring data for our Azure virtual machine

VM Insights Map feature

VM Insights also includes a Map experience that helps us visually identify active network connections between VMs, identify connection latency that's both inbound and outbound, and identify utilized TCP ports over a specific time range. Note that we used Windows Server 2022 virtual machines in this lab and that this OS version is not yet supported for the Map feature.

In addition, you will need to have processes and dependencies configured in your Data Collection Rules for the VM Insights to properly reflect Map features. More details on the Map feature can be found in this article: https://learn.microsoft.com/azure/azure-monitor/vm/vminsights-maps.

With that, we have learned how to monitor Azure Virtual Machine performance. Next, we will review what we have learned in this chapter and set the stage for the next chapter of this book.

Summary

In this chapter, we learned about the various ways we can monitor and alert ourselves to Windows Server performance and availability events. We learned how to use Performance Monitor and Data Collector Sets, how to monitor servers and event logs using Windows Admin Center, and how to monitor overall server health using System Insights.

We also learned how to deploy and configure Azure Monitor Agents to collect performance data and alert on predefined thresholds. Finally, we discussed how to use the Azure diagnostics extension in Azure VMs, as well as how to monitor performance using VM insights.

In the next chapter, we will learn how to effectively troubleshoot network connectivity for both on-premises and hybrid networking. We will learn how to utilize existing tools, as well as the **Azure Virtual Network troubleshooter** package, to gain additional troubleshooting insights.

19

Troubleshooting Windows Server On-Premises and Hybrid Networking

In this chapter, we will learn how to effectively troubleshoot network connectivity for both on-premises and hybrid networking. We will learn how to utilize existing tools to gain additional troubleshooting insights to achieve the AZ-801 exam's preparation objectives. We will focus on the requirements, best practices, and walkthroughs in this chapter that align with the AZ-801 exam objectives.

In this chapter, we will cover the following topics:

- Technical requirements and lab setup
- Troubleshooting hybrid network connectivity
- Troubleshooting on-premises connectivity

Technical requirements and lab setup

To successfully follow along and complete the tasks and exercises throughout this chapter and the following chapters in this book, we will need to ensure that the technical requirements from *Chapter 1, Exam Overview and the Current State of Cloud Workflows*, have been completed in full.

Let's begin by learning about some of the tools used for troubleshooting hybrid network connectivity, including where and how the tools can be used to quickly gather insights.

Troubleshooting hybrid network connectivity

It is fitting that we just wrapped up *Chapter 18* by learning how to *Monitor Windows Server Using Windows Server Tools and Azure Services*. As part of **Azure Monitor**, **Azure Monitor Network Insights** (or **Azure Monitor for Networks**) continues to grow with a deep set of troubleshooting tools and insights that help administrators review the health, metrics, and visual topology of their deployed network resources. The great part about Azure Monitor Network Insights is that no configuration is required to begin using the available features.

Azure Monitor Network Insights comprises network health and metrics, connectivity, traffic, a diagnostic toolkit, and topology monitoring components for your deployed network. As every environment differs in network design and architecture, monitoring and troubleshooting tools need to be as dynamic and intuitive as possible for both current architects and engineering teams, as well as new team resources to get all troubleshooting efforts on the same level.

Let's dig into each of these components to learn more about the ease of use, features, and functionality they bring to administrators.

Network health and metrics

Within **Azure Monitor**, the **Network Insights** overview page provides quick insights into various network resources and helps us identify the resources, counts, and locations, including the overall service health and any alerts affecting our resources. The dashboard page allows us to search and filter by subscription, resource group, or resource type while displaying a visual card or tile-based representation of the resources, available metrics, and alerts, as shown in *Figure 19.1*:

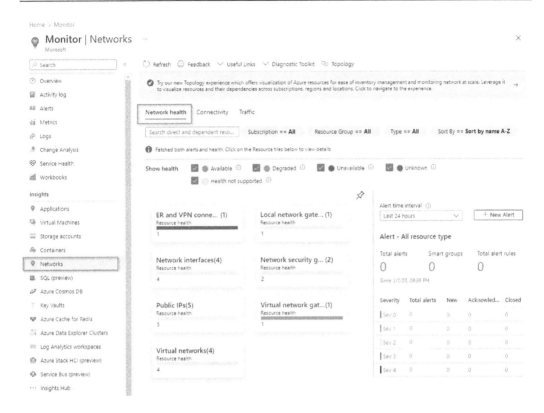

Figure 19.1 – Azure Monitor Network Insights overview page

Simply selecting a resource by clicking on a resource tile drills us into a view showing deeper metrics for the selected resource, as shown in *Figure 19.2*. Here, we are choosing the **Virtual network gateways** resource as an example:

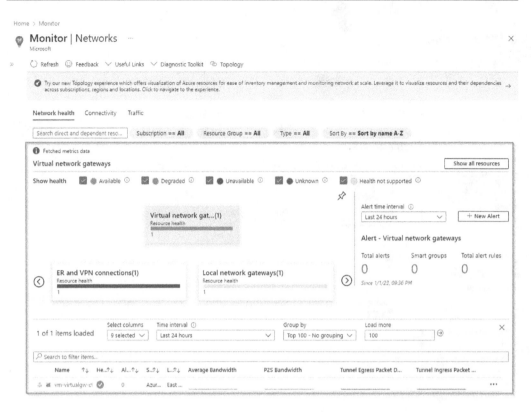

Figure 19.2 – Virtual network gateways drill-through metrics available when selecting the tile

Note that there is an available **Resource view** for the **Azure Virtual Gateway** resource, as evidenced by the additional details at the bottom of *Figure 19.2*. **Resource views** are smart in that they display appropriate metrics and visual details related to the selected resource. At the time of writing, the **Resource** view is only available for **Azure Application Gateway**, **Azure Virtual WAN**, and **Azure Load Balancer**.

Connectivity

The **Connectivity** tab that's part of **Network Insights** gives us a simple way to review all tests that have been configured for our resources using **Connection Monitor**. The default view in the **Connectivity** tab, as shown in *Figure 19.3*, shows **Sources** and **Destinations**, as well as the status details of each of the tests being run and included for the selected resources:

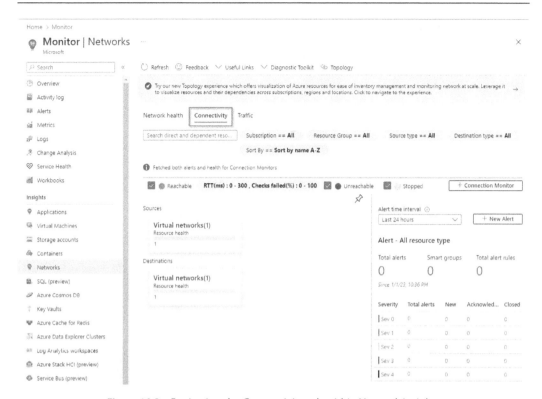

Figure 19.3 – Reviewing the Connectivity tab within Network Insights

Connection Monitor could easily take an entire chapter to cover, so we will dive into an overview of the tool as it will soon replace **Network Performance Monitor** in 2024 (more details available at this URL: `https://aka.ms/migrate-to-connection-monitor-from-network-performance-monitor`). The key objective is that these **Connection Monitor** tests and test groups allow for end-to-end connection monitoring via an approach where we identify source resources to run the tests (virtual networks, virtual machines, services), configure a test to run against these services (an HTTP over port 80 with a status of 200 as an example, as shown in *Figure 19.4*), and apply destination resources to run the tests against (which resource we want to check for availability):

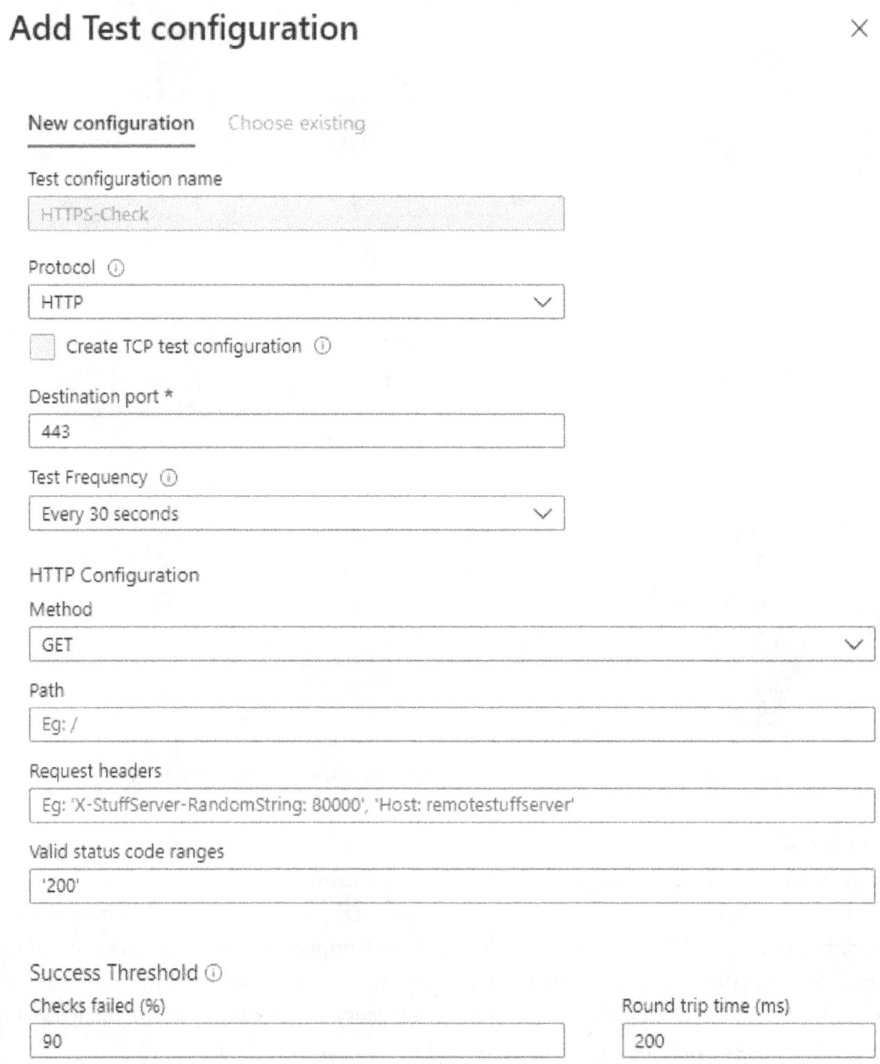

Add Test configuration   ×

New configuration   Choose existing

Test configuration name

```
HTTPS-Check
```

Protocol ⓘ

```
HTTP                                                          ∨
```

☐ Create TCP test configuration ⓘ

Destination port *

```
443
```

Test Frequency ⓘ

```
Every 30 seconds                                             ∨
```

HTTP Configuration

Method

```
GET                                                          ∨
```

Path

```
Eg: /
```

Request headers

```
Eg: 'X-StuffServer-RandomString: 80000', 'Host: remotestuffserver'
```

Valid status code ranges

```
'200'
```

Success Threshold ⓘ

Checks failed (%)   Round trip time (ms)

```
90                                         200
```

Figure 19.4 – Adding a test configuration to the Connection Monitor test group

A lengthy but worthwhile read on configuring and utilizing **Connection Monitor** can be found at the following URL: https://learn.microsoft.com/azure/network-watcher/connection-monitor-overview.

Traffic

The **Traffic** tab within **Network Insights** provides details on all **Network Security Groups** (NSGs) configured for both **Flow Logs** and **Traffic Analytics**, just **Flow Logs**, or no **Flow Logs** configured. The beauty of this tab is that you can search for either a selected NSG or an IP address, and all NSGs that are relevant to the search filter with the ability to view and edit the **Flow Log** and **Traffic Analytics** configurations, as shown in *Figure 19.5*:

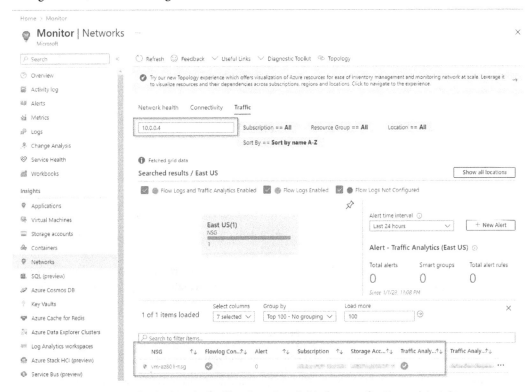

Figure 19.5 – Reviewing the Traffic tab and available features for Network Insights

An overview of what flow logging brings for both NSGs and Network Insights can be reviewed at the following URL: `https://learn.microsoft.com/azure/network-watcher/network-watcher-nsg-flow-logging-overview`.

Diagnostic Toolkit

The **Diagnostic Toolkit** is a one-stop shop for all the available network troubleshooting diagnostic tools and can be seen in *Figure 19.6* as an drop-down menu within **Azure Monitor Network Insights**:

Figure 19.6 – Introducing the Diagnostic Toolkit within Azure Monitor Network Insights

This menu allows you to quickly access the following tools, which are available for quick intuitive use (pro tip – all these tools are available as part of **Azure Network Watcher**, a topic we will cover later in this chapter):

- **Capture packets on virtual machines** or **Packet capture** within **Network Watcher**
- **Troubleshoot VPN** or **VPN troubleshoot** within **Network Watcher**
- **Troubleshoot connectivity** or **Connection troubleshoot** within **Network Watcher**
- **Identify next hops** or **Next hop** within **Network Watcher**
- **Diagnose traffic filtering issues** or **IP flow verify** within **Network Watcher**

Topology

The final network monitoring tool available within **Azure Monitor Network Insights, which is** also one of my favorites, is the **Topology** tab. This is an incredibly visual tool as it shows you which regions your resources are in and how they connect and interrelate with each other. The entire topology map is clickable, with hover tiles and interactive details!

As an example, I have taken a cross-section of my resource view to show a single **Azure Virtual Machine** virtual network interface with relationships to the virtual machine, the NSG, and the IP address resource, as shown in *Figure 19.7*:

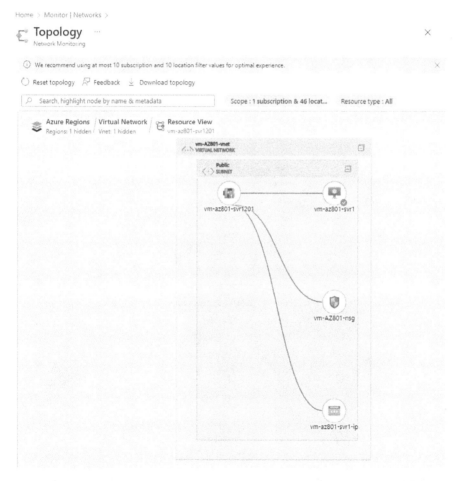

Figure 19.7 – Topology map within Network Monitoring showing resource relationships

Once you have completed your visual design, the topology can also be downloaded for additional documentation or visuals by quickly selecting **Download topology**. This gives you access to an *.svg file.

Effective routes

Available within the portal from the **Network Interface** resource view is a tool called **Effective routes**. When administering multiple virtual networks, you must also consider **Virtual Network (VNet)** peering to ensure that traffic is flowing between VNets. If you wish to have one resource in a VNet utilize a gateway in another VNnet, you will also need to consider enabling the **Use Remote Gateway** setting on the VNet peering configuration. Finally, if you are using **Azure Firewall** as a central control plane, you must also manage the network flow via route tables, which can be viewed per network interface from the **Effective routes** blade, as shown in *Figure 19.8*:

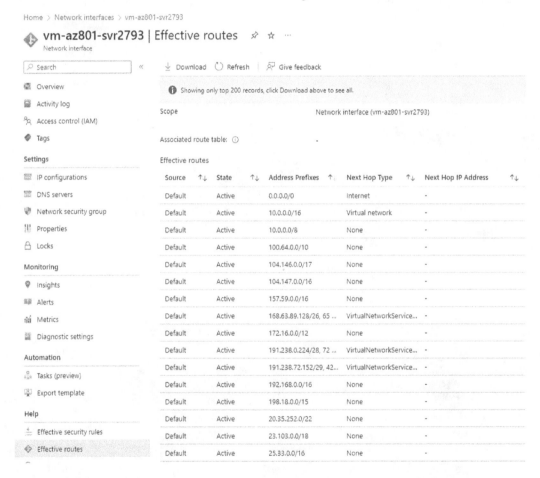

Figure 19.8 – Network interface Effective routes tool with the associated route table

Network Watcher

Network Watcher is a suite of tools that help us monitor and repair the overall network health of **Infrastructure-as-a-Service** (**IaaS**) resources such as **VNets**, **Application Gateways**, and **Virtual Machines**, and quickly enable or disable logging of resources within an Azure virtual network. When creating or updating a virtual network in your Azure subscription, Azure Network Watcher will be automatically enabled within the Virtual Network's region to assist in network and resource monitoring.

From within this single pane and dashboard, as shown in *Figure 19.9*, administrators can quickly monitor, diagnose, gather additional logging, and resolve issues in hybrid computing environments:

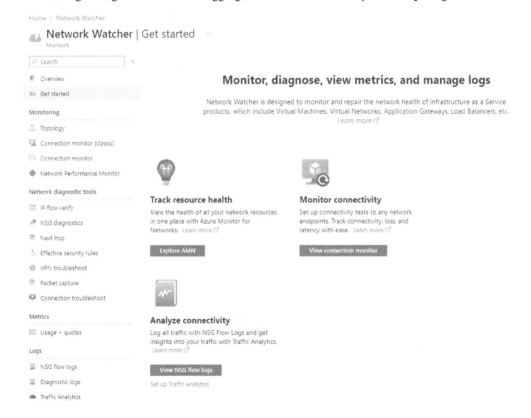

Figure 19.9 – Network Watcher dashboard showing available troubleshooting tools

Let's spend a few minutes reviewing all the available tools as it's extremely valuable to know of their existence, as well as the available options for each when it comes to diagnostics and troubleshooting.

Topology

The **Network Watcher | Topology** view allows you to see the available resources within a virtual network, including the relationships between the resources within the same resource group and region. The **Topology** view also identifies and visualizes the Azure resources associated with the resources within a virtual network (such as subnets, route tables, and IP addresses), as shown in *Figure 19.10*:

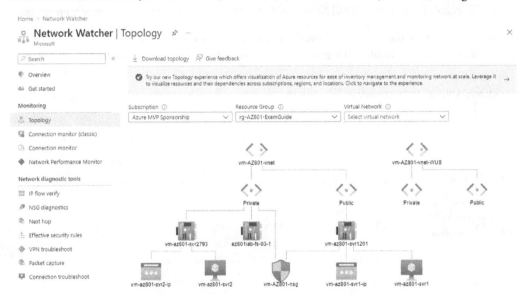

Figure 19.10 – Network Watcher Topology example giving more in-depth visualizations

IP Flow verify

The **IP Flow verify** tool quickly identifies connectivity issues on whether a packet is allowed or denied to or from an identified virtual machine. To complete the checks, you must enable logging of 5-tuple information for all traffic while considering the protocol being used, the direction of the traffic, the local IP address and local port, and the destination or remote IP address and remote port, as shown in *Figure 19.11*:

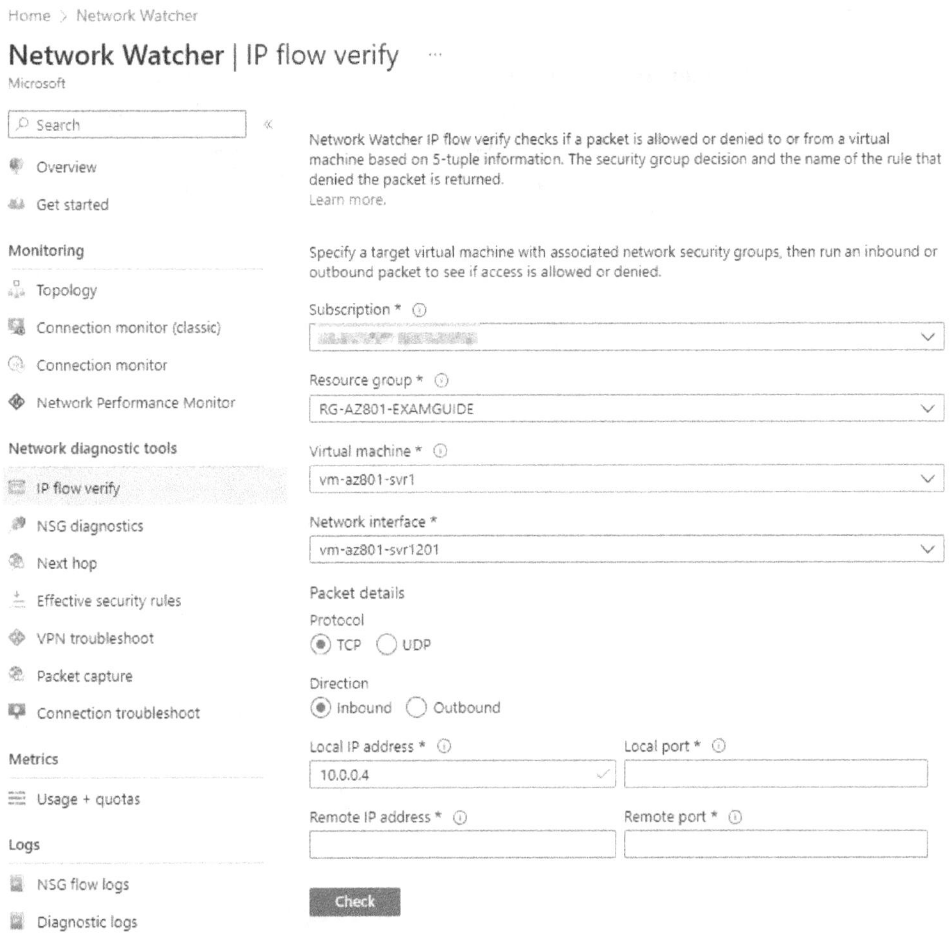

Home > Network Watcher

Network Watcher | IP flow verify ...
Microsoft

Search «

🔹 Overview

📧 Get started

Monitoring

🔹 Topology

🔹 Connection monitor (classic)

🔹 Connection monitor

🔹 Network Performance Monitor

Network diagnostic tools

📧 IP flow verify

🔹 NSG diagnostics

🔹 Next hop

🔹 Effective security rules

🔹 VPN troubleshoot

🔹 Packet capture

🔹 Connection troubleshoot

Metrics

📧 Usage + quotas

Logs

🔹 NSG flow logs

🔹 Diagnostic logs

Network Watcher IP flow verify checks if a packet is allowed or denied to or from a virtual machine based on 5-tuple information. The security group decision and the name of the rule that denied the packet is returned.
Learn more.

Specify a target virtual machine with associated network security groups, then run an inbound or outbound packet to see if access is allowed or denied.

Subscription * ⓘ

[   ] ⌄

Resource group * ⓘ

[RG-AZ801-EXAMGUIDE   ] ⌄

Virtual machine * ⓘ

[vm-az801-svr1   ] ⌄

Network interface *

[vm-az801-svr1201   ] ⌄

Packet details
Protocol
(●) TCP   ( ) UDP

Direction
(●) Inbound   ( ) Outbound

Local IP address * ⓘ   Local port * ⓘ

[10.0.0.4   ✓]   [   ]

Remote IP address * ⓘ   Remote port * ⓘ

[   ]   [   ]

[Check]

Figure 19.11 – Required fields and information for running the IP flow verify tool

The results that are returned during the checks will contain the following:

- The status of whether the check was allowed or denied
- The name of the network security group rule or rules that denied the packet

NSG diagnostics

Like **IP Flow verify**, **NSG diagnostics** provides more detailed information on configuration within your network, helping you to further investigate and troubleshoot security and connectivity issues. Details provided as results include all NSGs through which traffic is flowing, the rules that will be applied within each NSG, and the final allow/deny flow status, as shown in *Figure 19.12*:

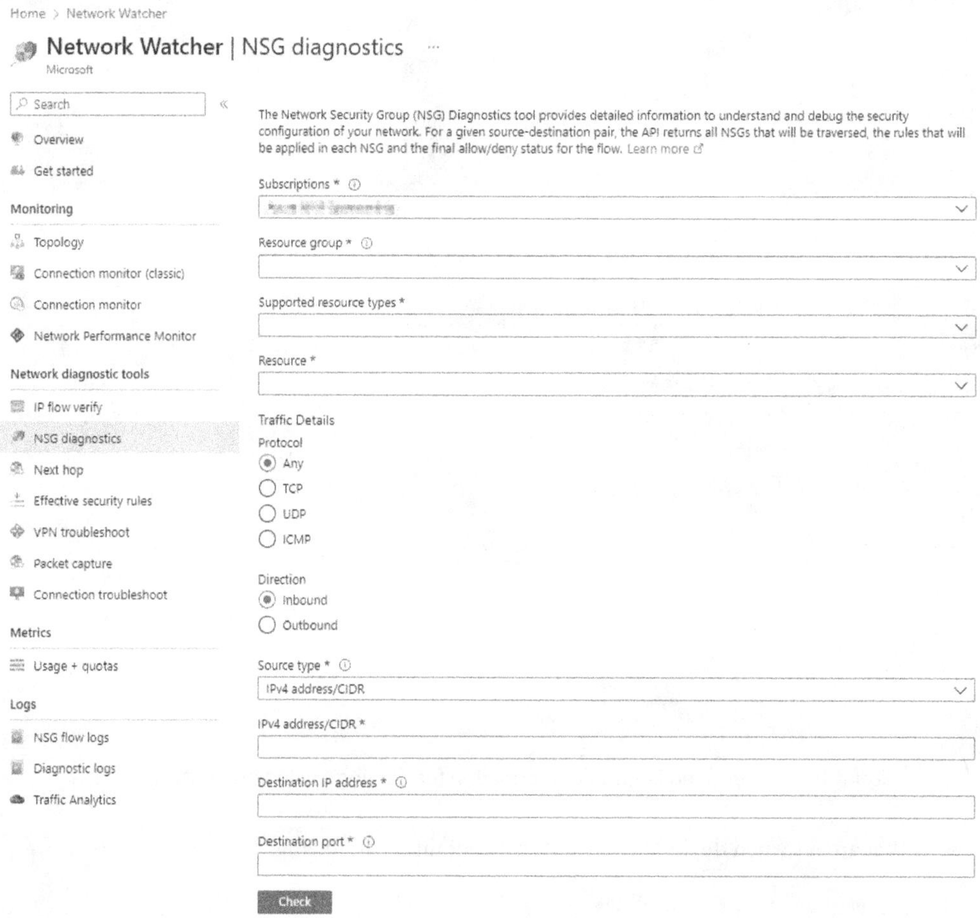

Figure 19.12 – Required fields and information for running NSG diagnostics

Next hop

Another small yet effective tool is the **Next hop** tool, which determines the next hop type and IP address of a given packet from a specified virtual machine and network interface, as shown in *Figure 19.13*. This resulting information tells you whether the traffic is being sent to the intended destination, or whether the traffic is not being sent anywhere. If the next hop is identified during the check, the associated route table is provided in the results:

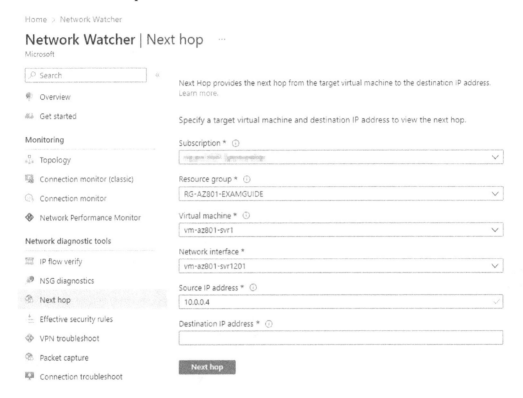

Figure 19.13 – The Next hop tool seeking a Virtual machine, Network interface, Source IP address, and Destination IP address

A quick overview of next hop values that may be returned is as follows:

- **VirtualAppliance**
- **VirtualNetworkInterface**
- **VirtualNetwork**
- **VirtualNetworkPeering**
- **VirtualNetworkServiceEndpoint**

- **Internet**
- **MicrosoftEdge**
- **None**

Additional details on the various next hop types can be found at the following URL: `https://learn.microsoft.com/azure/virtual-network/virtual-networks-udr-overview`.

Effective security rules

The **Effective security rules** tool returns all NSGs and rules that are associated with a NIC and subnet for a particular virtual machine. This gives quick insights into both the configuration and the currently active security rules for each of the NICs for a selected virtual machine. Using this view, we can quickly determine the associated NSG, as well as any destination prefixes, as shown in *Figure 19.14*:

Figure 19.14 – Effective security rules example indicating the associated NSG

VPN troubleshoot

Another simple yet powerful tool is the **VPN troubleshoot** tool from **Network Watcher**, which allows you to validate the status and health of your virtual network gateway or VPN connection. This action does take quite a while to complete and when finished, a troubleshooting status will be returned with a summary of details and potentially an **Action** plan suggesting ways to resolve any identified issues, as shown in *Figure 19.15*:

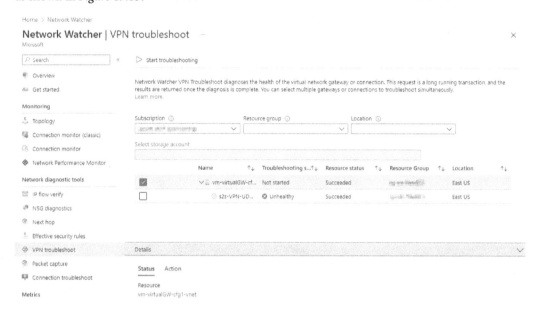

Figure 19.15 – The VPN troubleshoot tool providing logging information for a virtual network gateway

Packet capture

The **Packet capture** tool in **Network Watcher** is a lightweight extension that simply allows you to create packet capture sessions to log traffic to and from virtual machines within the environment, capturing the logs to file, storage account, or both. As shown in *Figure 19.16*, there are a good amount of packet capture configuration and filtering options, but this is meant to quickly capture and diagnose network anomalies, statistics gathering, and general application communication debugging:

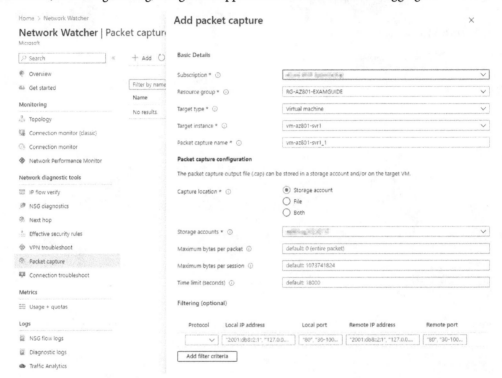

Figure 19.16 – Packet capture configuration and filtering options in Network Watcher

Connection troubleshoot

The **Connection troubleshoot** tool within **Network Watcher** gives administrators the ability to check or probe a TCP connection from a virtual machine to another virtual machine, a fully-qualified domain name, an IPv4 address, or a **Uniform Resource Identifier** (**URI**) on the internet, as shown in *Figure 19.17*. Results returned include the overall connection status, latency details, probes sent with a failure status, hops information (disclosing source to destination details), and fault types for the connection:

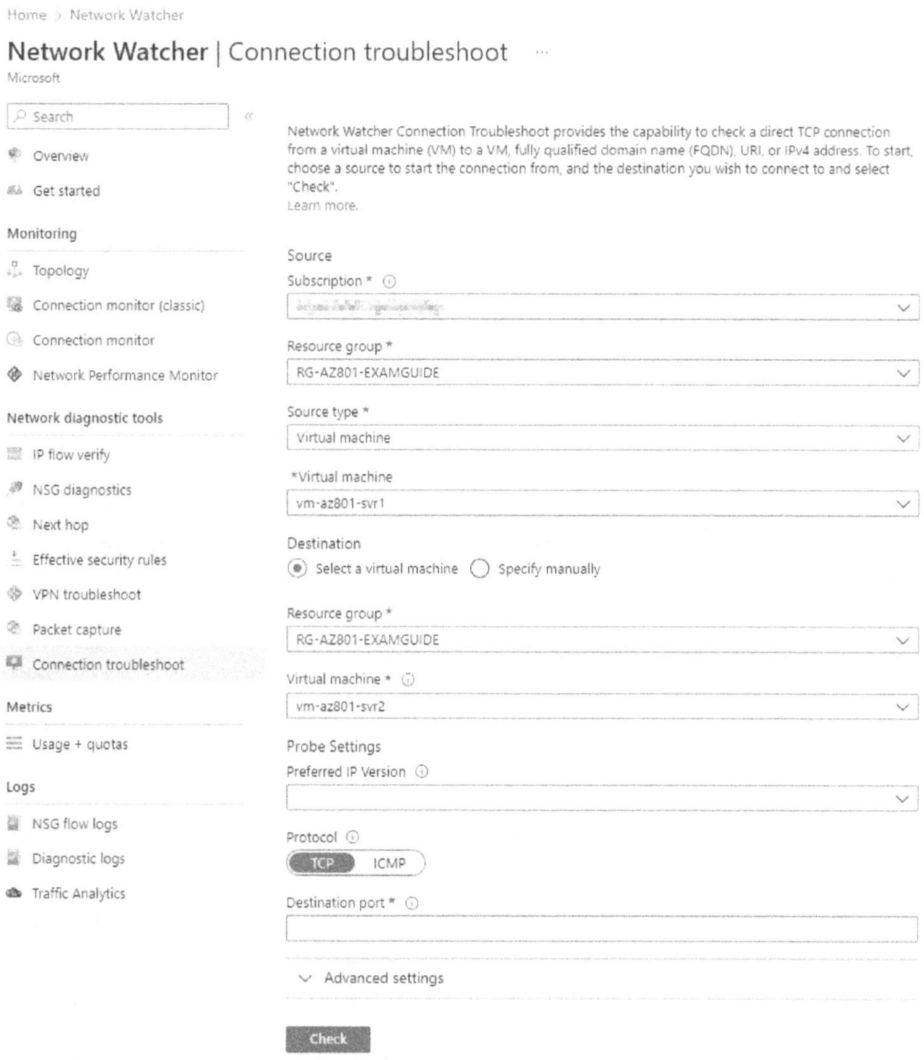

Figure 19.17 – Available check options for Connection troubleshoot in Network Watcher

The overall resulting fault types are classified as follows:

- **CPU** (high utilization)
- **Memory** (high utilization)
- **GuestFirewall** (traffic blocked due to VM firewall configuration)
- **DNSResolution** (resolution failed for the destination address)

- **NetworkSecurityRule** (name of rule blocking traffic is returned)
- **UserDefinedRoute** (dropped traffic due to either a system or user-defined route)

Now that we have had an overview of how to troubleshoot hybrid network connectivity using **Azure Monitor** and its related tools, let's learn about the tools we can use to troubleshoot on-premises network connectivity.

Troubleshooting on-premises connectivity

In general, **Azure VPN Gateway** enables you to create a secure connection from your on-premises network into your Azure network footprint. The computing environments and sets of requirements are unique to all businesses; different VPN devices have varying configuration settings that work to enable connectivity of your on-premises VPN device to the cloud.

That being said, the **Azure Network Watcher** troubleshooting API can help quickly you identify the root cause of your connectivity issues and provide insights into steps for resolution. It is recommended to connect to **Azure PowerShell** and utilize the `Start-AzNetworkWatcherResourceTroub leshooting` cmdlet to begin logging and diagnosing your VPN connection. After a few minutes, the command will complete, creating a `*.zip` archive of diagnostic logs for your review.

The various Azure VPN Gateway resource troubleshooting log files that can be written are as follows:

- The `ConnectionStats.txt` file, which contains connection status information, ingress and egress bytes, and the time that the connection was established
- The `CPUStats.txt` file, which contains memory and CPU details available at the time of the test
- The `IKELogs.txt` file, which contains any **Internet Key Exchange (IKE, IKEv1, or IKEv2)** activity for the established connection during monitoring
- The `IKEErrors.txt` file, which contains any IKE errors that were encountered during monitoring
- The `Scrubbed-wfpdiag.txt` file, which contains the logging of any dropped packets and/ or IKE/Authentication failures during monitoring
- The `Wfpdiat.txt.sum` file, which shows the events and any buffers processed during monitoring

> **An Azure CLI bug?**
>
> Note that if you are using the Azure CLI to run the troubleshooting command, both the storage account and the VPN Gateway must be in the same resource group. It is recommended that if the resources are in different resource groups, either the portal or PowerShell should be used for the troubleshooting operation.

A comprehensive list of all the Azure Resource Log files that are supported can be found at the following URL: `https://learn.microsoft.com/azure/azure-monitor/essentials/resource-logs-categories`.

With that, we have completed this chapter on how to troubleshoot on-premises network connectivity. We have also learned how to effectively troubleshoot network connectivity for hybrid networking. Next, we will review what we have learned in this chapter and set the stage for the next chapter of this book.

Summary

In this chapter, we learned how to effectively troubleshoot network connectivity for both on-premises and hybrid networking. We learned how to utilize existing tools to gain additional troubleshooting insights to achieve the AZ-801 exam's preparation objectives.

In the next chapter, we will learn how to effectively troubleshoot Windows Server virtual machine workloads in Microsoft Azure. This includes learning how to troubleshoot deployment and booting failures, VM performance and extension issues, disk encryption, storage, and overall VM connection issues. We will cover a wide range of tools that will help us successfully address the typical issues that are experienced when running Windows Servers in Microsoft Azure.

20

Troubleshooting Windows Server Virtual Machines in Azure

In this chapter, we will learn how to effectively troubleshoot Windows Server virtual machine workloads in Microsoft Azure. This includes learning how to troubleshoot deployment and booting failures, VM performance and extension issues, disk encryption, storage, and overall VM connection issues.

We will focus on the requirements, best practices, and hands-on walkthroughs in this chapter that align with the AZ-801 exam objectives. We will also cover a wide range of tools that will help you successfully address typical issues experienced when running Windows Servers in Microsoft Azure to achieve the AZ-801 exam's preparation objectives.

In this chapter, we will cover the following topics:

- Technical requirements and lab setup
- Troubleshooting deployment failures
- Troubleshooting booting failures
- Troubleshooting VM performance issues
- Troubleshooting VM extension issues
- Troubleshooting disk encryption issues
- Troubleshooting storage
- Troubleshooting VM connection issues

Technical requirements and lab setup

To successfully follow along and complete the tasks and exercises throughout this chapter and the following chapters in this book, we will need to ensure that the technical requirements from both *Chapter 1, Exam Overview and the Current State of Cloud Workflows*, and *Chapter 6, Securing Windows Server Storage*, have been completed in full.

Let's begin with a general overview of troubleshooting Azure VMs and some strategies for troubleshooting VM deployment issues in Azure.

Troubleshooting deployment failures

It's safe to say that we expect occasional issues to arise with any Azure Virtual Machines, just like we would encounter with on-premises hosted VMs. While the process for troubleshooting Azure-hosted VMs may have similarities to on-premises VMs, there are also key differences to keep in mind.

For instance, if a hosted VM fails to start, you do not have the same physical access to the device or the hosting environment as you would in an on-premises data center. Additionally, when accessing a hosted VM, you may need to consider and navigate just-in-time access and Azure Bastion. While these incredible features help secure access to your VMs, if they are not configured properly, VM connections for troubleshooting and remediation can quickly become difficult or nearly impossible to overcome.

When it comes to identifying VM deployment issues, there is a myriad of reasons why a deployment falls into the failed state. Let's review some of the more common deployment failures and what steps can be taken to correct the failures:

- An **allocation failure** occurs when the selected Azure region or cluster no longer has available resources for the VM (this could be CPU, RAM, disk, or when the requested VM size cannot be supported in that region). Additional troubleshooting techniques can be found at the following URL: `https://learn.microsoft.com/troubleshoot/azure/virtual-machines/allocation-failure`.

- An **authorization failure** occurs when the service principal or account that is being used for deployment does not have the appropriate permissions to complete the deployment activity. This could also be due to a resource provider registration error.

- A **conflict failure** occurs when a request has been made for an operation that is not allowed in the current state of the VM and is typically due to a VM being deallocated.

- A **deployment quota failure** occurs when your VM deployment request has exceeded the maximum allowed limit on either a service limit or a quota applied to a subscription, resource group, account, or additional scoping. In some cases, a quota increase can be administrator adjusted or requested from support (subscription limits and quotas information can be found at the following URL: `https://learn.microsoft.com/azure/azure-resource-manager/management/azure-subscription-service-limits`) within the Azure portal by visiting **Subscription**, then **Usage + quotas**, and finally selecting either the pencil icon for increasing the administrator quota or the user icon to **Create a new support request**, as shown in *Figure 20.1*:

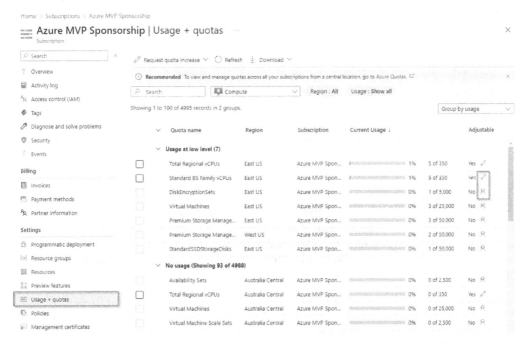

Figure 20.1 – Completing a quota increase or requesting an increase from Azure Support

- An **invalid template failure** occurs when errors are encountered in the template being used for deployment and are typically due to either a syntax error, an incorrect resource name, an improper parameter, or an unnecessary dependency being used in the template.

- A **provisioning failure** occurs when the operating system of the VM fails to load properly. This is typically due to a specialized VM image that has been captured as a generalized image where the error reports that the VM is stuck at the **out-of-box experience** (**OOBE**) screen during upload or receives a timeout error because it is marked as generalized during the capture process. The second error experience is related to the converse where a generalized VM image has been captured as a specialized image and gets stuck at the OOBE screen during upload primarily because the VM is running with the original source computer name, username, and/ or password and has been marked as specialized during the capture process.

- A **request disallowed by policy failure** occurs when there is an **Azure Resource Manager** (**ARM**) lock, Azure Policy, an incorrect region, type of resource, SKU, or other reason on the resource(s) that you are attempting to deploy and typically require a review of security policy definition and assignments to intended resources.

- A **stock-keeping unit** (**SKU**) **not available failure** occurs when either ARM or Bicep template files are expecting a resource that is currently not available in the intended Azure subscription region or availability zone.

- A **subnet is full failure** occurs when any IP addresses are no longer available in the subnet, thus requiring the existing subnet to be cleaned up, a different existing subnet to be used, or a completely new subnet to be created.

> **Additional resources for troubleshooting common Azure deployment failures**
>
> While we have reviewed a more concise list of frequently occurring failures, Microsoft has also published an ever-growing list of additional failure error codes and resulting details for troubleshooting, mitigation, and supporting information on errors received during virtual machine deployment for both Windows and Linux.
>
> These articles are available at `https://learn.microsoft.com/azure/azure-resource-manager/troubleshooting/common-deployment-errors` and `https://learn.microsoft.com/troubleshoot/azure/virtual-machines/troubleshoot-deployment-new-vm-windows`.
>
> Additional troubleshooting resources for Linux VMs can be found at the following URL: `https://learn.microsoft.com/troubleshoot/azure/virtual-machines/troubleshoot-deployment-new-vm-linux`.

In most of the cases we have covered, capturing activity or diagnostic logs and then adjusting your deployment before retrying the request or operation tends to resolve most of the failures encountered.

Now that we have covered some tips and tricks for troubleshooting Azure VM deployment errors, let's learn how to troubleshoot booting failures with Azure VMs.

Troubleshooting booting failures

Throughout this book, we have discussed the overall reliability of both the **Windows Server** operating system and **Microsoft Azure**. However, nothing is perfect and we do occasionally encounter a virtual machine that simply refuses to boot up properly. This could be due to a wide variety of reasons including, but not limited to, the following:

- Filesystem or disk stability or corruption
- Disk encryption (such as BitLocker or Azure Disk Encryption)
- Blue screen errors
- Post update processing
- *Critical service failed* error on startup or via blue screen error
- A Windows reboot loop

Let's begin by reviewing some of the built-in tools that can assist with identifying and potentially resolving a boot failure that is encountered due to the issues previously listed.

Reviewing Boot diagnostics

Boot diagnostics is enabled by default on **Azure Virtual Machines** and can be accessed using the following steps:

1. Within the **Virtual Machines resources** area in the Azure portal, select `vm-az801-svr1` from the list.
2. Scroll down to the **Help** blade section, then select **Boot diagnostics**.

3. By default, you will see the last screen shown to the console under the **Screenshot** tab, as shown in *Figure 20.2*. You also have the option of selecting **Download screenshot** to capture a copy of the latest screenshot:

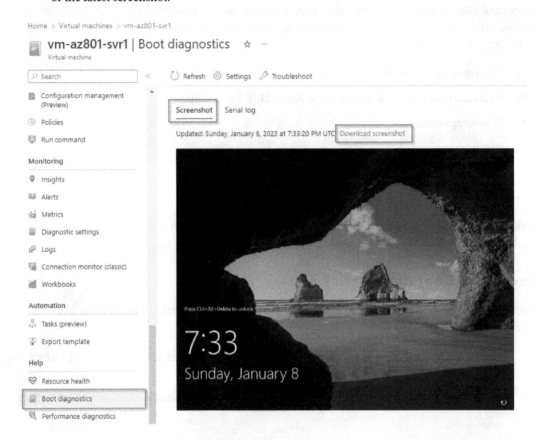

Figure 20.2 – Example of the Boot diagnostics screenshot feature within Azure Virtual Machines

4. Selecting the **Serial log** tab displays the output from the serial log for our VM, indicating that the VM has started successfully, as shown in *Figure 20.3*:

Figure 20.3 – Example of the serial log output from our Virtual Machine

5.   Finally, there is the **Troubleshoot** tool within the **Boot diagnostics** page, whereby you can automatically troubleshoot and recommend mitigation steps, as shown in *Figure 20.4*:

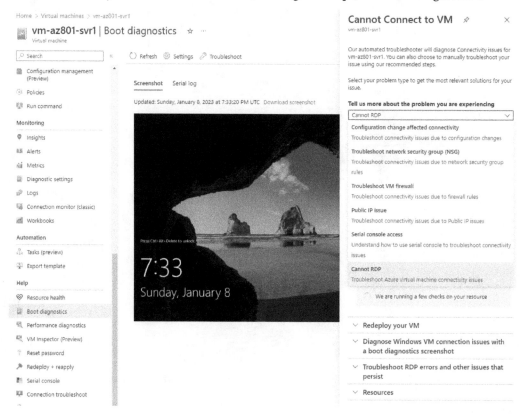

Figure 20.4 – Using the automated troubleshooter feature within Boot diagnostics

Automatic options that can be run as part of the troubleshooting tool are **Configuration change affected connectivity**, **Troubleshoot network security group (NSG)**, **Troubleshoot VM firewall**, **Public IP issue**, **Serial console access**, and **Cannot RDP**. This completes the steps for this exercise on **Boot diagnostics**.

Reboot loops on Azure VMs

This may or may not appear on the exam, but it is valuable to know that you may experience boot loops on an Azure VM. The following article discusses three reasons why this may happen and gives extensive details on how to resolve the issue in each of the three cases: `https://learn. microsoft.com/troubleshoot/azure/virtual-machines/troubleshoot- reboot-loop`.

Now, let's move on to another tool called **Azure Serial Console**, which is used for boot troubleshooting and diagnosis.

Working with Azure Serial Console

If you encounter a case where you cannot create a new connection to a VM after startup, Azure Serial Console becomes your new best friend for troubleshooting. Let's complete a walkthrough to see how we can use this feature to get additional information about the status of our VM:

1. To utilize this troubleshooting feature, we must have **Enable with managed storage account** or **Enable with custom storage account** enabled. This can be completed from within the **Boot diagnostics** settings for our VM and is configured as shown in *Figure 20.5*:

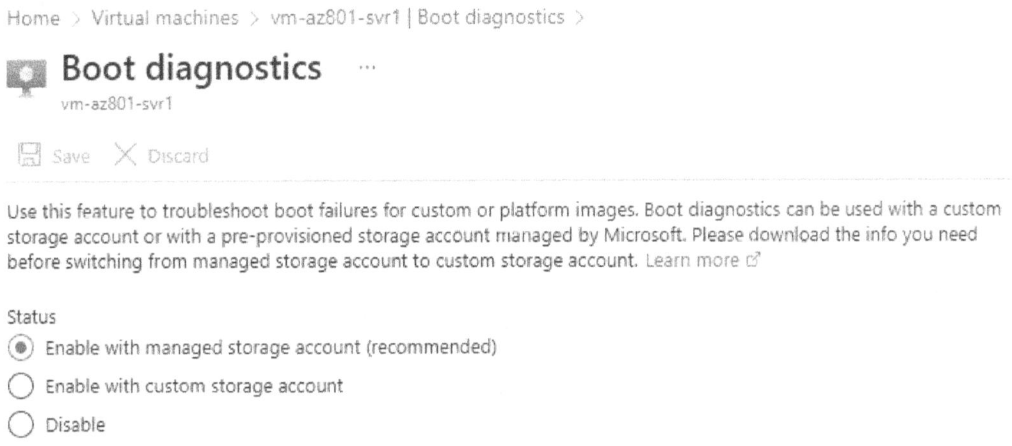

Home > Virtual machines > vm-az801-svr1 | Boot diagnostics >

🖥️ Boot diagnostics   ...
vm-az801-svr1

💾 Save   ✕ Discard

Use this feature to troubleshoot boot failures for custom or platform images. Boot diagnostics can be used with a custom storage account or with a pre-provisioned storage account managed by Microsoft. Please download the info you need before switching from managed storage account to custom storage account. Learn more ↗

Status

⦿ Enable with managed storage account (recommended)

◯ Enable with custom storage account

◯ Disable

Figure 20.5 – Enabling the Boot diagnostics feature with storage account configuration

2. Since we have not yet connected to our Azure VM via console, we will need to establish a new user account and credential for the Virtual Machine. To do this, select **Reset password** under the **Help** blade, enter vmadmin for **Username**, then type Packtaz801guiderocks in both the **Password** and **Confirm password** fields. Then, click the **Update** button to create the account on the VM, as shown in *Figure 20.6*:

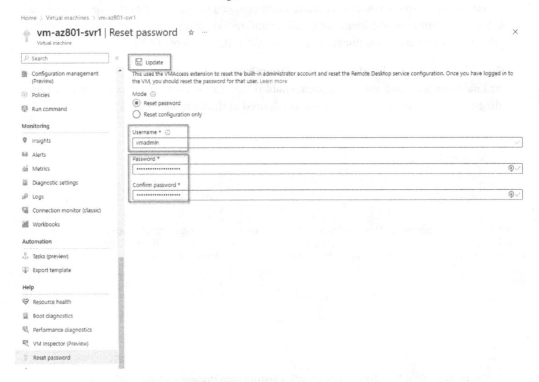

Figure 20.6 – Setting a new admin user and credential for our VM

3. Next, under the **Help** section for our VM, select **Serial console**. Once connected, you should see a screen similar to what is shown in *Figure 20.7*:

Home > Virtual machines > vm-az801-svr1 >

vm-az801-svr1 | Serial console ...
Virtual machine

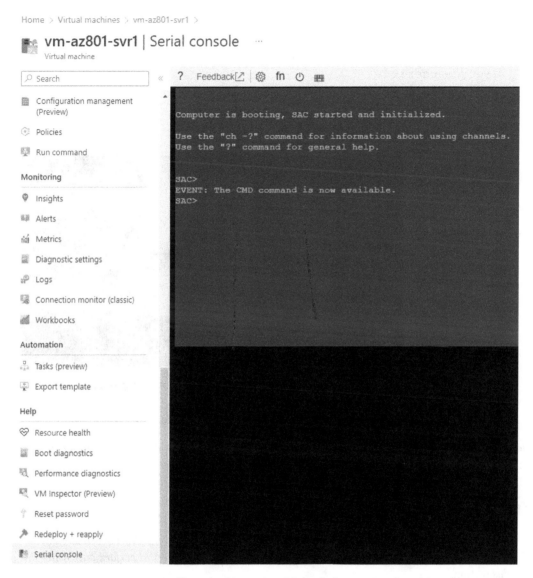

Figure 20.7 – Initial launch of Azure Serial Console for our virtual machine

4. Enter cmd at the prompt, then press the *Esc* key on your keyboard, followed by the *Tab* key, to switch to the newly created **CMD** console channel.

5. When prompted, enter the new username you just created, hit *Enter* for domain (if prompted), and supply a password of Packtaz801guiderocks. Then, hit *Enter*.

6. Once authenticated, type `powershell` and hit *Enter*, noticing that you can now utilize the familiar set of PowerShell tools to investigate and troubleshoot the VM, as shown in *Figure 20.8*. The command that was used in the example is `Get-EventLog -LogName system -Source user32`:

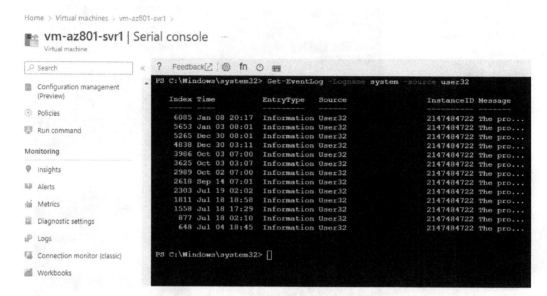

Figure 20.8 – Running PowerShell cmdlets to troubleshoot boot/startup issues

7. This completes the steps for this exercise on Azure Serial Console.

Now that we have covered using Azure Serial Console to gain troubleshooting access to an Azure VM, let's move on to troubleshooting and diagnosing VM performance issues.

Troubleshooting VM performance issues

We covered how to troubleshoot VM performance issues in *Chapter 18, Monitoring Windows Server Using Windows Server Tools and Azure Services,* in the *Monitoring Azure VM performance using VM Insights* section.

However, there is yet another diagnostics tool available within the **Help** section of an individual Windows or Linux VM called **Performance diagnostics**. This tool installs a VM extension named **PerfInsights** on the VM.

The following steps will help you set up the tool and run **Performance diagnostics**:

1. Select our `vm-az801-svr1` virtual machine from within the Azure portal.

2. Navigate to the **Help** section of the VM, and then select **Performance diagnostics**.

3. You have the option of changing the **Settings** properties, where you can change the storage account. For this exercise, we will keep the default settings, so select the **Install performance diagnostics** button, as shown in *Figure 20.9*:

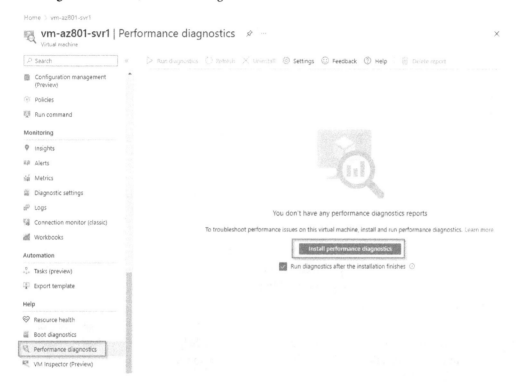

Figure 20.9 – Installing the performance diagnostics VM extension on an Azure VM

4. Here, we have the option of selecting one of the four performance diagnostics, as shown in *Figure 20.10*:

Figure 20.10 – Selecting an analysis scenario to run for our exercise

The additional details outlined for each of the scenarios are given here:

I.   **Quick performance analysis** checks against known issues and baselines against best practices, and collects general diagnostic data within a few minutes. This analysis scenario can be run on both Windows and Linux VMs.

II. **Performance analysis** gives more configuration options on top of the checks available under the **Quick performance analysis** option. It can be configured anywhere from 30 seconds to 15 minutes for a performance trace and helps troubleshoot high CPU, RAM, and disk utilization on VMs. This analysis scenario can be run on both Windows and Linux VMs.

III. **Advanced performance analysis** contains all the checks under **Performance analysis** and allows us to select one or more of the additional traces, as shown in *Figure 20.11*. Note that this analysis scenario can only be run on Windows:

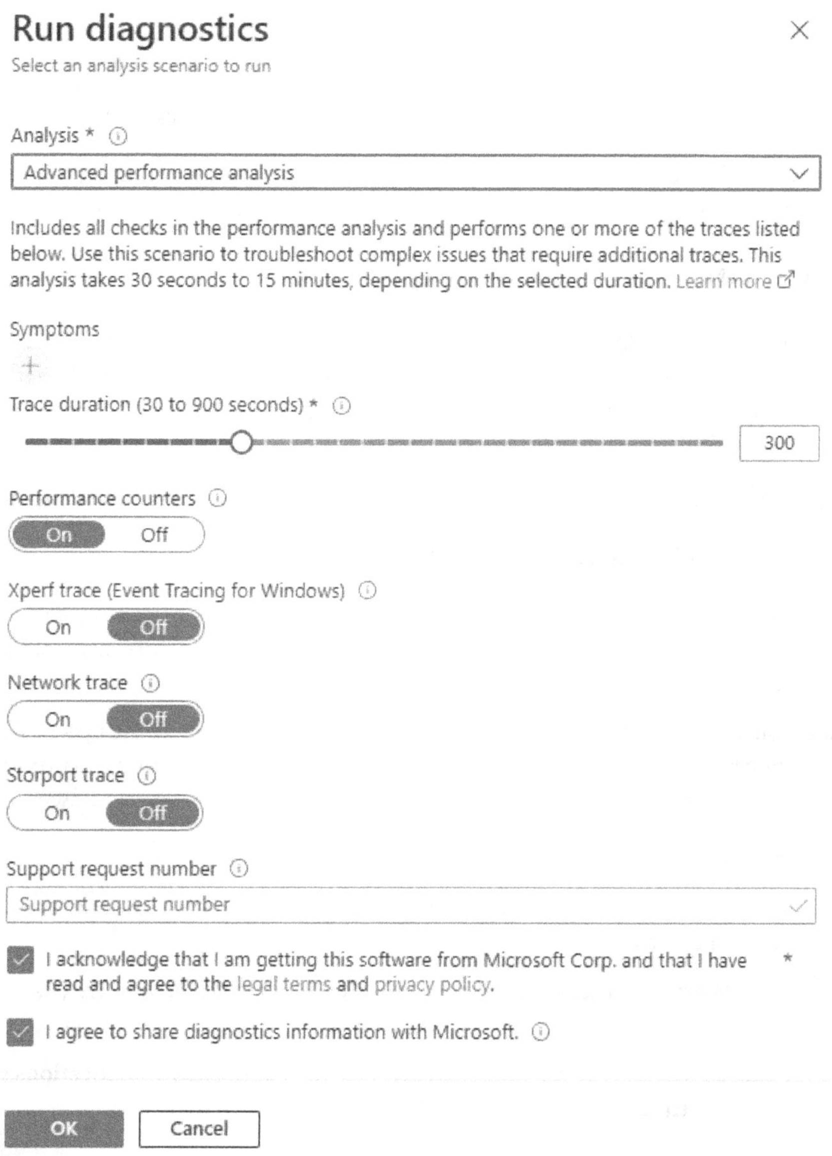

Run diagnostics   ✕

Select an analysis scenario to run

Analysis * ⓘ

Advanced performance analysis	⌄

Includes all checks in the performance analysis and performs one or more of the traces listed below. Use this scenario to troubleshoot complex issues that require additional traces. This analysis takes 30 seconds to 15 minutes, depending on the selected duration. Learn more ⌐

Symptoms

+

Trace duration (30 to 900 seconds) * ⓘ

━━ ━━━ ━━ ━━ ━━ ━━ ━━●━ ━━ ━━ ━━ ━━ ━━ ━━ ━━ ━━ ━━ ━━ ━━ ━━ ━━ | 300 |

Performance counters ⓘ

(**On**   Off)

Xperf trace (Event Tracing for Windows) ⓘ

(On   **Off**)

Network trace ⓘ

(On   **Off**)

Storport trace ⓘ

(On   **Off**)

Support request number ⓘ

Support request number	✓

☑ I acknowledge that I am getting this software from Microsoft Corp. and that I have   *
  read and agree to the legal terms and privacy policy.

☑ I agree to share diagnostics information with Microsoft. ⓘ

OK		Cancel

Figure 20.11 – Display of the available traces from Advanced performance analysis

IV.   **Azure Files analysis** also includes all checks under **Performance analysis** and allows additional network trace and **SMB** counters to help troubleshoot the performance of **Azure files**. It can be configured anywhere from 30 seconds to 15 minutes for a performance trace. Note that this analysis scenario can only be run on Windows.

5. Once any of the diagnostic reports have been created, they can simply be reviewed or downloaded for up to 30 days from within the diagnostics report list.

6. This completes the steps for this exercise on **Performance diagnostics**.

Now that we have covered using **Performance diagnostics** to gain troubleshooting access to an Azure VM, let's move on to troubleshooting and diagnosing VM extension issues.

Troubleshooting VM extension issues

For those who have used Azure VM extensions, the installation process is seemingly magical at times where you either enable a feature, run a script, or simply press a button and the VM extension is installed on your Azure VM. However, for those less-than-magical moments, let's discuss a few quick ways to troubleshoot VM extension failures.

The following list of tips and tricks should be used to troubleshoot and resolve any VM extension failures:

- Ensure that the **VM Agent** (or Azure Guest Agent) is installed for a VM:

  - Additional troubleshooting for the Windows VM Agent is located here: `https://learn.microsoft.com/troubleshoot/azure/virtual-machines/windows-azure-guest-agent`

  - Additional troubleshooting for the Linux VM Agent is located here: `https://learn.microsoft.com/troubleshoot/azure/virtual-machines/linux-azure-guest-agent`

- Ensure that the extension is supported for the version of your Azure VM

- Review the VM extension logs (for Windows) stored in `C:\WindowsAzure\Logs\Plugins directory`

- Review the VM extension settings and status file (for Windows) stored in `C:\Packages\Plugins directory`

- View the extension status for the VM, available from the **Extensions + applications** selection within the VM **Settings** blade, as shown in *Figure 20.12*:

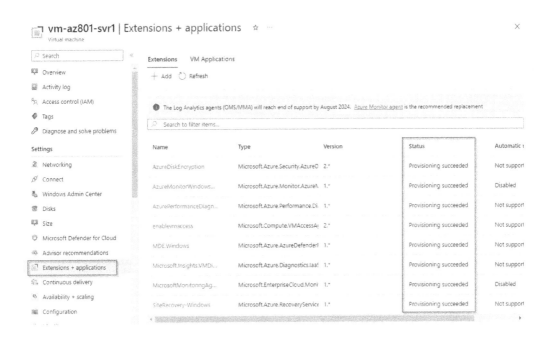

Figure 20.12 – Viewing the extension status of an Azure VM

- Rerun or uninstall the extension on the VM, as shown in *Figure 20.13*, where **AzPerfDiagExtension** failed to install on the VM:

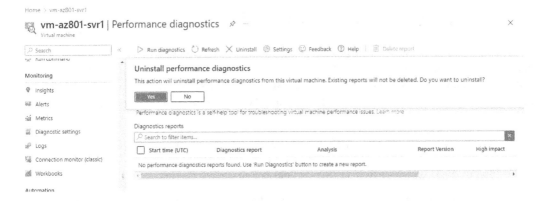

Figure 20.13 – Uninstall the performance diagnostics VM extension

- Utilize the VM **Activity log** to determine the root cause of VM extension failures, selecting **Activity log** from within a selected VM, as shown in *Figure 20.14*:

Figure 20.14 – Reviewing VM extension details within the VM Activity log

Now that we have covered using troubleshooting VM extensions for an Azure VM, let's move on to troubleshooting VM disk encryption issues.

Troubleshooting disk encryption issues

We covered what **Azure Disk Encryption** using **BitLocker** means to **Windows Server** in *Chapter 6*, in the section - *Enabling Storage Encryption Using Azure Disk Encryption*, where the operating system drive is unlocked during startup. That said, if a VM fails to start as expected during the boot phase, there may potentially be an issue with BitLocker, and the screenshots from the boot diagnostics will assist with troubleshooting.

Methods of recovery may be determined based on one of the following messages sent to the console screen:

- **Plug in the USB driver that has the BitLocker key**

- **Enter your recovery key – Load your recovery key from a USB device**

- **Enter the password to unlock this drive [ ] Press the Insert Key to see the password as you type**

- **You're locked out! Enter the recovery key to get going again (Keyboard Layout: US) The wrong sign-in info has been entered too many times, so your PC was locked to protect your privacy. To retrieve the recovery key, go to** `https://windows.microsoft.com/recoverykeyfaq` **from another PC or mobile device. In case you need it, the key ID is XXXXXXX. Or, you can reset your PC.**

Typically, for almost all the on-screen messages as indicated, an administrator can and should consider stopping the running VM, deallocating the VM, and starting the VM back up so that the device can attempt to retrieve the **BitLocker Recovery Key** stored within the Azure Key Vault.

Additionally, if there is a recent backup of the VM that can be restored, the backup version of the VM may also fix the inherent boot problem.

More advanced recovery techniques for BitLocker recovery

There are many scenarios and environmental designs, so it's difficult to account for every restore scenario. However, Microsoft does offer an incredibly detailed set of processes that can help you recover or decrypt an encrypted disk or recover your disk when utilizing a **Key Encryption Key** (KEK) scenario for protecting **BitLocker Encryption Keys** (BEKs) and **Content Encryption Keys** (CEKs).

The article referencing these more advanced techniques is detailed at the following URL: `https://learn.microsoft.com/troubleshoot/azure/virtual-machines/troubleshoot-bitlocker-boot-error`.

Now that we have provided an overview of troubleshooting disk encryption for an Azure VM, let's move on to troubleshooting VM storage issues.

Troubleshooting storage

The available tiers of Azure storage, called Standard and Premium, allow you to store Azure VM disk files. Generally, the disks exist as `.vhd` files and are stored as optimized page blobs. Do remember that **Standard storage** has the equivalent performance of **hard disk drives** (HDDs), whereas **Premium storage** gives superior (and even **Ultra disk**) performance equivalent to the latest **solid-state drive** (SSD) technologies on the market.

In addition, there is quite a bit of consideration regarding managed versus unmanaged disks. Unmanaged disks have been traditionally designed and made available for low-priority, testing, disaster backup, or proof-of-concept workloads and are only available in HDDs. This approach of unmanaged disks also has a higher administrative overhead with planning, naming, availability and resiliency options, and storage constraints regarding support and service maximums.

For managed disks, there are far more available benefits:

- Offers server-side and **Azure Disk Encryption**
- More resiliency for disks within the same Azure VM availability set
- More granular, disk-level **Role-Based Access Control** (RBAC)
- The ability to convert from an unmanaged to a managed disk (by stopping and deallocating the VM or all VMs within an availability set)

- A higher number of disks per subscription
- Less administrative effort as Azure manages the VM disk placement and reduces the overall storage management complexity

Now that we have set the stage, some of the common issues for storage issues within your Azure VMs include, but are not limited to, the following:

Issue	Resolution
Azure disk encryption errors	See the *Troubleshooting disk encryption issues* section. Also, consider any network components that may prevent or disrupt your connectivity, such as proxy requirements, firewall requirements, or NSG settings or rules. Finally, periodically review the platform-level encryption settings to ensure compliance and proper VM extension settings for encryption.
Storage errors when deleting storage resources	Attempt to identify any blob storage attached to VMs, consider deleting the VMs with the attached OS disk, detach any or all the data disks from the remaining VMs, then retry the deletion operation.
Inability to extend an encrypted operating system volume within Windows	This is most likely caused by a disk that has the wrong partition ID assigned to it and thus presents an issue with booting the system partition. Potential steps to resolve this issue start with assigning a larger disk SKU to the OS disk in the portal, then extending the system reserved volume into the new unallocated space on disk, creating a new boot volume in the remaining unallocated space, then deleting the system reserved volume and safely extending the Windows volume as originally intended.
Unable to create an Azure VM by uploading a VHD file	This is typically due to non-compliance with a 1 MB alignment rule and can be resolved by first installing a Hyper-V role on the Windows Server, then converting the disk into a fixed-size VHD.

> **Deprecation of Azure unmanaged disks by September 30, 2025**
>
> Note that even though you should know the difference between managed versus unmanaged disks in Azure, unmanaged disks are on the deprecation list for 2025. For additional information on the reasons why, including migration strategies, be sure to check out the following article: `https://learn.microsoft.com/azure/virtual-machines/unmanaged-disks-deprecation`.

Now that we have covered troubleshooting storage issues for an Azure VM, let's move on to troubleshooting VM connection issues.

Troubleshooting VM connection issues

Sure, **Windows Admin Center** and **Azure Cloud Shell** solve most management and administrative concerns, but there are still valid reasons that administrators must rely on remote connectivity to Azure VMs. Two of those approaches utilize **Remote Desktop Protocol** (**RDP**) over TCP port 3389 and **Secure Shell** (**SSH**) over TCP port 22 for remote management and connectivity.

From an RDP connectivity perspective, Azure has introduced additional validation for the **Connect** setting available on Azure VMs running Windows. As you can see in *Figure 20.15*, Azure not only crafts the RDP file for connecting to your VM but also pre-validates whether your connectivity is sufficient and recommended:

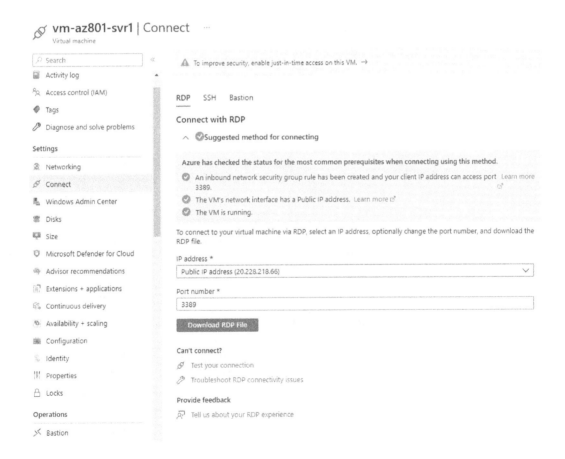

Figure 20.15 – Reviewing the Connect with RDP page for remote connections

From this same screen, an administrator can choose from **Test your connection** and **Troubleshoot RDP connectivity issues** to run the automated validation checks. If one of these checks fails, you will want to consider reviewing the following:

- Confirm that this is the expected remote connection type for this VM

- Validate that any **Network Security Group** rules are not blocking RDP traffic

- If traffic is indeed permitted, either reset the virtual machine's NIC or restart the VM

Similar configuration options and connectivity validation are available for **SSH**, as shown in *Figure 20.16*.

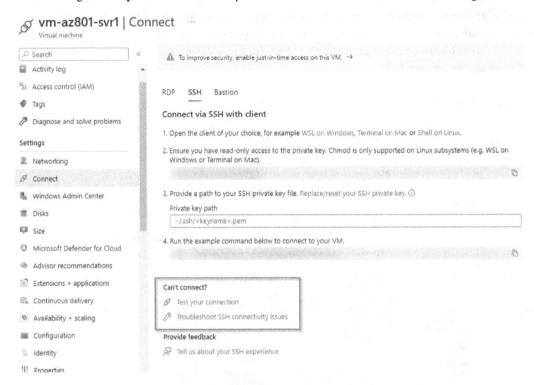

Figure 20.16 – Testing and troubleshooting SSH connections in Azure

However, both previous options require opening remote administrative access to the internet, thus opening up inherent vulnerabilities to your workloads and your business. Microsoft has created a solution called **Azure Bastion** that provides a secure connection to Azure VMs remotely using a modern web browser with **HTML5** over the latest TLS protocols.

> **Azure Bastion for secure administration**
>
> In a nutshell, Azure Bastion establishes what's called a Bastion host within the Azure virtual network. The administrative users then connect to the Azure portal using HTML5 and TLS in a modern browser and select the destination VM for connection. The Bastion host is then responsible for brokering the RDP/SSH connection to the VM via privileged administrative workstations or "jump stations."
>
> For more details on the benefits, scale, architecture, and overall cost of the Azure Bastion feature set, be sure to check out the following URL: `https://learn.microsoft.com/azure/bastion/bastion-overview`.

With that, we have completed this chapter on how to troubleshoot VM connection issues in Azure. Next, we will review what we have learned in this chapter and set the stage for the next chapter of this book.

Summary

In this chapter, we learned how to effectively troubleshoot Windows Server virtual machine workloads in Microsoft Azure. We learned how to troubleshoot deployment and booting failures, VM performance and extension issues, disk encryption, storage, and overall VM connection issues. We also learned about a wide range of tools that will help us successfully address typical issues experienced with running Windows Servers in Microsoft Azure, consistent with the AZ-801 exam objectives.

In the next chapter, we will learn about valuable tools that help you troubleshoot and recover from typical issues experienced when running **Active Directory Domain Services** workloads. We will discuss how to enable and use the Active Directory Recycle Bin to restore deleted objects, how to use **Directory Services Restore Mode** to recover a corrupt **Active Directory** database and/or corrupted objects, and how to recover the SYSVOL folder and files necessary for running Active Directory services. We will then learn about tools available to determine Active Directory replication issues and how to resolve them, how to troubleshoot hybrid authentication issues, and how to troubleshoot overall issues that are encountered when running Active Directory on-premises.

21

Troubleshooting Active Directory

In this chapter, we will learn about valuable tools that help you troubleshoot and recover from typical issues experienced when running **Active Directory Domain Services** (**AD DS**) workloads. We will discuss how to enable and use AD Recycle Bin to restore deleted objects, how to use **Directory Services Restore Mode** (**DSRM**) to recover a corrupt AD database and/or corrupted objects, and how to recover the SYSVOL folder and files necessary for running AD services.

We will focus on requirements, best practices, and hands-on walkthroughs in this chapter that align with the *AZ-801* exam objectives. We will then learn about the tools available for determining AD replication issues and how to resolve them, how to troubleshoot hybrid authentication issues, and how to troubleshoot general issues encountered when running AD on-premises.

In this chapter, we will cover the following topics:

- Technical requirements and lab setup
- Restoring objects from AD Recycle Bin
- Recovering an AD database using DSRM
- Recovering SYSVOL
- Troubleshooting AD replication
- Troubleshooting hybrid authentication issues
- Troubleshooting on-premises AD

Technical requirements and lab setup

To successfully follow along and complete the tasks and exercises throughout this chapter and the following chapters in this book, we will need to ensure that the *Technical requirements* section from *Chapter 1, Exam Overview and the Current State of Cloud Workflows*, has been completed in full.

When it comes to on-premises and even hybrid authentication and authorization, AD DS provides the necessary user, group, device, policy, and identity management tools for all the objects within your infrastructure. Knowing how to effectively restore and recover objects and services for AD DS, troubleshoot on-premises authentication issues, and troubleshoot hybrid authentication and authorization issues are a must.

This becomes even more important when organizations have users and devices connecting to hybrid services and **Software-as-a-Service (SaaS)** apps such as **Microsoft 365**. The integration of both on-premises and cloud identity management systems quickly becomes a blur with services such as **single sign-on (SSO)**, password protection, or **self-service password reset (SSPR)** happening either exclusively on-premises, exclusively in the cloud, or somewhere in between.

This chapter is focused on troubleshooting AD DS no matter where the identity control plane is being managed and will help you learn about details, tips, and tricks for both recovery and troubleshooting using common and new tools as part of this hybrid world.

Restoring objects from AD Recycle Bin

We have all been there before – a support representative calls and goes on to share that an AD object has been accidentally deleted either manually or automatically. The discussion quickly turns to raising the following questions for restoration:

- What type of AD object has been deleted and are there any attributes that might be lost due to the deletion?

- When has the object been deleted? AD garbage collection runs every 12 hours, cleaning up the AD database and any aging or tombstoned objects that have exceeded a default lifetime of 180 days, so this is important to determine during the restore request process.

- Where has the object been deleted from and to where should it be restored? The **Organizational Unit (OU)** is very important, as the restored object should be placed back into its original location to restore any inherited permissions, policies, or other AD attributes necessary to protect the object.

The msDS-deletedObjectLifetime and tombstoneLifetime attributes

The default value of a deleted object lifetime is set to 180 days even though both attribute values are set to Null by default (for additional details, be sure to check out this URL: `https://learn.microsoft.com/windows-server/identity/ad-ds/get-started/adac/advanced-ad-ds-management-using-active-directory-administrative-center--level-200-#BKMK_EnableRecycleBin`). That said, if a requirement is identified that forces a change to these values, the PowerShell commands can be run to update both values as required.

For example, if the requirement was set to retain deleted objects for 7 years, we would need to use 2555 days as the value in the following PowerShell commands.

Setting the tombstoneLifetime to 365 days:

```
Set-ADObject -Identity "CN=Directory Service,CN=Windows
NT,CN=Services,CN=Configuration,DC=AD,DC=az801,DC=com" -
Partition "CN=Configuration,DC=AD,DC=az801,DC=com" -Replace:@
{"tombstoneLifetime" = 2555}
```

Setting the deleted object lifetime to 365 days:

```
Set-ADObject -Identity "CN=Directory Service,CN=Windows
NT,CN=Services,CN=Configuration,DC=AD,DC=az801,DC=com" -Partition
"CN=Configuration,DC=AD,DC=az801,DC=com" -Replace:@{"msDS-
DeletedObjectLifetime" = 2555}
```

There are a few techniques that can be used to restore (or reanimate) the identified object. Still to this day, as has been true for quite some time, the `ldp.exe` tool can be used to connect, bind, and ultimately restore objects from AD using a deprecated predefined control called **Return deleted objects**, as shown in *Figure 21.1*. However, this roughly 18-step process is lengthy and, while well documented, still trips up even the most experienced AD engineers.

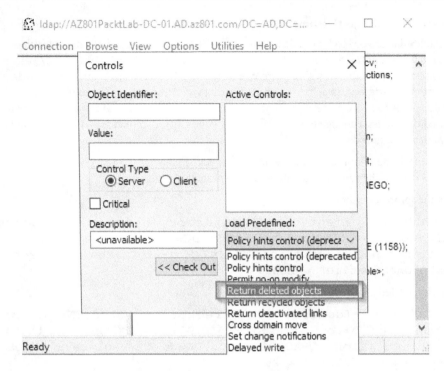

Figure 21.1 – Utilizing ldp.exe to return and restore a deleted item from AD

The preferred and simplified approach to restoring deleted objects with a full authoritative restore process is to enable and utilize the **Active Directory Recycle Bin** feature. This tool doesn't require pulling data from backups, restarting AD DS services, rebooting domain controllers, or attempting to identify and restore all link-valued and non-link-valued AD attributes for the restored object. This provides a consistent and easy approach to restoration in any environment, so let's learn how to enable and use **Active Directory Recycle Bin**. Note that to enable this feature, the forest functional level of the AD DS environment must be at **Windows Server 2008 R2** or higher:

1. From a virtual machine connection to `AZ801PacktLab-DC-01`, open **Active Directory Administrative Center** from the **Start** menu.

2. Select **AD (local)** and, from the **Tasks** menu on the right, then select **Enable Recycle Bin ...**, as shown in *Figure 21.2*.

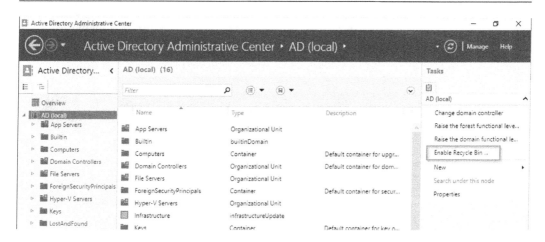

Figure 21.2 – Enabling AD Recycle Bin from Active Directory Administrative Center

3. Say **Yes** to the **Are you sure you want to perform this action? Once Recycle Bin has been enabled, it cannot be disabled.** message, and then say **OK** to the **Please refresh AD Administrative Center now AD DS has begun enabling Recycle Bin for this forest. The Recycle Bin will not function reliably until all domain controllers in the forest have replicated the Recycle Bin configuration change.** prompt.

4. Once **Active Directory Administrative Center** has been refreshed, navigate to the **AD (local)** > **Users** container and select the object named **RonHD** from the list. Right-click on the account and select **Delete** from the menu and then **Yes** to confirm, as shown in *Figure 21.3*.

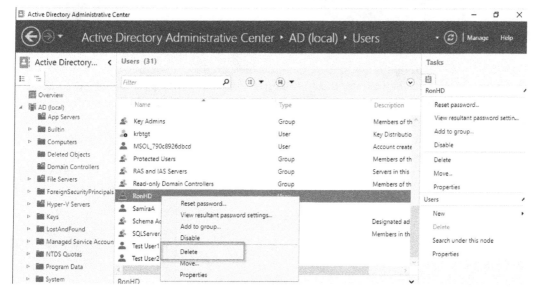

Figure 21.3 – Deleting a user from AD.az801.com as an example in AD Recycle Bin

5. Next, navigate to **Deleted Objects**, select the recently deleted user, **RonHD**, from the list, and select **Restore** to complete the authoritative restoration, noting that **Last known parent** shows the AD OU from where the object originated, as shown in *Figure 21.4*.

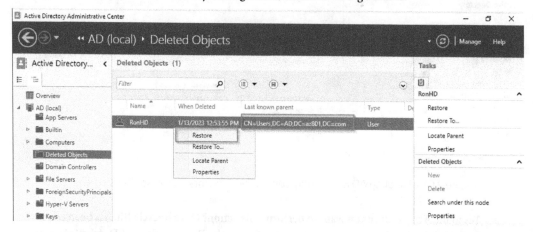

Figure 21.4 – Restoring a deleted object from AD Recycle Bin to the original OU

6. Return to the **Users** OU and confirm that the **RonHD** user object has been successfully restored. This completes this Active Directory Recycle Bin exercise.

Let's now move on to learning about how to recover an AD database using the tools readily available within the **Windows Server** operating system.

Recovering an AD database using DSRM

DSRM is a boot option that is like safe mode but is only available to Windows Server domain controllers, allowing an administrator to authoritatively restore or repair a corrupted AD database (identified as the `Ntds.dit` file). Typically, if there is an error raised on a domain controller that cannot be easily resolved, DSRM must be used to restart the domain controller in an offline mode, allowing the administrator to recover system files and work with the collection of database files located in `%SystemRoot%\NTDS` by default.

This collection of files comprising an AD database results in the following:

AD file	File description
Ntds.dit	This is the main database file that contains all AD objects and directory partitions.
Edb*.log (one or multiple files)	These files are considered the AD DS transaction logs for the database, tracking changes to the AD DS database.
Edb.chk	This file is considered a database check file and determines the place where transactions have been successfully written to the database, as well as identifying what transactions are waiting to be processed and written to the database.
Edbres00001.jrs and Edbres00002.jrs	These files are unique in that they reserve additional transaction log space in the event that a domain controller runs out of disk space. If this situation occurs, AD DS recoups the space from these two files, processes the remaining transactions in the waiting queue, and then safely shuts down AD services while dismounting the database to avoid corruption.

Table 21.1 – Collection of files comprising an AD database

Let's review the following walkthrough to gain familiarity with all the available tools to manage an AD database in a failed state due to one of these various failure scenarios:

- Cleaning up and removing failed AD domain controllers from the environment

- Authoritative restoration of a deleted object(s) in AD

- Managing the AD snapshots for a domain controller

- Forcible transfer of the **Flexible Single Master Operation** (**FSMO**) role or roles for a domain controller

- Completing AD database maintenance to recover from corruption by checking database integrity, moving the files to a new location, or completing cleanup/compression of the database files to restore performance

- Maintenance of the administrator password for DSRM (a change or rollover)

Booting a domain controller into DSRM requires access to a console, as the *F8* key needs to be pressed during startup to enter **Advanced Boot Options**, and more importantly, to be able to select **Directory Services Repair Mode**, as shown in *Figure 21.5*.

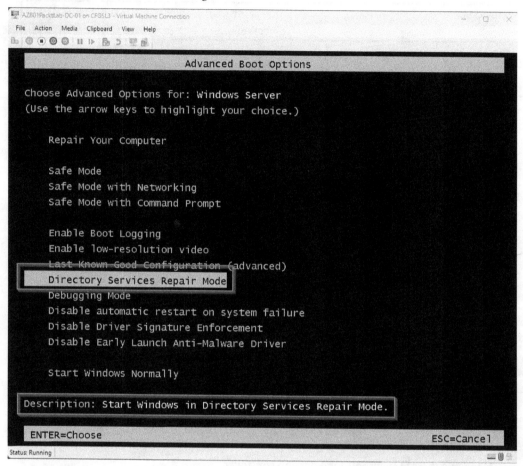

Figure 21.5 – Booting a domain controller into DSRM

Remotely restarting a domain controller into DSRM

If you do not have console access to the domain controller, it is more than likely you have **Remote Desktop Protocol (RDP)** access to the DC and can utilize the `msconfig.msc` user interface tool to configure the boot configuration and startup options for the DC. For more details on this remote DSRM process, be sure to review the following article: `https://learn.microsoft.com/previous-versions/windows/it-pro/windows-server-2008-r2-and-2008/cc794729(v=ws.10)`.

From this point, as we first learned in *Chapter 2*, *Securing the Windows Server Operating System*, a DSRM password must be set for AD, should always be protected (using a hardened password that is not reused elsewhere in the environment), and should only be used for DSRM maintenance and restoration efforts. When the domain controller is booted into DSRM mode, know that this is considered an AD DS **Stopped** state, meaning that the domain controller is still a member server and will get **Group Policy** applied to it. However, it will not be able to service authentication requests while it remains in this state. Note that other domain controllers within the AD DS environment will continue to host authentication requests while one domain controller is in DSRM mode, but replications to that domain controller in the DSRM state will fail until the domain controller has been rebooted and restored to the AD DS **Started** state.

For this walkthrough, suppose the environment did not have **Active Directory Recycle Bin** enabled and our user **RonHD** from earlier in this chapter was accidentally deleted from AD. If we have recent backups and snapshots of the AD environment, we could utilize DSRM and `ntdsutil.exe` to authoritatively restore the deleted object, as shown in *Figure 21.6*, then complete a reboot on the domain controller to complete the restoration process, and return to the AD DS **Started** state to host authentication requests for the domain controller again.

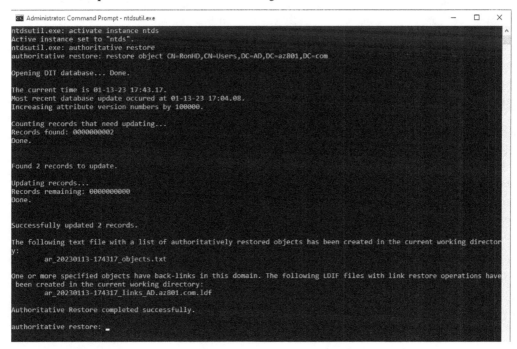

Figure 21.6 – Using ndtsutil.exe to complete an authoritative restore example

AD snapshots provide a historical capture of the state of a directory service on a domain controller at a certain time. Many organizations utilize a simple scheduled task and a script to create snapshots at a certain cadence as another method of backing up and allowing AD objects to be recovered. The `ntdsutil.exe` tool can also mount these snapshots, browse the contained objects within the snapshot, and ultimately export objects for restoration purposes.

One important thing to note is that if this were in a production/live environment, you would absolutely want to disable the inbound replication while performing any restoration activities (be sure to review this URL for additional replication configuration options: `https://learn. microsoft.com/previous-versions/windows/it-pro/windows-server-2003/ cc811569(v=ws.10)`). As with any of these processes, they can be quite involved based on the environment, the design, and additional complexities, so the following article is a recommended read to familiarize yourself with additional commands and restoration scenarios: `https://learn. microsoft.com/troubleshoot/windows-server/identity/retore-deleted- accounts-and-groups-in-ad`.

For the full set of available commands and switches surrounding `ndtsutil.exe`, be sure to check out the following article: `https://learn.microsoft.com/previous-versions/ windows/it-pro/windows-server-2012-r2-and-2012/cc753343(v=ws.11)`.

Let's now learn how to recover AD `SYSVOL` using the tools readily available within the Windows Server operating system.

Recovering SYSVOL

First, `SYSVOL` is a folder that resides on every domain controller within a domain, is located at `C:\ Windows\SYSVOL` by default, and should be moved to another location only during promotion of the domain controller. `SYSVOL` contains folders, files, and junction points that utilize **Distributed File System Replication (DFS-R)** to keep files and folders synchronized between domain controllers. More details on the recommendations of how to place both the databases and `SYSVOL` can be found at the following URL: `https://learn.microsoft.com/azure/architecture/reference- architectures/identity/adds-extend-domain#vm-recommendations`.

The `SYSVOL` folder is typically accessed through a namespace share such as `\\ad.az801.com\ sysvol` or via the local domain controller share path of `\\AZ801PacktLab-DC-01\sysvol`. Additional folders and shares are listed as such:

- `\SYSVOL\domain`
- `\SYSVOL\staging\domain`
- `\SYSVOL\staging areas`
- `\SYSVOL\domain\Policies`

- `\SYSVOL\domain\scripts`

- `\SYSVOL\SYSVOL`

When it comes to **Group Policy Objects (GPOs)**, there is a **Group Policy container** that is represented as a GUID value and uniquely identifies the object within AD, and contains the basic properties of the GPO. The second component is the **Group Policy template**, which stores all the GPO's configured settings and changes and this collection of files is stored at the following location: `%SystemRoot%\SYSVOL\Domain\Policies\GPOGUID`. It is important to know that the Group Policy container is replicated throughout the AD DS environment via a directory replication agent, whereas the Group Policy template is replicated using DFS-R. Since these two objects are replicated at different intervals, it is possible for objects to be out of sync for brief periods of time, but the issue quickly self-corrects.

Whether intentional or accidental, when items are deleted from `SYSVOL`, they will be deleted on every domain controller after replication has completed. There are at least two approaches that can be taken for restoration purposes, so let's review each high-level process.

Let's complete a system state restoration to restore `SYSVOL`:

1. Identify the location of the current system state backup for the domain controller that needs to be recovered.

2. Ensure that the account completing the restore is part of either the **Backup Operators** group or the **Administrators** (built-in) group.

3. Issue the following command to kick off the restoration process: `wbadmin -authsysvol`.

4. Monitor progress and replication to the domain controller replication partners.

Let's complete authoritative restoration by recovering a domain controller to a point-in-time backup:

1. On the domain controller you are using for the authoritative restore process, open **Active Directory Users and Computers** (enabling **Advanced Features** and showing **User, Contacts, Groups, and Computers as containers**), as shown in *Figure 21.7*.

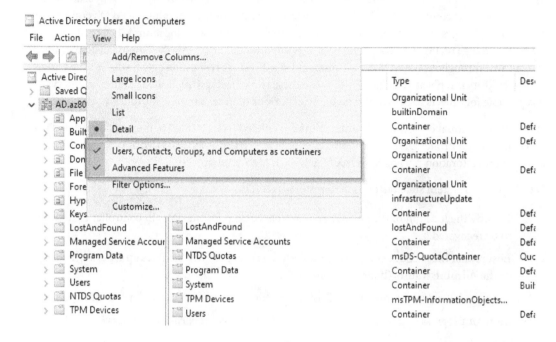

Figure 21.7 – Enabling Advanced Features and Users, Contacts, Groups, and Computers as containers

2. Expand **Domain Controllers**, then the domain controller that has been restored > DFSR-LocalSettings > **Domain System Volume**, then right-click on the **SYSVOL Subscription** object, and select **Properties**.

3. Select the **Attribute Editor** tab, then locate msDFSR-Options (you may need to select the **Filter** button and deselect **Show only attributes that have values**), and change the value to 1, as shown in *Figure 21.8*.

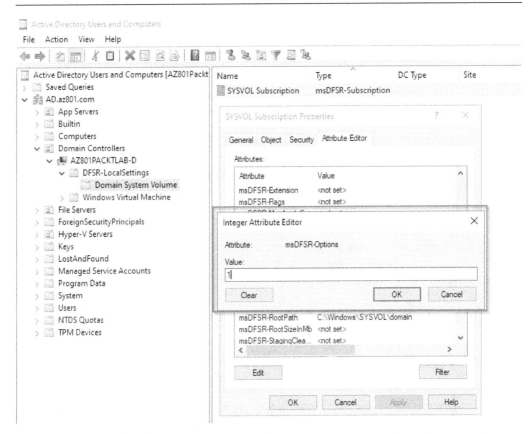

Figure 21.8 – Changing the msDFSR-Options attribute to complete an authoritative restoration

4. This will mark the domain controller as authoritative and the SYSVOL replication and repair will begin behind the scenes.

Last-resort SYSVOL rebuild approach to recovery

Microsoft provides steps for a quicker and more detailed recovery approach where downtime (completed after-hours) and a lack of authentication mechanisms are not a concern, while also requiring all domain controllers to be restarted in tandem. Again, this is considered a last-resort approach due to the downtime incurred by this process. For more details on this approach, be sure to check out the following article: https://learn.microsoft.com/troubleshoot/windows-server/group-policy/rebuild-sysvol-tree-and-content-in-a-domain.

Let's continue with more depth on troubleshooting AD DS replication as part of the Windows Server operating system.

Troubleshooting AD replication

Let's begin with a quick overview of AD DS regarding its structure and defining what replication does for AD DS. Every domain controller in an AD DS environment contains a replica of several partitions (or naming contexts), and each directory partition contains objects that have a specific replication scope and purpose for the AD DS environment.

In general, AD replication achieves the replication of objects via an attribute-level change monitor to complete granular updates of attributes without collisions or conflict. In addition, AD replication also builds an autogenerated replication topology of additional replication partner DCs, ensuring that none of the partners are more than two to three hops apart from each other.

The four AD logical directory partitions are defined as follows:

- A **configuration** partition that represents the overall logical structure of the forest and provides information on the physical topology and includes the services, sites, and subnets of the AD DS environment. This is a mandatory replication partition.

- A **domain** partition that represents all the AD DS objects in the domain such as users, groups, computers, and GPO containers. This is a mandatory replication partition.

- A **schema** partition that represents the attributes and object classes for the entire directory, including custom attributes and classes. This is a mandatory replication partition.

- An **application** partition that represents the ability to limit replication of certain application-specific data to a controlled subset of domain controllers. This is an optional replication partition.

One of the more frequent problems to occur in AD environments is that of replication issues. Whether there are scheduled or unexpected changes in configuration, networking, DNS issues, hardware failures, firewall configuration, or other contributing factors, both inbound and outbound replication failures can cause a myriad of issues and lead to inconsistencies for the following objects:

- Domain controllers, including both the replication topology and replication schedule, leading to delays and large delta synchronization times

- Users, computers, and devices generally leading to inconsistent attribute updates

- Passwords and password changes, leading to leading to inconsistent password replication across the environment and potential account lockouts

- Security groups, distribution groups, and group membership changes

- GPO setting changes resulting in inconsistencies across the environment

The hardest part of troubleshooting AD replication is systematically working to identify a root cause not only for the issue at hand but also ensuring that the issue doesn't occur again to provide availability and resiliency.

Let's discuss a general rule-of-thumb approach to troubleshooting and fixing replication issues in an AD DS environment:

- Be on the lookout for replication and authentication health within an AD environment on a daily cadence. Establish daily status reports, active monitoring and notification of anomalies, and utilize replication tools.

Some of the tools available that can be utilized for troubleshooting, analysis, and resolving AD replication errors include, but are not limited to, the following:

- The Event Viewer on any domain controller. AD events appear in many of the event log channels and can give great insights on where to begin your troubleshooting efforts. Do know that for some of the logs, there may be an additional audit logging requirement that needs to be completed at the domain level (via Advanced Audit Policy Configuration).

- Use **Active Directory Sites and Services** to easily determine the replication partners for any domain controller, as well as force replication from a listed partner domain controller (can be inbound or outbound replication).

- Use the built-in `repadmin.exe` tool to monitor replication times and replication states, as well as resolve a myriad of issues. The `repadmin /replsum` and `repadmin /showrepl` commands are two great starting points, as shown in *Figure 21.9*. A complete list of all of the `repadmin.exe` commands can be found at the following URL: `https://learn.microsoft.com/previous-versions/windows/it-pro/windows-server-2012-R2-and-2012/cc770963(v=ws.11)`.

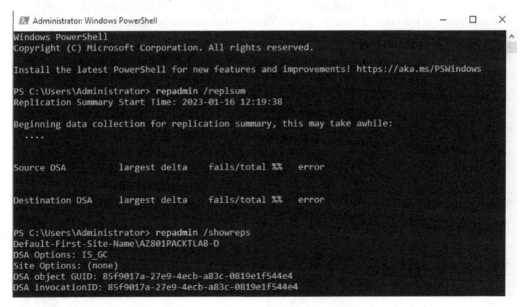

Figure 21.9 – Using repadmin.exe to complete troubleshooting tasks for AD DS

- The **Active Directory Replication Status** tool can be used to review and analyze the replication status and can be downloaded from the following URL: `https://www.microsoft.com/download/details.aspx?id=30005`.

- The **Microsoft Support and Recovery Assistant tool (SARA) for AD DS** can be used to troubleshoot and resolve issues based on a question and answers and recommendations and can be downloaded from the following URL: `https://outlookdiagnostics.azureedge.net/sarasetup/SetupProd_ADReplication.exe`.

- **New**: Use **Virtual Agent** to quickly identify and potentially resolve some of the most common AD replication issues. **Virtual Agent** is located at the following URL: `https://vsa.services.microsoft.com/v1.0/?partnerId=7d74cf73-5217-4008-833f-87a1a278f2cb&flowId=DMC&initialQuery=adrepl`.

- Ensure that there are no hardware failures or disruptions in network and site connectivity between sites and branch offices.

- Ensure that appropriate ports are open between domain controllers. TCP port `135` is used for RPC and is necessary for replication to succeed.

- Ensure that all domain controllers holding at least one of the five specific FSMO roles are available and if not, consider moving the role to an available domain controller to restore functionality for the entire environment.

Operations master roles in AD

Some domain controllers are special and hold a very specific operations master role. The FSMO roles are defined as follows, and need to be available for certain tasks to be successfully performed:

The **Primary Domain Controller** (PDC) emulator [0] master not only receives urgent password changes but is also the master time source for the domain. Among other responsibilities, this role also ensures that when a GPO is edited, the PDC emulator master holds the edited copy.

The **Relative ID** (RID) master [1] is responsible for allocating blocks of available IDs and those are delegated to each domain controller within the domain to build **security descriptors** (SIDs) for creating new AD objects.

The **infrastructure** master [2] is responsible for maintaining object references both inside of the domain as well as other domains.

The **schema** master [3] is responsible for all updates and modifications to the schema, even though schema changes are infrequent.

The **domain naming** master [4] is responsible for adding or removing domains within a forest.

Note that each forest has only one schema master and one domain naming master while each AD DS domain has one RID master, one infrastructure master, and one PDC emulator.

Let's now learn how to troubleshoot hybrid authentication issues using Windows Server and Microsoft Azure tools.

Troubleshooting hybrid authentication issues

As we learned in *Chapter 3, Securing a Hybrid Active Directory (AD) Infrastructure*, there are four main AD DS integration points with Azure AD when using **Azure AD Connect Sync** to achieve cloud-scale identity and single control plane management. However, it's not always as simple as installing an agent or application; you must ensure that the environment you are about to synchronize contains few errors and is well prepared and documented.

One of the tools provided by Microsoft to overcome some of the common issues to prepare your environment for synchronization is called the **IdFix** tool. The **IdFix** tool helps to identify duplicated and formatting issues within your on-premises directory, investigating **AD** objects and recommending remediation steps for identified anomalies prior to synchronization. Running the **IdFix** tool can be done directly from **GitHub** via this URL – `https://raw.githubusercontent.com/Microsoft/idfix/master/publish/setup.exe` – with additional documentation on the tool available at the following URL – `https://microsoft.github.io/idfix/`.

Once Azure AD Connect has been established in your environment, Azure AD Connect offers a troubleshooting feature from within the **Azure AD Connect** tool. To launch and utilize this tool, open **Azure AD Connect**, select the **Configure** button (noting that while this tool is open, all synchronization tasks are suspended), select **Troubleshoot**, and click the **Next** button. Finally, select the **Launch** button to open the PowerShell-based troubleshooting tool, as shown in *Figure 21.10*.

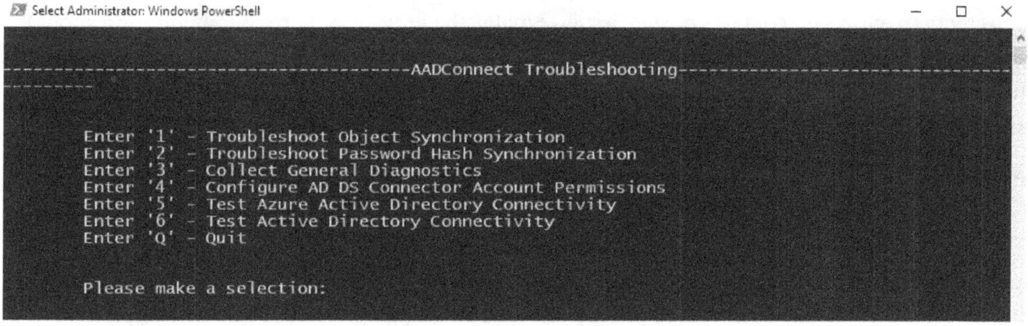

Figure 21.10 – Running the AADConnect Troubleshooting tool

Additionally, simply opening the **Azure AD Connect** application or **Synchronization Service Manager** can also aid in troubleshooting. For instance, there may be times when the server has been recently patched or rebooted and, upon boot up, the **Azure AD Connect** service did not start properly, as shown in *Figure 21.11*.

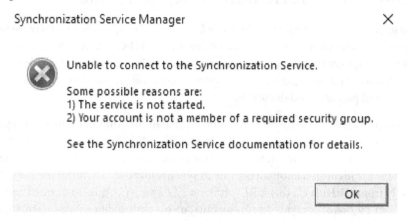

Figure 21.11 – AD Connect synchronization service failed to start properly

To troubleshoot, it's best to review the `services.msc` entry for **Microsoft Azure AD Sync** to ensure that the service has an expected status. Changes to the hosting server (an account no longer having permission to start the service or another policy issue encountered on the device) should also be reviewed to ensure they are not preventing the startup of the service. Using the **Azure AD Connect** application, the **Refresh directory schema** task is also beneficial, as it checks both cloud and on-premises connectivity to ensure no conflicts or sync issues. Finally, the firewall and proxy configuration should also be reviewed to ensure connectivity to the Microsoft cloud services. Additional details on the setup and requirements can be found at the following URL: `https://learn.microsoft.com/azure/active-directory/hybrid/how-to-connect-health-agent-install`.

For synchronization issues, additional troubleshooting can be completed by reviewing the **Operations** tab within **Synchronization Service Manager** for the **Azure AD Connect** host. This can be combined with PowerShell to review currently configured sync schedules, as well as manually run synchronization cycles, as shown in *Figure 21.12*, where we use `Get-ADSynScheduler` to identify the current schedule and issue `Start-ADSyncSyncCycle -PolicyType Initial` to complete a full synchronization pass.

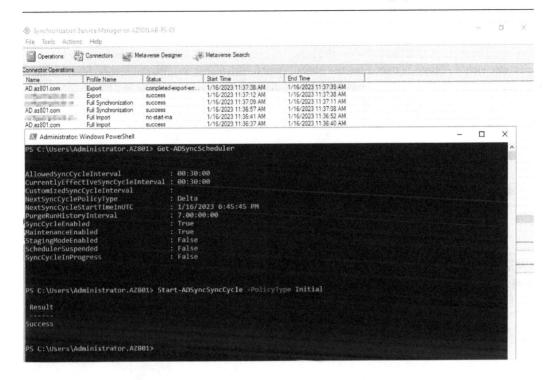

Figure 21.12 – Using Synchronization Service Manager operations and PowerShell for troubleshooting

For customers on Azure AD Premium P1 or P2, the **Azure AD Connect Health** feature is also available for additional monitoring and troubleshooting your environment. Most likely, this was done for you automatically during installation; if not, the `Register-AzureADConnectHealthSyncAgent` PowerShell command will also walk you through the registration process. Additional details on the reasons for using Azure AD Connect Health for both Azure AD and AD FS can be found at the following URL: `https://learn.microsoft.com/azure/active-directory/hybrid/whatis-azure-ad-connect#why-use-azure-ad-connect-health`.

Accessed from `https://portal.azure.com/#view/Microsoft_Azure_ADHybridHealth/AadHealthMenuBlade/~/QuickStart`, administrators have the ability to navigate rich dashboards that show the overall sync health status, number of errors, recent sync and sync failure heartbeats, and a myriad of other information, as shown in *Figure 21.13*:

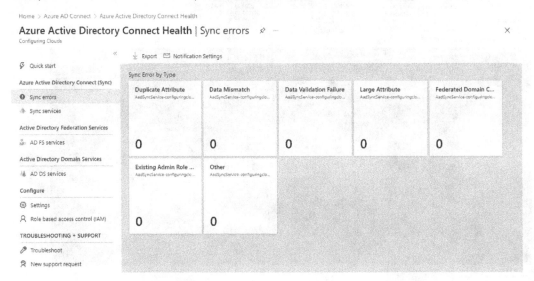

Figure 21.13 – Navigating the Sync errors dashboard in Azure AD Connect Health

In addition, the **Troubleshoot** blade also provides some quick remedial steps to help with troubleshooting some common sync issues, as shown in *Figure 21.14*.

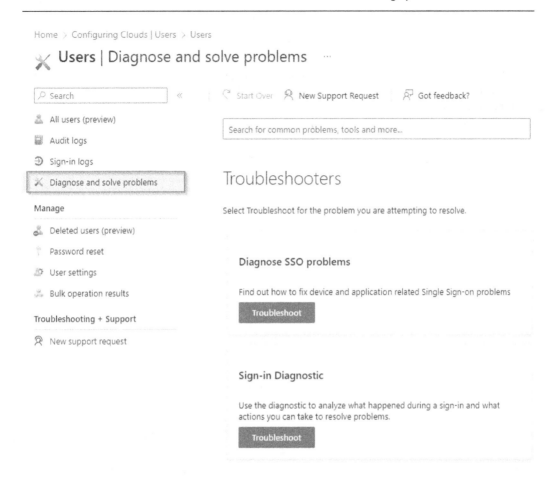

Figure 21.14 – Using the Diagnose and solve problems troubleshooting tool within Azure AD

Beware and be cautious of Azure Conditional Access policies

While Conditional Access policies are incredibly powerful, you must ensure that the on-premises AD Sync service account is not subjected to MFA requirements that prevent the synchronization, troubleshooting, and installation of the sync or health agents. This includes Conditional Access policies for users as well as devices. Be sure that the user is exempted or excluded from these policies and ensure that the change is documented as an exception to your organizational policies and standard operating procedures.

Finally, it's important to always remember that Azure AD sign-in logs are available for your troubleshooting efforts. Simply reviewing **Sign-in logs** from the Azure portal for a user or users can quickly point you in additional directions simply from a failed attempted sign-in or Conditional Access policy that failed authentication. For example, the figure in *Figure 21.14* shows what happens when **On-Premises Directory Synchronization Service Account** is affected inadvertently by a new **Conditional Access** policy introduced into the Azure tenant.

Figure 21.15 – Reviewing sign-ins and failures from the Users blade in Azure

Let's now learn how to troubleshoot on-premises AD issues using Windows Server tools.

Troubleshooting on-premises AD

When it comes to troubleshooting on-premises AD, there are a ton of tools, built-in logs, and general commands for troubleshooting and resolving domain controller issues. However, we will focus on the key tools that help to uncover another layer of detail that logs simply do not provide when troubleshooting:

- As discussed previously, `repadmin.exe` is still a great friend to have in your corner for on-premises troubleshooting.

- `Dcdiag.exe` gives you access to a tool that runs tests across domain controllers and can be selectively targeted to specific DCs that have reported issues or faults. The tests include replication checks, Netlogon checks, advertising of roles, intersite checks and balances, FSMO checks, `SYSVOL` replication, and most importantly, DNS tests, which are the root cause of many an outage.

- **Task Manager**, `MSInfo32.exe`, and `AutoRuns.exe` can be used to identify any superfluous applications that may be running during startup, applications that are consuming excessive amounts of RAM or CPU, and other system anomalies.

- Network validation within **Network Monitor** or other third-party network capture tools identifies whether any data or network packets are being dropped or sent to an unexpected location.

- You can use tools to assist in the metadata cleanup of a forced removal of a domain controller, also called a tombstoned DC. Additional details surrounding the graphical UI tools and steps for remediation can be found at the following URL: `https://learn.microsoft.com/windows-server/identity/ad-ds/deploy/ad-ds-metadata-cleanup`.

- You can review the Event Viewer and the following set of logs for domain controllers to review any AD DS deployment issues (for `dcpromo` processes) or forest/domain upgrades:

  - `%systemroot%\debug\dcpromoui*.log`

  - `%systemroot%\debug\dcpromo*.log`

  - `%systemroot%\debug\adprep\{datetime}\adprep.log`

  - `%systemroot%\debug\adprep\{datetime}\dspecup.log`

- **New**: Use **Virtual Agent** to quickly identify and potentially resolve some of the most common on-premises domain controller issues. **Virtual Agent** is located at the following URL: `https://vsa.services.microsoft.com/v1.0/?partnerId=7d74cf73-5217-4008-833f-87a1a278f2cb&flowId=DMC&initialQuery=31806257`.

As with any troubleshooting efforts, domain controllers are no different in the sense that you use the following process to identify, troubleshoot, repeat, and resolve any issues encountered:

1. Identify the issue and document date and time, the criticality of event, and the services affected.

2. Determine whether any changes were recently made in the environment.

3. Begin working through a review of the available logs, the output of troubleshooting tools, and configuration to identify the root cause, and repeat the troubleshooting review steps as necessary to identify the cause.

4. Implement the recommended or prescribed fixes to resolve the issue.

One additional tool that can be utilized for troubleshooting and monitoring on-premises AD environment is called **Azure AD Connect Health agent for Azure AD DS**. The link for installation of the agent is located at `https://go.microsoft.com/fwlink/?LinkID=820540` and requires the following for communication:

- Azure AD Premium P1 or P2

- A **Hybrid Administrator** or **Global Administrator** role in **Azure AD**

- The **Azure AD Connect Health** agent installed on necessary domain controllers

- Outbound connectivity to Azure service endpoints

- Firewall opened for TCP port `443` (TCP port `5671` is no longer needed with the most recent version of the agent)

- PowerShell version 5.0 or newer installed on the machine

- If using a proxy, the `Set-AzureAdConnectHealthProxySettings` PowerShell cmdlet can be used for configuration (as outlined at this URL: `https://learn.microsoft.com/azure/active-directory/hybrid/how-to-connect-health-agent-install#configure-azure-ad-connect-health-agents-to-use-http-proxy`)

- Once established with Azure, health statistics will be sent to the portal and monitoring details will be sent to an insights dashboard showing details on the following:

  - The name of the forest and its functional level

  - Details on the FSMO operational master roles holder(s)

  - Operational and availability alerts

  - LDAP successful binds/sec monitoring across all DCs for the last 24 hours

  - NTLM authentication/sec monitoring across all DCs for the last 24 hours

  - Kerberos authentication/sec monitoring across all DCs for the last 24 hours

This dashboard can be accessed at this URL – `https://portal.azure.com/#view/Microsoft_Azure_ADHybridHealth/AadHealthMenuBlade/~/AddsServicesList` – and can be seen from the following example shown in *Figure 21.16*.

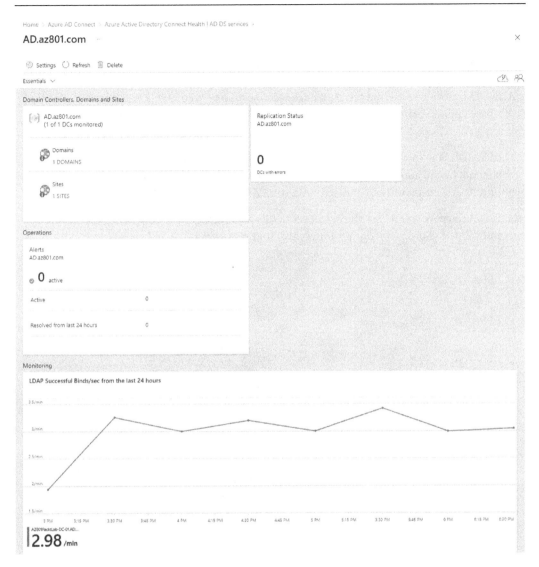

Figure 21.16 – Azure AD Connect Health for AD DS services showing available insights

We have completed this section on how to troubleshoot on-premises AD. Next, we will complete a review of everything we have learned in this chapter and set the stage for the next chapter of this book.

Summary

In this chapter, we learned about the valuable tools that help you troubleshoot and recover from the typical issues experienced when running AD DS workloads. We also discussed how to enable and use Active Directory Recycle Bin to restore deleted objects, how to use DSRM to recover a corrupt AD database and/or corrupted objects, and how to recover the SYSVOL folder and files necessary for running AD services.

We then learned about tools available to determine AD replication issues and how to resolve them, how to troubleshoot hybrid authentication issues, and how to troubleshoot the general issues encountered when running AD on-premises, consistent with the *AZ-801* exam objectives.

In the next chapter, we will review the objectives in this guide using a question-and-answer approach to test and validate your knowledge prior to scheduling your exam.

22

Final Assessment and Mock Exam/Questions

Throughout this book, you have learned about the basic, intermediate, and advanced techniques of designing, managing, and migrating on-premises, hybrid, and cloud workloads. You have also learned how to prepare yourself for the AZ-801: Configuring Windows Server Hybrid Advanced Services exam objectives with immersive hands-on exercises so that you may successfully pass the exam.

This final chapter can be used as additional reinforcement for the objectives reviewed in this guide, giving you a question-and-answer approach to test and validate your knowledge before you schedule your exam.

In this chapter, you can expect the following sections:

- Mock exam practice questions
- Mock exam answers and supporting chapter references
- Lab environment housekeeping
- Summary

Mock exam practice questions

I encourage you to treat this as a true exam experience – try to approach it in a closed-book manner, set a countdown timer for 185 minutes, and truly focus on this assessment. Going through this assessment, you are aiming for 90% or higher, which means you should be ready to schedule and take this AZ-801 exam.

So, what are you waiting for? Let's begin the 70-question assessment!

1. When setting up a new Domain Controller, `SafeModeAdministratorPassword` is required to complete the Domain Controller installation. `SafeModeAdministratorPassword` is used for which of the following features listed?

    A. **Microsoft Azure Backup Server (MABS)**

    B. **Directory Services Restore Mode (DSRM)**

    C. Active Directory Safe Mode

    D. Active Directory Troubleshooting Mode

2. Your business has an Azure VM that runs Windows Server and has a new line of business application installed that requires images to be blocked when accessing image files on a file share or UNC path on the network. What feature should be configured for the VM?

    A. Microsoft Defender SmartScreen

    B. Exploit Protection

    C. Microsoft Defender Application Control

    D. Windows Defender Credential Guard

3. Your organization has decided to begin piloting the use of Microsoft Defender for Endpoint. Which of the following onboarding deployment methods is not supported?

    A. Microsoft Endpoint Configuration Manager

    B. Mobile Device Management/Microsoft Intune

    C. Scripted or manual update directly to registry keys on a device

    D. A local script that can be used for up to 10 devices

4. Which of the following is not a requirement for enabling Windows Defender Credential Guard on a device?

    A. Virtualization-based security support

    B. Secure boot

    C. 64-bit capable CPU

    D. UEFI lock

5. Your organization is strongly considering a new hybrid authentication approach that reduces complexity, allows self-service password management in the cloud, and is highly available via the use of lightweight agents installed on-premises. The solution must also ensure that the on-premises passwords are never stored in the cloud in any form while still requiring on-premises AD connectivity for both users and devices. Which of the following hybrid authentication concepts is the best answer?

   A. **Pass-through Authentication (PTA)**

   B. **Seamless Single Sign-On (SSSO)**

   C. **Active Directory Federation Services (ADFS)**

   D. **Password hash synchronization (PHS)**

6. When installing the Azure AD Connect application to a host, the recommended AD attribute to be used as the source anchor is which of the following?

   A. `userPrincipalName`

   B. `objectGUID`

   C. `ms-DS-ConsistencyGuid`

   D. `samAccountName`

7. When configuring **Self-Service Password Reset (SSPR)**, the administrator is required to delegate privileges to the user being used for synchronization, selecting a specified set of AD properties. Which of the following properties should not be delegated to the sync account?

   A. Reset Password (under permissions)

   B. Write lockoutTime (under properties)

   C. Write pwdLastSet (under properties)

   D. Read lockoutTime (under properties)

8. When creating an Authentication Policy Silo, which of the following tools can be used to complete the task?

   A. CMD prompt

   B. Active Directory Users and Computers

   C. Active Directory Administrative Center

   D. ADSIEdit

9. When onboarding servers to Azure Arc, which of the following options is not a way to successfully install and onboard the Azure Connected Machine agent?

    A. Add servers with Microsoft Endpoint Manager Configuration Manager

    B. Add a single server

    C. Add multiple servers

    D. Add servers with Azure Migrate

10. True or False: When sending data to Log Analytics and Microsoft Sentinel, the data collection rule can be created in a different geographic region than the region the Log Analytics workspace resides.

11. When completing data manipulation and querying results within Microsoft Sentinel, which language is used?

    A. **Structured Query Language (SQL)**

    B. **Kusto Query Language (KQL)**

    C. **Windows Query Language (WQL)**

    D. Microsoft Power Query

12. Which of the following choices is not documented as one of the advanced protection capabilities within Microsoft Defender for Cloud?

    A. **Adaptive Network Hardening (ANH)**

    B. Vulnerability assessment and management features for devices

    C. **Just-in-time (JIT)** virtual machine access

    D. Vmware host hardening for unmanaged virtual machines and containers

13. When establishing a new firewall security rule for IPSec connections to ensure that only authenticated machine accounts can connect to Domain Controllers, which of the following is the best practice approach to ensure the rule can be evaluated before finalizing and hardening the connections?

    A. Require authentication for inbound and outbound connections

    B. Require authentication for inbound connections and request authentication for outbound connections

    C. Request authentication for inbound connections only

    D. Request authentication for inbound and outbound connections

14. When creating Windows Defender Firewall rules for a device, which of the following is not a way to create and manage firewall rules?

    A. Group Policy

    B. Using wf.msc on an individual device

    C. Utilizing a VBScript

    D. Onboarding the device(s) into Microsoft Defender for Business

15. When establishing Advanced Key Exchange settings for IPSec specifically for use within data integrity and encryption, which of the following algorithms cannot be selected?

    A. AES-CBC-128

    B. AES-CBC-256

    C. SHA-256

    D. SHA-512

16. When implementing connection security rules within Windows Defender Firewall, which of the following is not one of the main types of connection security rules?

    A. Isolation

    B. Authentication exemption

    C. Server-to-client

    D. Server-to-server

17. When it comes to managing firewall rules within Azure **Network Security Groups** (NSGs), there are six default rules. Which of the following are considered default rules?

    A. DenyAllInbound

    B. AllowVnetOutbound

    C. AllowInternetOutbound

    D. AllowInternetInbound

18. When specifying public network access for an Azure Storage account, which of the following options is not a configurable setting?

    A. Enabled from all networks

    B. Enabled from Internet only

    C. Enabled from selected virtual networks and IP addresses

    D. Disabled

19. When creating an outbound security rule for an NSG, which of the following is not a required property?

    A. Destination port ranges

    B. Priority

    C. Source service tag

    D. Service

20. Your organization wants to ensure that certain security regulations are established for BitLocker encrypted drives. To ensure that a user is forced to supply a PIN when booting up and utilizing a laptop managed by the organization, which of the following options should be configured as a protector for the drive?

    A. `StartupKeyProtector`

    B. `TpmAndPinAndStartupKeyProtector`

    C. `TpmAndPimProtector`

    D. `TpmAndStartupKeyProtector`

21. When configuring a cluster quorum witness for a Windows Server Failover Cluster, which of the following cannot be used?

    A. A standard file share witness

    B. A USB drive attached to a network switch

    C. A symbolic link

    D. A cloud witness using an Azure Storage account

22. When deploying a Windows Server cluster, which of the following does not meet the prerequisites?

    A. An account that's a member of the local Administrators group on each server

    B. One or more network adapters with static IP addresses per server for management, and separate adapters for other traffic

    C. At least two servers with supported hardware

    D. Windows Server 2012 R2 or later installed on each server

23. When considering Windows Server Failover Cluster Storage options, which of the following cannot be used for failover cluster shared storage?

    A. **Storage Spaces Direct (S2D)**

    B. **Cluster Shared Volumes (CSV)**

C. **Write Anywhere File Layout (WAFL)**

D. **Scale-Out File Server (SOFS)**

24. When analyzing cluster readiness before completing the cluster-aware updating configuration, which of the following tools cannot be used to complete a readiness check?

    A. The `Test-CAUSetup` PowerShell cmdlet

    B. Server Manager

    C. The Analyze Cluster updating readiness link within **Cluster-aware updating**

    D. The readiness tool within Windows Admin Center

25. True or False: Applying updates using the Cluster-Aware updating tool in Windows Admin Center is as easy as selecting the **Check for updates** button and the **Install** button.

26. Which of the following is not a cluster update state during the cluster-aware updating process?

    A. Starting

    B. Waiting

    C. Completing

    D. Staging

27. During the Cluster OS Rolling Upgrade process for a server migrating from Windows Server 2019 to Windows Server 2022, which of the following is not a required step for a successful upgrade to occur?

    A. The cluster node does not need to have storage and network reconfigured

    B. The cluster workload data and configuration must be backed up

    C. The cluster node must be paused, and all roles drained from the node

    D. The cluster database must be backed up

28. To achieve the recommended network connectivity for a high-performance cluster for Storage Spaces Direct (where four or more nodes are required), which of the following is not a recommended component of the design?

    A. 10 Gbps network interface(s)

    B. 25 Gbps (or faster) network interface(s)

    C. **Network interface cards (NICs)** that are **remote-direct memory access (RDMA)** capable with iWARP or RoCE

    D. Two or more network connections from each node recommended to improve both redundancy and overall performance

29. Which of the following options are available when considering an upgrade to your Storage Spaces Direct cluster?

    A.   Complete a clean OS install where virtual machines are in the stopped state

    B.   In-place upgrade on cloned virtual machines while they are in the stopped state

    C.   In-place upgrade while all virtual machines are in the running state

    D.   In-place upgrade while all virtual machines are in the stopped state

    E.   Complete a clean OS installation while virtual machines are running

30. True or False: In an S2D Converged mode, the compute and storage resources reside on the same clusters to support large-scale deployments and workloads.

31. Azure Backup supports all but which of the following backup scenarios?

    A.   Azure VM with disk deduplication for any backup component

    B.   Azure VM backup using VM extension

    C.   Using agent extension to back up Azure VM that's running Linux OS

    D.   On-premises backup to **Microsoft Azure Backup Server (MABS)**

32. Which of the following options is not considered to be a snapshot consistency model?

    A.   Application-consistent snapshots

    B.   Crash-consistent snapshots

    C.   OS-consistent snapshots

    D.   Filesystem-consistent snapshots

33. True or False: When adding a new retention category such as a weekly, monthly, or yearly backup point, this new retention category will be applied to all existing and future recovery points.

34. Which of the following is not a required component for establishing Azure Site Recovery for your environment?

    A.   An Azure Recovery Services Vault within your subscription

    B.   Access to the Microsoft 365 IP ranges for authentication and use of the Azure Active Directory (Azure AD) service tag-based NSG

    C.   Appropriate credentials for Azure, requiring both the Virtual Machine Contributor and the Recovery Contributor RBAC roles

    D.   The Site Recovery Provider and the MABS agent

35. For the Configuration server role in Azure Site Recovery, the configuration server cannot be which of the following?

    A. VMware machine (either an appliance or a PowerShell script run on an existing virtual machine to complete appliance configuration)

    B. A server with System Center **Virtual Machine Manager** (**VMM**) with Azure Site Recovery provider registered

    C. A server with the Microsoft Endpoint Manager Configuration Manager with Azure Site Recovery Provider registered

    D. Hyper-V sites/hosts with the Azure Site Recovery provider installed and registered

36. When creating a recovery plan, which of the following is not an available failover option?

    A. Hyper-V to Azure

    B. VMware to Azure

    C. XenServer to Azure

    D. Azure to Azure

37. When creating an authorization entry in the Hyper-V Replica configuration for a primary replica server, which of the following is not required?

    A. Specifying the primary server as a short hostname

    B. Specifying the primary server as a **fully-qualified domain name** (**FQDN**)

    C. Specifying the default location to store the replica files

    D. Specifying the trust group

38. The available options for replication of virtual machines from a secondary Hyper-V replica server include which of the following?

    A. Test failover

    B. Extend replication

    C. Failback

    D. Failover

39. True or False: When using certificate-based authentication in a Failover Cluster, all the cluster nodes and the **Client Access Points** (**CAPs**) supporting the Hyper-V Replica role need to have appropriate machine certificates issued and installed.

40. When establishing the Storage Migration Service, which of the following are prerequisites for the service?

    A. An identified source server

    B. An orchestrator server

    C. A Windows workstation or Windows Server running **Windows Admin Center** (**WAC**) components version 2103 or higher

    D. A non-administrator migration account is to be used on the source and orchestrator servers

41. When setting up a new Storage Migration Service job, which of the following source devices cannot be selected for use?

    A. NetApp FAS array

    B. IBM Storage servers

    C. Linux servers

    D. Windows servers and clusters

42. When specifying a destination for the transfer data step in Storage Migration Service, which of the following is not an option that can be used as a destination?

    A. Use an existing server or VM

    B. Create a new Azure VM

    C. Copy data to a network share

    D. Don't transfer files from the server

43. When configuring and integrating Azure File Sync, which of the following is not a prerequisite?

    A. Windows Admin Center 1904 or later

    B. A source file server running Windows 2008 R2 or newer

    C. An Azure subscription with existing Storage Sync Service, Azure Storage account, and Azure File Share resources created

    D. A Windows Admin Center gateway connection registered to Azure

44. When using the Azure Migrate appliance to collect performance data, which of the following details are collected?

    A. RAM utilization

    B. CPU utilization

C. Free disk remaining

D. Per-virtual network adapter information on network utilization

45. True or False: When selecting **Discover using appliance** and acknowledging that the servers intended for migration are virtualized with Hyper-V, as part of the Azure Migrate Appliance installation, there is no need to generate a project key to register the appliance as it will automatically register with Azure.

46. As part of the Azure Migrate Discovery and Assessment tools, which of the following resources cannot be assessed within the Azure portal?

A. Azure VM

B. Azure SQL

C. On-premises web server

D. **Azure VMware solution (AVS)**

47. When establishing the target or intent configuration for Azure Migrate replication settings for a migration project, which of the following is not a valid option for the replicate configuration?

A. Replication Storage account

B. Virtual network

C. Subnet

D. Azure region

48. True or False: The Web Deploy application can be used not only to export applications but also to migrate the application contents and configuration to a new destination host.

49. When migrating Hyper-V hosts to a newer OS version, which of the following migration options does not require additional hardware?

A. Cross-version live migration

B. In-place upgrade

C. Export and Import

D. Copy Cluster Roles Wizard

50. When completing a **Remote Desktop Services (RDS)** migration, which of the following options provides the correct order of upgrade?

A. RD Session Hosts, RD Connection Broker(s), RD Web Access Host(s), RD Virtualization Host(s)

B. RD Connection Broker(s), RD Session Hosts, RD Web Access Host(s), RD Virtualization Host(s)

    C.   RD Connection Broker(s), RD Session Hosts, RD Virtualization Host(s), RD Web Access Host(s)

    D.   RD Web Access Host(s), RD Session Hosts, RD Connection Broker(s), RD Virtualization Host(s)

51. When planning for an application migration using the Azure App Service Migration Assistant, which of the following URLs can be used to complete an automated assessment?

    A.   `https://aka.ms/MigrateIIS`

    B.   `https://portal.azure.com`

    C.   `https://admin.microsoft.com`

    D.   `https://appmigration.microsoft.com`

52. True or False: The Azure App Migration Assistant will only assess the IIS applications you have on a server and provide only ARM templates for migration and importing into Azure.

53. When attempting to migrate an IIS workload into a container, which of the following tools can be used to complete the export and container configuration?

    A.   Web Deploy

    B.   Msdeploy

    C.   Image2Docker

    D.   Export-WindowsImage

54. When using the Active Directory Migration Tool, which of the following tasks cannot be completed?

    A.   Migration of user accounts

    B.   Migration of computer accounts

    C.   Migration of group accounts

    D.   Migration of user profiles from an older version of Windows to a newer one

55. True or False: Server Manager can be used to remove Active Directory Domain Services cleanly from a Domain Controller by selecting **Demote this domain controller**.

56. Which of the following is not considered a requirement for upgrading or migrating an existing forest from Windows Server 2016 to Windows Server 2022?

    A.   Ensure that DNSSEC has been established for all DNS zones in the existing environment

    B.   Ensure you have a system state backup covering boot files, the registry, and the SYSVOL folder

    C.   Patch the servers to the latest available before migration to avoid migration blockers

    D.   Ensure that the environment is already running DFS-R instead of FRS for SYSVOL folder replication

57. When using the Active Directory Migration Tool, which of the following can the Reporting Wizard provide as insights?

    A. Report on migrated user accounts

    B. Report on expired accounts

    C. Report on account name conflicts

    D. Report on unused Group Policy Objects

58. True or False: Data Collector Sets can be created using either Performance Monitor or Windows Admin Center.

59. When using System Insights as part of Windows Server 2019 and later, which of the following is not a default capability?

    A. CPU capacity forecasting

    B. Networking capacity forecasting

    C. Certificate life cycle and expiration forecasting

    D. Volume consumption forecasting

60. When deploying Azure Monitor Agent for use, which of the following approaches can not be used for installation and deployment?

    A. Azure VM extension installation

    B. Windows client installer

    C. Microsoft Intune's proactive remediation

    D. Windows Admin Center (via gateway registration)

61. Using **Network Insights** within Azure Monitor, which of the following resources are not available as tracked metrics on the Network Health dashboard?

    A. Private IPs

    B. Network interfaces

    C. Local network gateways

    D. Network security groups

62. When using the **Topology** view for troubleshooting and overview, what file type can the topology be exported in?

    A. `.pdf`

    B. `.docx`

    C. `.svg`

    D. `.png`

63. When troubleshooting the Azure VPN Gateway, which of the following log files is not written as part of the diagnostic logs gathering?

    A. `ConnectionStats.txt`

    B. `IKELogs.txt`

    C. `CPUPerf.txt`

    D. `IKEErrors.txt`

64. When attempting to deploy an Azure VM, you receive a warning that the deployment request has exceeded the maximum allowed limit on either a service or another restriction on a subscription, resource group, or account. Which of the following failure types can be associated with this error?

    A. Deployment quota

    B. Conflict failure

    C. Authorization failure

    D. Allocation failure

65. When using the Azure VM performance diagnostics tool, which one of the following is not an available analysis option?

    A. Quick performance analysis

    B. Boot analysis

    C. Advanced performance analysis

    D. Azure Files analysis

66. When troubleshooting Azure VM extension issues, how can you troubleshoot and resolve any VM extension failures?

    A. Ensure that the VM Agent (or Azure Guest Agent) is installed for a VM

    B. Review the logs in the `C:\WindowsAzure\Logs\Plugins` directory

    C. Uninstall and reinstall the VM extension from the **Extensions + applications settings** blade for the VM

    D. Use **Add/Remove programs** to remove the VM extension

67. True or False: If an Azure VM is having trouble booting up due to a BitLocker issue, two troubleshooting steps to take are to review the boot diagnostics screenshot and enable the serial console to begin reviewing the startup issues.

68. When presented with a restore opportunity for an AD DS environment, which of the following answers is the best for completing the restore task as quickly as possible?

    A. Using `ldp.exe` to search for and restore a deleted item

    B. Utilizing DSRM to mount and restore the user from a snapshot

    C. Utilizing the Active Directory Recycle Bin within Active Directory Administrative Center

    D. Completing an authoritative restore from backup media for the Domain Controller

69. When reviewing the collection of database files located in `%SystemRoot%\NTDS` for Active Directory database recovery purposes, which of the following is not a file that exists for AD DS?

    A. `Ntds.dit`

    B. `Edb.mdb`

    C. `Edb.log`

    D. `Edb.chk`

70. When using Azure AD Connect in your organization for cloud identity plane synchronization and management, which of the following selections is not considered a requirement for Azure AD Connect Health Agent connectivity to the cloud?

    A. Outbound connectivity to Azure service endpoints

    B. Azure private link

    C. Firewall opened for TCP port 443

    D. Azure AD Premium P1 or P2

Mock exam practice answers

We recommend that you review the following answers after attempting to answer the assessment questions. Check your answers and review the sections within the chapters for additional clarification:

1. The answer is B. *Chapter 2, Securing the Windows Server Operating System.*

2. The answer is B. *Chapter 2, Securing the Windows Server Operating System.*

3. The answer is C. *Chapter 2, Securing the Windows Server Operating System.*

4. The answer is D. *Chapter 2, Securing the Windows Server Operating System.*

5. The answer is A. *Chapter 3, Securing a Hybrid Active Directory (AD) Infrastructure.*

6. The answer is C. *Chapter 3, Securing a Hybrid Active Directory (AD) Infrastructure.*

7. The answer is D. *Chapter 3, Securing a Hybrid Active Directory (AD) Infrastructure.*

8. The answer is C. *Chapter 3, Securing a Hybrid Active Directory (AD) Infrastructure.*

9. The answer is A. *Chapter 4, Identifying and Remediating Windows Server Security Issues Using Azure Services.*

10. The answer is False. *Chapter 4, Identifying and Remediating Windows Server Security Issues Using Azure Services.*

11. The answer is B. *Chapter 4, Identifying and Remediating Windows Server Security Issues Using Azure Services.*

12. The answer is D. *Chapter 4, Identifying and Remediating Windows Server Security Issues Using Azure Services.*

13. The answer is D. *Chapter 5, Secure Windows Server Networking.*

14. The answer is C. *Chapter 5, Secure Windows Server Networking.*

15. The answer is D. *Chapter 5, Secure Windows Server Networking.*

16. The answer is C. *Chapter 5, Secure Windows Server Networking.*

17. The answer is A, B, and C. *Chapter 6, Secure Windows Server Storage.*

18. The answer is B. *Chapter 6, Secure Windows Server Storage.*

19. The answer is D. *Chapter 6, Secure Windows Server Storage.*

20. The answer is C. *Chapter 6, Secure Windows Server Storage.*

21. The answer is C. *Chapter 7, Implementing a Windows Server Failover Cluster.*

22. The answer is D. *Chapter 7, Implementing a Windows Server Failover Cluster.*

23. The answer is C. *Chapter 7, Implementing a Windows Server Failover Cluster.*

24. The answer is B. *Chapter 8, Managing Failover Clustering.*

25. The answer is True. *Chapter 8, Managing Failover Clustering.*

26. The answer is D. *Chapter 8, Managing Failover Clustering.*

27. The answer is A. *Chapter 8, Managing Failover Clustering.*

28. The answer is A. *Chapter 9, Implementing and Managing Storage Spaces Direct.*

29. The answer is B. *Chapter 9, Implementing and Managing Storage Spaces Direct.*

30. The answer is False. *Chapter 9, Implementing and Managing Storage Spaces Direct.*

31. The answer is A. *Chapter 10, Managing Backup and Recovery for Windows Server.*

32. The answer is C. *Chapter 10, Managing Backup and Recovery for Windows Server.*

33. The answer is False. *Chapter 10, Managing Backup and Recovery for Windows Server.*

34. The answer is D. *Chapter 11, Implementing Disaster Recovery Using Azure Site Recovery.*

35. The answer is C. *Chapter 11, Implementing Disaster Recovery Using Azure Site Recovery.*

36. The answer is C. *Chapter 11, Implementing Disaster Recovery Using Azure Site Recovery.*

37. The answer is A. *Chapter 12, Protecting Virtual Machines by Using Hyper-V Replicas.*

38. The answer is C. *Chapter 12, Protecting Virtual Machines by Using Hyper-V Replicas.*

39. The answer is True. *Chapter 12, Protecting Virtual Machines by Using Hyper-V Replicas.*

40. The answer is A, B, and C. *Chapter 13, Migrating On-Premises Storage to On-Premises Servers or Azure.*

41. The answer is B. *Chapter 13, Migrating On-Premises Storage to On-Premises Servers or Azure.*

42. The answer is C. *Chapter 13, Migrating On-Premises Storage to On-Premises Servers or Azure.*

43. The answer is B. *Chapter 13, Migrating On-Premises Storage to On-Premises Servers or Azure.*

44. The answer is C. *Chapter 14, Migrating On-Premises Servers to Azure.*

45. The answer is False. *Chapter 14, Migrating On-Premises Servers to Azure.*

46. The answer is C. *Chapter 14, Migrating On-Premises Servers to Azure.*

47. The answer is D. *Chapter 14, Migrating On-Premises Servers to Azure.*

48. The answer is True. *Chapter 15, Migrating Workloads from Previous Versions to Windows Server 2022.*

49. The answer is B. *Chapter 15, Migrating Workloads from Previous Versions to Windows Server 2022.*

50. The answer is C. *Chapter 15, Migrating Workloads from Previous Versions to Windows Server 2022.*

51. The answer is D. *Chapter 16, Migrating IIS Workloads to Azure.*

52. The answer is False. *Chapter 16, Migrating IIS Workloads to Azure.*

53. The answer is C. *Chapter 16, Migrating IIS Workloads to Azure.*

54. The answer is D. *Chapter 17, Migrating an Active Directory Domain Services (AD DS) Infrastructure to Windows Server 2022 AD DS.*

55. The answer is True. *Chapter 17, Migrating an Active Directory Domain Services (AD DS) Infrastructure to Windows Server 2022 AD DS.*

56. The answer is A. *Chapter 17, Migrating an Active Directory Domain Services (AD DS) Infrastructure to Windows Server 2022 AD DS.*

57. The answer is D. *Chapter 17, Migrating an Active Directory Domain Services (AD DS) Infrastructure to Windows Server 2022 AD DS.*

58. The answer is True. *Chapter 18, Monitoring Windows Server Using Windows Server Tools and Azure Services.*

59. The answer is C. Chapter 18, Monitoring Windows Server Using Windows Server Tools and Azure Services.

60. The answer is C. *Chapter 18, Monitoring Windows Server Using Windows Server Tools and Azure Services.*

61. The answer is A. *Chapter 19, Troubleshooting Windows Server On-Premises and Hybrid Networking.*

62. The answer is C. *Chapter 19, Troubleshooting Windows Server On-Premises and Hybrid Networking.*

63. The answer is C. *Chapter 19. Troubleshooting Windows Server On-Premises and Hybrid Networking.*

64. The answer is A. *Chapter 20, Troubleshooting Windows Server Virtual Machines in Azure.*

65. The answer is B. *Chapter 20, Troubleshooting Windows Server Virtual Machines in Azure.*

66. The answer is D. *Chapter 20, Troubleshooting Windows Server Virtual Machines in Azure.*

67. The answer is True. *Chapter 20, Troubleshooting Windows Server Virtual Machines in Azure.*

68. The answer is C. *Chapter 21, Troubleshooting Active Directory.*

69. The answer is B. *Chapter 21, Troubleshooting Active Directory.*

70. The answer is B. *Chapter 21, Troubleshooting Active Directory.*

Lab environment housekeeping

As a rule of thumb, be kind to your devices and your cloud computing environments and remove any unused or no longer needed resources when you have completed your training and evaluation for this exam guide. While the virtual machines stored on your laptop may only be using up storage, resources allocated in the Azure or Microsoft 365 cloud environments can incur additional unexpected costs if they're not properly deleted or removed.

To completely remove all local Hyper-V virtual machines

Follow these steps:

1. Open **Hyper-V Manager**.
2. Select all virtual machines that begin with the name `AZ801...` and then right-click and select **Delete...**.
3. Open **Virtual Switch Manager**, select any virtual switches that begin with the name `AZ801`, and select **Remove** one by one.
4. Navigate to the `C:\AZ801PacktLab` folder located on your device and delete the folder and its entire contents.

To completely remove all Azure resources for your AZ801 exam tenant

Follow these steps:

1. Open a browser to `https://portal.azure.com` and sign in using the Global Administrator account you used for the exercises in this book.

2. Select or search for **Resource Groups**, then search for `AZ801` to identify any cloud **Resource Groups** and accompanying **Resources** that are no longer in use.

3. Select the `rg-AZ801-ExamGuide` **Resource Group**.

4. Select **Delete Resource Group** from the menu bar. When prompted, ensure that you select the checkbox for **Apply force delete for selected Virtual machines and Virtual machine scale sets** and then type the name of the Resource Group – that is, `rg-AZ801-ExamGuide`.

5. Note that if you have completed all the exercises in this guide, there will be anywhere from 40-50 resources to be deleted and this operation will require considerable time to complete.

6. Select **Delete** at the bottom of the page, grab a cup of tea or coffee, and wait for the resources to be removed. There may be a few resources that require additional steps to confirm they've been deleted, so please pay close attention to the individual service or resource instructions on a case-by-case basis.

Summary

This completes your assessment and preparation for the AZ-801: Configuring Windows Server Hybrid Advanced Services exam. I do hope that you had as much fun learning this content as I did creating it. Good luck with your continued successes in your career and certification journeys!

CG

Index

Symbols

Packtpub.com

Subscribe to our online digital library for full access to over 7,000 books and videos, as well as industry leading tools to help you plan your personal development and advance your career. For more information, please visit our website.

Why subscribe?

- Spend less time learning and more time coding with practical eBooks and Videos from over 4,000 industry professionals

- Improve your learning with Skill Plans built especially for you

- Get a free eBook or video every month

- Fully searchable for easy access to vital information

- Copy and paste, print, and bookmark content

Did you know that Packt offers eBook versions of every book published, with PDF and ePub files available? You can upgrade to the eBook version at packtpub.com and as a print book customer, you are entitled to a discount on the eBook copy. Get in touch with us at customercare@packtpub.com for more details.

At www.packt.com, you can also read a collection of free technical articles, sign up for a range of free newsletters, and receive exclusive discounts and offers on Packt books and eBooks.

Other Books You May Enjoy

If you enjoyed this book, you may be interested in these other books by Packt:

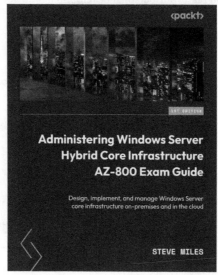

Administering Windows Server Hybrid Core Infrastructure AZ-800 Exam Guide

Steve Miles

ISBN: 978-1-80323-920-0

- Deploy and manage AD DS on-premises and in cloud environments
- Implement and manage hybrid core infrastructure solutions for compute, storage, networking, identity, and management
- Discover expert tips and tricks to achieve your certification in the first go
- Master the hybrid implementation of Windows Server running as virtual machines and containers
- Manage storage and file services with ease
- Work through hands-on exercises to prepare for the real world

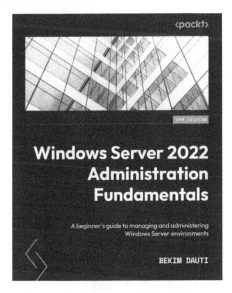

Windows Server 2022 Administration Fundamentals

Bekim Dauti

ISBN: 978-1-80323-215-7

- Grasp the fundamentals of Windows Server 2022
- Understand how to deploy Windows Server 2022
- Discover Windows Server post-installation tasks
- Add roles to your Windows Server environment
- Apply Windows Server 2022 GPOs to your network
- Delve into virtualization and Hyper-V concepts
- Tune, maintain, update, and troubleshoot Windows Server 2022
- Get familiar with Microsoft's role-based certifications

Packt is searching for authors like you

If you're interested in becoming an author for Packt, please visit authors.packtpub.com and apply today. We have worked with thousands of developers and tech professionals, just like you, to help them share their insight with the global tech community. You can make a general application, apply for a specific hot topic that we are recruiting an author for, or submit your own idea.

Share Your Thoughts

Now you've finished *Configuring Windows Server Hybrid Advanced Services Exam Ref AZ-801*, we'd love to hear your thoughts! Scan the QR code below to go straight to the Amazon review page for this book and share your feedback or leave a review on the site that you purchased it from.

https://packt.link/r/1804615099

Your review is important to us and the tech community and will help us make sure we're delivering excellent quality content.

Download a free PDF copy of this book

Thanks for purchasing this book!

Do you like to read on the go but are unable to carry your print books everywhere?

Is your eBook purchase not compatible with the device of your choice?

Don't worry, now with every Packt book you get a DRM-free PDF version of that book at no cost.

Read anywhere, any place, on any device. Search, copy, and paste code from your favorite technical books directly into your application.

The perks don't stop there, you can get exclusive access to discounts, newsletters, and great free content in your inbox daily

Follow these simple steps to get the benefits:

1. Scan the QR code or visit the link below

https://packt.link/free-ebook/9781804615096

2. Submit your proof of purchase
3. That's it! We'll send your free PDF and other benefits to your email directly

www.ingramcontent.com/pod-product-compliance
Lightning Source LLC
Chambersburg PA
CBHW060635060326
40690CB00020B/4411